Color TV Servicing

Color TV Servicing

Newt Smelser

Nelson-Hall nh Chicago

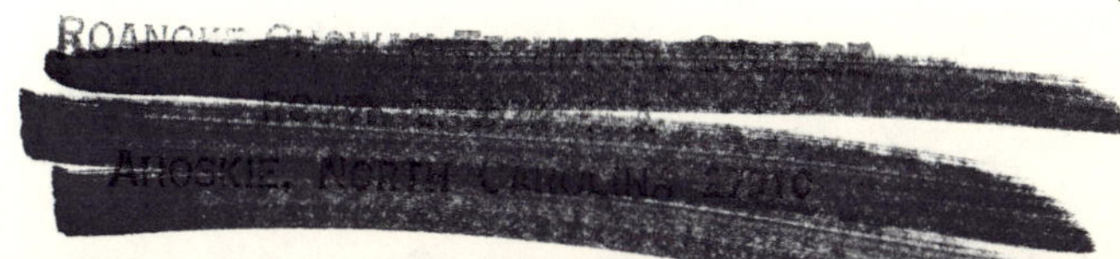

Library of Congress Cataloging in Publication Data

Smelser, Newt. 1930–
    Color TV servicing.

    1.   Color television—Repairing.   I.   Title.
TK6670.S58   621.388′874   80-26523
ISBN 0-88229-549-7 (cloth)
ISBN 0-88229-770-8 (paper)

Manufactured in the United States of America

10   9   8   7   6   5   4   3   2   1

Contents

Preface

IN THE 1940s and even into the early 1950s anyone who wanted to learn radio and television repairing could get a few magazines on the subject month by month and in time pick up considerable knowledge. Today electronics and the electronics industry are so complex that magazines only hit the highlights, assuming that the reader already has basic and perhaps even advanced knowledge in a particular area. It is impossible to learn the complicated circuits, technical terms, and use of expensive equipment through apprenticeship if you are completely unfamiliar with electronics, as very few TV shop repairmen would have the patience to teach this skill on the job. Color TV repair is definitely not something that you can "pick up"; it takes study and more study. What then is a suitable method of imparting TV repair knowledge to the beginner?

A good technical school is certainly one approach to learning complicated TV systems. In a good school the instructors should recently have done servicing of color circuits themselves. Further, they must be good teachers, for it takes particular patience, organization, and skill to teach TV service.

The instructor and pupils must have modern TV sets to test, observe, trace integral circuits, and adjust. Working with snap lead parts on a circuit board is good for fundamentals, but later only working with actual TV receivers will be truly useful. A reference library should be available and students should be encouraged to make good use of it.

Finally, the textbook for the course must be modern, factual, and easy to understand. One great problem with TV texts available has been that they were written for pre-engineering training, and parts of them are quite hard for the student to grasp. Certainly we need engineers, but we also need skilled technicians, and technicians do not need to know a great deal of math, theory, or atomic structure. Instead, the TV technician needs to know how a TV works and how to make it work when it doesn't. This book is designed to teach just that.

With a simplified text, good instruction, and plenty of equipment for practical experience, a TV school can do wonders for a pupil in a reasonable amount of time. This book has been carefully planned so that it can be the central unit text for a TV servicing course. Because some schools do not have laboratory facilities available to every class all of the time, much of the beginner's lab work included here can be completed at the pupil's desk.

The book can be used in the traditional way, with study first, then the experiments. It would be more effective, however, to try the innovation of having pupils read and work the projects as they come up in the text. The instructor would have to go around the room helping pupils as needed with the reading and with the work involved. Finally, the instructor should plan a discussion covering the content of each chapter or portion of a chapter, and follow with periodic exams.

In the later chapters, after pupils have completed the basic experiments and have mastered the basic information about circuit elements, it would be well to have the students study a chapter or a section of a chapter, have an instructor's lecture, discuss the material, then have a lab lesson based on finding and measuring in actual sets the voltage, ohms, and so on of the parts just studied. From Chapter 9 on this would probably be the most useful approach.

Not everyone cares to or can afford to go to a good TV school. Therefore, this text has been prepared so that it can be used in high school classes or other programs for younger people. Many persons, however, want to learn

TV servicing by just studying the book. Again, I have tried to avoid complicated material, and the book is designed to be self-teaching to some degree. If you study it carefully and conscientiously, actually do the experiments, and study some color TVs, you definitely can and will learn color TV servicing by the proper use of this book alone. For study you can look at your own set, friends' sets, and so on for variety.

When the text refers you back to a previous schematic or drawing, be sure that you go back to it. The pictorial illustrations that show how the parts really are arranged in the set are extremely important. The student who compares schematics with pictorial parts layout illustrations is quickly orienting himself so he will recognize actual TV circuits. Many TV books deal only with schematics, and resemblance to actual TV sets is almost nonexistent. You must be able to read schematics, but you must also be able to find the parts in the actual set. This book is unique in that with the schematic, or at least usually in the same chapter, is a pictorial drawing showing how the parts really look or are laid out.

It is quite an effort to bridge the gap between the fundamentals of electricity and electronics and the far more complicated technology of a color TV, whether learned in a course or from a book. Usually, along the way, the pupil picks up too much engineering, spends too much time on knowledge he will not use, and has too little time to learn TV repair thoroughly.

The beginning chapters of this book may seem like radio study for radio's sake. Closer examination will reveal, however, that with each project the beginner learns several valuable facts about circuits he will later experience in TV study. The book is carefully designed to that end. Even if you have made a radio or two and assume you know everything up to Chapter 9, don't skip these early chapters; you may find the later work hard and wonder why. It is important to do the work in the first eight chapters, as they teach very important fundamentals that make analysis of TV circuits easier. My intent has been to make the fundamental chapters interesting, even fun to work through. Many people don't think they would enjoy making transmitters and receivers until they start. They learn to their surprise that it is great fun to connect a few parts together and find that you can actually receive programs.

It may seem that the book occasionally strays into training for amateur radio and for broadcast work. My experience as a TV school instructor, however, has shown that some TV repair students make a career of broadcasting. The study is somewhat parallel, so this book gives a little introduction to broadcasting. If you understand transmitters, it helps fix ideas regarding carriers and modulation methods in your mind. These come up in color TV, as you will observe later.

Amateur radio work will not put bread on the table, but many a TV serviceman is also a ham operator. It makes a good companion hobby to the vocation. It is well worth thinking about learning amateur radio as an additional skill. The experience will aid you as a TV serviceman.

This book is built on the support principle. Until you understand a chapter perfectly, it is poor policy to proceed to the next. You must understand every part, as all circuits are interdependent. Fundamentals in early chapters are vitally important for understanding TV stages in later chapters. If you are studying the book alone, as a home project, the quizzes at the ends of chapters are an excellent means of checking your mastery of the chapter and seeing how well you comprehended the material. The numbered summary statements are worth almost memorizing. Note that I say "almost." TV repair cannot be learned by memory. You must *understand* the material, not just be able to repeat it to someone.

By the time you have finished the book you

will have been introduced to several modern sets. Your training should continue, however. Get every manufacturer's service manual you can find for color TV. Look at every *Sams Photofact* you can lay hands on. Continue to learn day by day; that is the best way to become a good TV serviceman. TV repair certainly needs competent servicemen—it has enough poor ones.

Since this book is designed for beginners, the language sometimes may appear a bit simplistic. It is not my intent to talk down to the reader, but rather to be sure that simple concepts are objectively introduced. Many basic ideas are repeated chapter after chapter. This redundancy is intentional and is necessary for all but advanced TV students who already know black-and-white servicing. I do not start with magnets, electromagnets, and static charges, because the beginning radio experiments are more interesting and more practical to the circuitry cause. Most readers will have done numerous experiments in electricity and magnetism in high school general science or even in elementary school.

The solid state circuits used for lab work were planned mainly because TV is rapidly moving in that direction. Even so, the serviceman has tubes to contend with, and liberal information on tubes is given in the book. I did not feel obliged, however, to introduce tubes into the laboratory work, because knowledge can be gained with TV work, and using tube circuits would have more than doubled the cost of the equipment necessary.

As a former TV station transmitter engineer, TV shop owner, microwave technician, and instructor in TV servicing, I am aware of the educational needs of a beginner in TV servicing study. If the student is willing to invest time and patience, this book, properly used, should train him to perform TV repairs successfully.

WARNING

Use great care when handling a picture tube. Rough handling may cause it to implode. Do not nick or scratch glass or subject it to undue pressure. Wear heavy gloves and safety goggles for protection. Discharge picture tube by shorting the anode to chassis ground. Avoid long exposure to unshielded areas of picture tube when the set is operating to avoid X-ray radiation. Avoid electrical shock when working on live circuits: use plastic tweezers or insulated screwdrivers to examine possible loose components; hold voltmeter test leads only by the insulated handles; and use only the special high voltage probe to measure voltages at horizontal output, high voltage rectifier, or picture tube anodes.

---

1. Audio Amplifiers

IN 1947, when much of our country found itself with a new home entertainment medium, little did we realize how much effect TV would have on our lives, our entertainment, and our pocketbooks. Radio had been a rather cheap means of communication, and most people who plunked down over a hundred dollars for their first TV fully expected the new mechanism to produce good pictures and sound for many a year before it had to go to the shop. Such was not the case. TV by its very makeup, with complicated, interdependent circuits, is subject to frequent breakdown. By now most citizens are aware of this fact and accept it. Your work as a technician who returns these ailing monsters to proper working order not only is a valuable service to others, but can be very profitable for you.

Because TV sets are so complicated, you must study the subject of repair carefully, without interference, and with as much practical, inside-the-set experience as you can get. The first six lessons or chapters in this book may seem to be radio centered. You will make radio sets for broadcast band and shortwave. Then you will make a low-power transmitter. All this has a very serious purpose. It gives you experience in working with transistors so you can find troubles in them and their associated parts. Later, when you study TV sets, you will find many of these parts, but you will be able to separate, in your mind, each transistor and its functions. Therefore, it is imperative that you build all the sets designated as pro-

jects and pay particular attention to each point stressed as you build so you will understand the follow-up TV circuits. Also, if you ever decide to have your own shop, some of the circuits can be used for inexpensive pieces of test equipment for TV repair.

In these beginning experiments you will use a 6-volt lantern battery that is not dangerous and has no shock hazard in itself. Later, as you explore the parts of a TV receiver, the equipment presents a considerable possibility of painful, though usually not serious, shock. It is important to learn good shop practices from the start to avoid shocks. Touch high-voltage circuits only with the testing probe of the meter, not with your hand or finger. Usually touching the metal of the chassis of a TV with the power on is safe if the set is operating normally and if you are seated or standing in a dry area. Special attention must be given to the high-voltage system of a TV set, and we will explore this later. The picture tube of a TV set has a very high vacuum drawn on it, and if broken will often produce a loud report and send glass flying in all directions. Always handle a picture tube with extreme care.

As you work through this book, refer to the list of parts needed for all constructions in Appendix A. You can buy them as needed or get the entire set at once to avoid delays. This overall plan of circuit building has been carefully planned so that it takes a very limited outfit of parts, and the building of one radio or other device goes a long way toward building the next.

For Chapter 1 you will need 2 audio transistors, 3 resistors, 3 capacitors, the 6-volt battery, some hookup wire, and a ground. You will also need an earphone or a set of two earphones with at least 1,000 ohms resistance. The 8-ohm earphones that come with a transistor radio will not work in these circuits, nor will crystal earphones. Be sure you have the type that are called 1,000 ohm for a single phone, or 2,000 ohm for a headset of two (one for each ear).

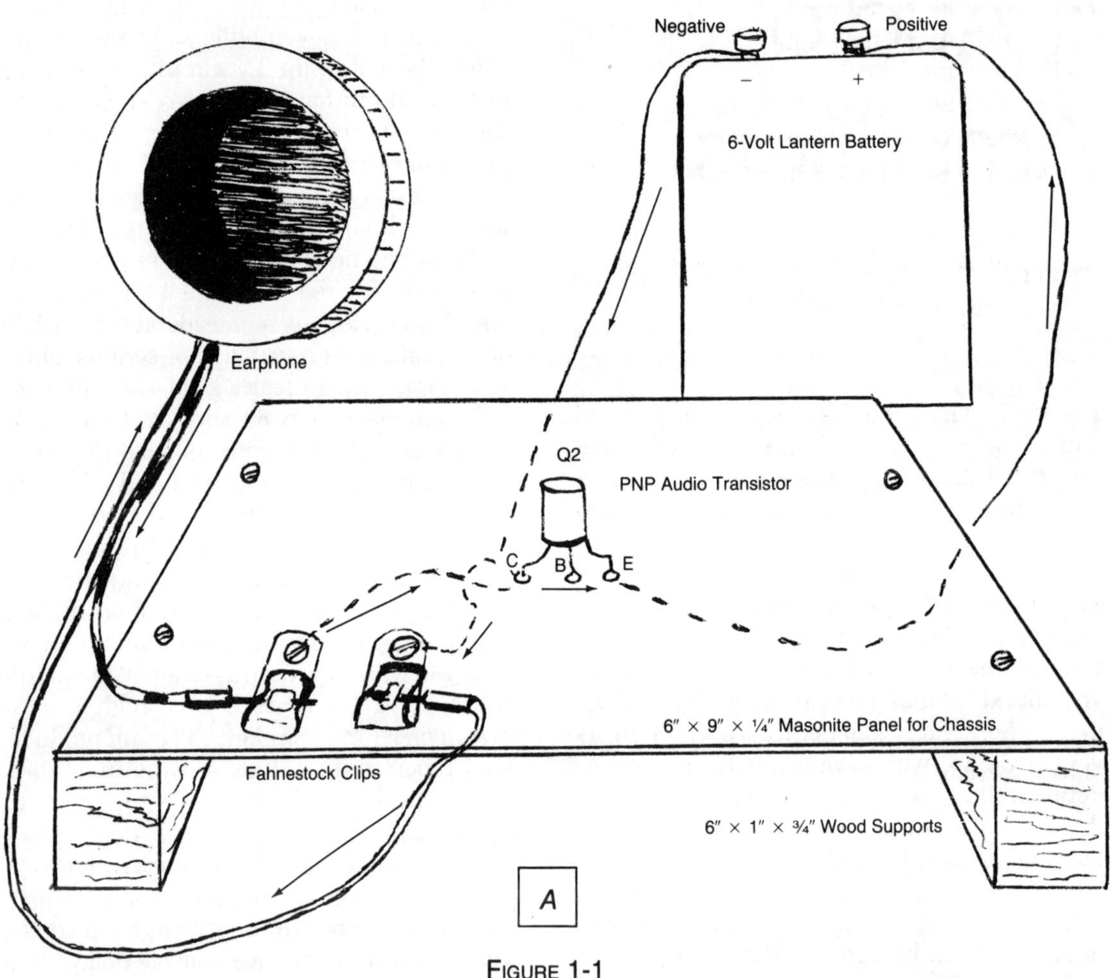

FIGURE 1-1

SERIES CIRCUITS

For your first hookup you will make the base that will be used in all the following radio circuits. Cut a piece of masonite 6 by 9 inches by ¼ inch thick for the chassis. Attach it with wood screws to two supports 6 by 1 inches by ¾ inch or wider if necessary. Consult Figure 1-1 A. Use a drill or icepick to make holes for the fahnestock clips that will attach the earphone to the rest of the circuit. These clips are secured near the left side of the chassis as shown. Each fahnestock clip is held in place with a metal screw inserted from above and secured with a nut below the chassis. Wires that are drawn with a broken line are below the chassis, and two wires are secured to the nuts below the fahnestock clips. This makes an electrical connection between the earphone and the rest of the amplifier or radio.

The audio transistor, Q2, is mounted by drilling three small holes in a triangular arrangement to accommodate its leads. An audio transistor is a kind of transistor that is used to amplify music or voice. It is the type that would be found in a guitar amplifier or in the part of a radio after the broadcast signal is removed. An audio transistor is designed only for pitches audible to the human ear. Parts of a radio operate at higher frequencies and require radio frequency transistors. Place the leads down into the holes very carefully, as these leads are quite easily broken.

In the next part of this chapter you will learn how a transistor amplifies, but for now you merely need to learn which part is which. Figure 1-1 *C* shows three types of transistors turned upside down with the leads coming out. Chances are you will use one like the one shown at the top. The little tab is your starting point. Some transistors don't have the tab, and if it is absent arrange the transistor so that all three leads are to left of center, and then you can begin. The bottom left lead is called the *emitter;* the next lead clockwise is the *base;* and the third is the *collector.*

Arrange your transistor on the chassis a bit to the left of center and so that the emitter will be to the right and the collector to the left. Twist connecting wires to the leads under the chassis and attach the collector side to the left fahnestock clip nut, and attach the emitter lead to a wire that will connect to the positive side of the battery. So far the base connects to nothing. A negative lead connects the battery to the other fahnestock clip (right clip). The transistor bottom lead digrams shown in Figure 1-1 *C* are usually correct, but some transistors have the collector in the middle or another arrangement.

It is best to look at a schematic of any TV set you attempt to repair and note the small drawings of the transistor leads nearby. See Figure 12-2 (small drawings near right center) and note that Q2 on that circuit has the emitter in the middle. Figure 12-8 also shows some unusual arrangements. Always check before you assume you know the leads.

Hookup wire usually has a plastic or rubber coating that must be removed at the ends where connections are made. Other places the coating, called *insulation,* is left on. Enameled wire that will be used for coils must be scraped clean at connecting places, but the enamel must stay on to separate turns of the coils except at connecting points. Illustrations sometimes show a little half-circular bend or loop in a wire (such as the one in the line from the negative battery to the right fahnestock clip). These are drawn only to show that the wires are not connected; it is not necessary to make the loop when you put the components together. Note the schematic drawing of Figure 1-1 *B.* This shows every electrical part of 1-1 *A* but does not show the chassis, clips, and so on.

Try as soon as possible to learn the schematic symbols for the parts, as these symbols are commonly used in TV servicing. Note the schematic parts of a transistor. The part with the arrow is the emitter. The line on a side by itself represents the base. The other line is the collector. Schematics don't always show the base at the left and the emitter at the bottom. Watch out for differences. The transistor we are using is a PNP type. Its arrow always points in toward the transistor. If the arrow points away from the transistor it is an NPN type (see Figure 1-2 *E*).

Be sure the negative side of the battery is connected to the right clip and the positive side of the battery to the emitter. A PNP transistor must always have the positive (+) battery in its emitter circuit, and the negative (−) battery in its collector circuit.

Now complete the circuit with the earphone by connecting the left side fahnestock clip with the earphone tip. Leave the right side connection off for now. Listen in the phone. Do you hear anything? Without the right side connected you can not hear any noise, as you do not have a complete circuit. If a circuit is broken anywhere, it will not work. Now touch the phone tip to the fahnestock clip. You should hear a click as the circuit is completed. This is a series circuit. In a series circuit all members must be working or none can work, and the same current must flow through every part of the circuit.

Disconnect the right fahnestock clip from the phone tip now and save your battery, and let's trace the circuit. Current always flows

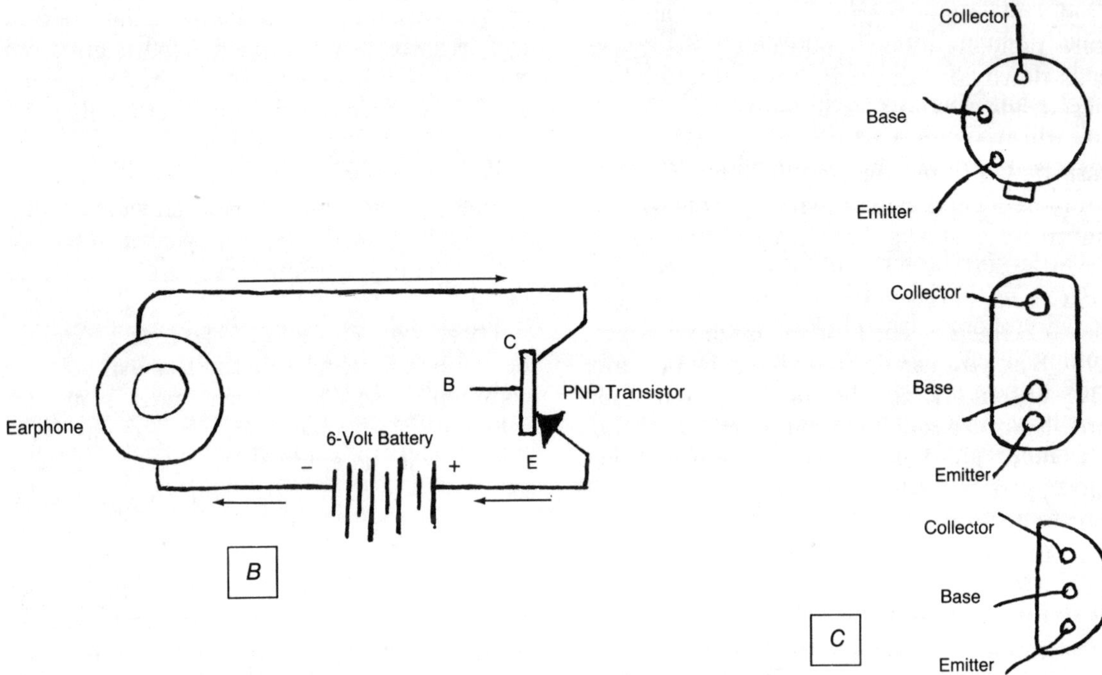

FIGURE 1-1 (*cont.*)

from negative to positive. It leaves the negative side of the battery and goes under the chassis on the wire up through the fahnestock ·clip and through the phone cord to the earphone. After traveling through the earphone it comes back and goes through the left fahnestock connection to the collector of the transistor, through the transistor, out the emitter, and back to the positive side of the battery. Connect the right tip and fahnestock clip connection and note the click again. Try disconnecting at the left clip and at each battery terminal, reconnecting before disconnecting at the next place. As you can observe, no matter where the circuit is broken, it is broken all over as far as current flow is concerned.

The voltage is supplied by the battery. Voltage is a measure of the amount of electricity for a circuit or within a circuit. Current is the flow of electricity through the circuit. Resistance is the force offered by circuit elements such as the earphone and transistor to keep the current from going too strongly and wearing out the battery in a short time. Voltage is measured in volts. Current is measured in amps, milliamps (a thousandth of an amp), or microamps (a millionth of an amp). An amp of current is a heavy amount, like that used by a starter motor in an automobile. Radios and TV sets use smaller currents, such as milliamps in the collector-to-emitter circuit and microamps in the base-to-emitter circuits in a PNP transistor.

Resistance is measured in ohms. The more ohms of resistance there are in a circuit the less current can flow in that circuit. The earphone changes pulses of electricity to sound waves, as we shall see later. The transistor takes weak pulses and produces strong pulses of current. This is called *amplification*. The transistor amplifies, or makes the signal stronger. The wires and connectors offer very little resistance. They merely carry the current between useful resistance objects in the circuit.

Another circuit element called a *resistor* offers resistance by piling up electrons. It is usually a long, thin cylinder with wire leads at each end. Figure 1-3 *A* shows three resistors in a circuit. They are labeled R1, R2, and R3. You will study these resistors fully later in this chapter. For now, though, remember that earphones and transistors offer resistance to a current flow in a series circuit.

The lantern battery supplies 6 volts of *direct current* to the circuit you just worked with. Direct current always flows in one direction, from negative battery, out into the circuit, and back to positive battery. Again note the current flow arrows. This is repetitious, but it is extremely important. Later you will study alternating current, which flows first in one direction through a circuit, stops, reverses itself, goes in the opposite direction, reverses itself, and goes back in the first direction, many times a second.

Alternating current is the type that is supplied to your house by the electric company. We label it *AC* for short. Batteries supply direct current or *DC*. Current is actually a movement of tiny electrical particles called *electrons* from atom to atom through wires or other electrical components. These tiny electrons are negative ($-$) charges of electricity, and they hop from atom to atom as long as there is a "push" of more electrons from behind to cause a current to flow.

Electrons readily jump from atom to atom in good conducting materials such as copper and aluminum; they move fairly well in iron but hardly at all in glass, rubber, cloth, or dry wood. The materials that pass them well are called *conductors*, and wires are usually made of good conducting material such as copper. Materials that prevent or restrain electron movement are called *insulators*. The chassis of your experimental workshop is made of masonite, which is a good insulator. Otherwise it would short out the emitter-to-collector part of the transistor where the wires go through the chassis. In a radio or TV with a metal chassis, all parts have to be fastened to terminal boards with insulators from the metal of the chassis.

The 6 volts that are applied to the circuit you made must be dropped across the circuit. Part of the voltage will be dropped across the earphone and part across the transistor, but both voltages must add up to 6 volts because that is what is supplied. A resistance always drops a voltage across it in proportion to its resistance. The more resistance, the more voltage will be dropped across it. Of course, the greater the resistance, the less the current that will flow in the circuit.

Assume that you have a single earphone of 1,000 ohms and that the transistor you have for Q2 offers 500 ohms from collector to emitter. The total resistance is thus 1,500 ohms. Since the earphone offers two-thirds of the total resistance it would develop two-thirds of the voltage across it. Why would more voltage pile up there? Since more resistance is offered, the tiny electrons are piling up in the earphone trying to fight their way through. Other electrons are pushing them from behind, but the resistance of the earphone is holding them back. Where electrons pile up they produce a voltage, because voltage is nothing more than a pileup of electrons at one place and a shortage of them at another. Where they enter the resistance—the earphone in this case—electrons are piled up in great numbers. The opposite side of the earphone has a shortage, as the electrons there have been able to flow on toward the transistor. In the circuit you have created you should be able to measure 4 volts across the earphone and 2 volts across the transistor.

If you like to fool around with math, you will find in Appendix F the formula of Ohm's law. In short, it states that if you divide the resistance of a circuit element into the voltage, you will get the current. For this whole circuit we have 1,500 ohms, because resistances add up in a series circuit. Now divide 1,500 into 6, the voltage, and you get 0.004 amps. Since this is 4/1,000 amps we call it 4 milliamps. That is the current flow in the entire circuit, for the current does not vary within a series circuit. True, where the electrons are piling up at the entrance to the earphone and at the collector of the transistor, there is a slowdown, but still 4 milliamps of current is pushing through there.

Suppose you are using a headset of two earphones. Usually these offer 2,000 ohms of resistance. Now suppose you are using a high-resistance transistor that offers 20,000 ohms between collector and emitter. In this case the transistor instead of the phones will offer most of the resistance. Altogether, 22,000 ohms is offered to the circuit. With 6 volts, the current flow would be approximately 0.000273 amps or 273 microamps. This is a rather small current, yet it is strong enough to produce a click in the earphones. Multiplying current times resistance gives voltage. The transistor offers 20,000 ohms and the current is always the same in any part of a series circuit, so 0.000273 times 20,000 gives about 5.46 volts dropped across the transistor. This leaves only about half a volt across the phones, since only 6 volts is supplied to all the circuit.

Now you will see what the voltages are in the circuit you have made. For this part, you will need a voltmeter. If you can borrow one it would be well to use it for these experiments, or if you are taking this as a school subject one will be furnished. In Chapter 8 you will learn how to make an effective high-impedance voltmeter-ohmmeter, and if you don't have access to one before then it might be well to repeat the exercises that require one.

Use a voltmeter in DC mode set on a range that will measure more than 6 volts. Reconnect your series circuit. First measure the voltage dropped across the earphone or earphones, whichever you have. Put the black lead of the meter $(-)$ to the right fahnestock clip, as it is the most negative point. Touch the red lead of the meter $(+)$ to the left fahnestock clip, as this is closer to the positive part of the battery. (It is closer electrically, although not necessarily physically.) What voltage do you read?

Now touch the black lead of the meter where you had the red one. Put the red lead to the positive battery terminal. This measures across the transistor. (Of course, you could have put the black lead at the collector and the red at the emitter, and in TV servicing that would be the way to measure this voltage.) Record your voltages; if all goes well they should add up to the 6 volts applied. Measure across the battery terminals with the red lead to $(+)$ and the black lead to $(-)$. You should get 6 volts. Minor variations may occur, because no meter is perfect, but in general the voltages across the earphone and transistor should equal the supply voltage.

Connect the voltmeter across the transistor. You can secure the negative (black) lead to the collector with a small alligator clip or, lacking this, connect to the left fahnestock clip. Connect the red lead of the meter either to the emitter if you can or to the positive battery terminal. This frees your hands to do other things. Disconnect the right fahnestock clip from the earphone tip. What happens to the voltage across the transistor? Since you broke the series circuit all current ceased to flow. This, in turn, ended the pileup of electrons, there is no voltage. With this earphone lead still off, measure across the battery again. Yes, there is voltage stored in the battery. No current has to flow for voltage to be in the battery.

If you have an ohmmeter you can measure the resistance across the earphone and transistor. Always measure resistances with the power source removed. In this case disconnect the battery. The ohmmeter supplies its own voltage with batteries. Any external voltage will damage the meter and possibly the circuit under test as well.

Put your meter on as high an ohm scale as it has. Low scales will damage the transistor, as they send too great a current into the circuit. R × 100 is a high setting. R × 1 is low. Set on R × 100 and remember to add two zeros to whatever number it reads. You should be able to measure the 1,000 or 2,000 ohms across the earphone. If the leads on the ohmmeter are placed one way from emitter to collector on the transistor, you will get a high reading. If they are reversed you will get a lower reading. Usually the lower reading is correct. Remember, these ohms are ohms of resistance. A transistor with less than 500 ohms (a lot less) from collector to emitter is shorted and useless. Certain high-power auto radio and CB transistors are exceptions, but usually a good transistor will have 500 ohms or more from collector to emitter.

By now you should have certain basic thoughts well in mind:

1. Direct current, DC, travels from negative supply source, battery or other, through the circuit to positive supply source. It always travels in one direction.

2. A series circuit is arranged so that all parts are continuous one after another.

3. Current is the same value in amps or milliamps in all parts of a series circuit.

4. Any break in the circuit stops current flow in all parts of the circuit, and voltages cease except at the source. (This is true of series circuits only.)

5. Every circuit element offers some resistance to current flow. The larger the resistance, the less the current that can flow.

6. In series circuits all resistances add up to produce the total resistance to current in the circuit.

7. Across each resistance will be voltage drop in proportion to the amount of resistance, more voltage drops for more resistance.

8. Voltages dropped across circuit elements will add up to the supply voltage.

9. Insulators prevent electric current flow. Conductors carry current flow.

10. Transistors are fragile at the leads. Large currents also can damage them.

How a Transistor Works

By now you should know this information and how to read the leads of transistors and how to read a schematic of the parts you have studied thus far. It does not hurt, though it seems to, to review the material, and it might be well to do so before continuing.

Although you have used a transistor as a resistive unit we have not yet explored a transistor's internal workings. Consult Figure 1-2 throughout this discussion. Figure 1-2 *A* is a greatly enlarged cutaway view of a typical transistor. Most of the material inside the plastic or metal shell is a soft jell that helps keep the device from overheating. In the middle are the actual working elements, and usually they are smaller than shown. First we will consider a PNP transistor like the one you have worked with.

The terms *emitter, base,* and *collector* are very unfortunate. The emitter, for example, does not emit or give off anything. With a PNP transistor the collector doesn't collect anything either. Still, we are stuck with the terms. All three elements are small pieces of germanium, silicon, or some other material that is neither a good insulator nor a good conductor of electricity. The material must carry a current, but not too well. Charges are placed in these elements by doping them with impurities. In a PNP transistor the emitter is made positive, the base negative, and the collector positive. Get it? PNP.

From the cutaway view you can see that the emitter and collector are reasonably thick. The base is the very thin wafer between them. Any current that goes from collector to emitter must go through this base. Now the next statement is one you must remember, as it applies to everything in television or other electronics: *Unlike (meaning "opposite") charges of electricity attract; current will flow from a negative ($-$) charge to a positive ($+$) charge.* It will never flow the opposite way. *Like charges of electricity repel.* Put a ($-$) to a ($-$) or a ($+$) to a ($+$) and nothing happens except that they try to get away from each other. *Current will not flow between them.*

Current enters the collector. It can go in because a negative source is always offered to the collector. Negative current flows easily into a positively doped collector, but it stops there because positive can't flow to the negative base and is repelled by it (like charges repel). A blockage results. This is the resistance of the transistor discussed earlier. Obviously some current breaks the barrier of the thin base and goes on to the emitter, or we would not have had a complete circuit in the earlier experiments. This current is small, however.

A current could be put in at the base, negative of course, and it would flow easily out the emitter and to its other side of source. Current can enter a negative base and go easily to a positive emitter. Such a current is often very small, such as that from a radio station. See Figure 1-2, *B* and *D*. This would not be a good radio, but it could work with a very strong signal, if you had a long antenna and were a few blocks from the station.

The light arrows show direction of current flow. The heavy arrows indicate larger currents such as milliamps. The light arrows indicate small currents in microamps, such as would come through the air from the radio station. The arrow on the transistor is opposite current flow. This is another thing we are stuck with in electronics. In the early days of radio it was believed that current flowed from positive to negative. Thus, the arrows showed current as the scientists then thought it flowed. Now, with the electron theory, it is believed that current flows from negative to positive, and tubes, transistors, meters, and so on all prove this to be true. Since the arrows in circuits have been drawn showing current going from positive battery toward negative since 1920, however, we are stuck with this incorrect representation. Within a circuit, always assume that current goes into the head of the arrow rather than in the direction the arrow points.

The parts within the dashed vertical lines may seem strange to you. Do not worry about them yet. They represent the antenna, ground, and tuning mechanism. A radio station current, feeble though it is, is going up

and into the base of the transistor in Figure 1-2 *B*. Since base to emitter represents negative to positive the current flows freely. This little current causes a breakaway in the barrier voltage at the base. You might think of it as somewhat like a vacuum. As the current goes by and down, it opens the way for current to jump across from collector to emitter easily. Since a battery of several volts is applied across the circuit, a large current flows from the battery, through the earphone, through the collector to emitter and back to the battery. This will happen as long as the little base-to-emitter current from the radio station is flowing upward, in down the base to emitter, and back to the tuned circuit. Since the earphone is in series with collector to emitter current it makes the earphone work and music is produced. If louder music is to play at one place, more current will come in at the antenna tuner system from the statin.

But now the station is between notes or at the end of the record (Figure 1-2 *D*). The station current reverses itself. It tries to go from emitter to base, but that would be positive to negative, so the current can't flow in the transistor. Since this little current quit breaking down the barrier at the base, the collector current no longer goes to the emitter in any great amount, so the earphone is silent. It would click if you attached and unattached a battery as in the early experiments, but you would not hear any music at this point, as very little current would flow through the earphone. You can see that a small base-to-emitter current allows a large collector-to-emitter current to flow. Since any changes in volume, pitch, quality, and so on in the tiny current affect the large current through the earphone directly in proportion, the transistor amplifies the signal. Even cheap transistors have a current amplification factor of 20, and some even amplify the current 100 times. Compare this with early radio tubes that had an amplification of 8 and were considered very good in their time.

NPN transistors are connected just the opposite of PNP types in terms of the supply voltage. NPN types are used a great deal in TV circuits, but it is quite common to have a mixture of both types in the same part of the TV set, as you will see much later.

Note Figure 1-2 *E*. You might say that we put an NPN transistor in place of the PNP and turned the battery around. The collector of an NPN transistor must have a positive voltage supply and the emitter must have a negative supply. The arrow points away from the transistor to indicate that it is NPN. This is the emitter arrow, of course. The emitter-to-base current from the radio station or wherever must now flow upward from emitter to base rather than the opposite, as it did with a PNP type. When the small current is flowing properly a large current is able to go from emitter to collector and out through the earphone and to the positive battery. In this case it would be more appropriate to think of the small current opening the barrier between base and collector as a slopover rather than a vacuum type of opening. As the current goes from emitter to base, some electrons jump over into the collector area by speed alone, not by the charges. This opens the barrier as long as the condition continues. This slopover will vary with the amount of emitter-to-base current and with its pitch, quality, and so on.

In either a PNP or an NPN transistor, the small base-emitter current affects the larger emitter-collector current. Though the directions of flow are opposite, the same general idea applies. A small input current affects a much larger output current, and the radio signal is amplified. Learn the symbols for PNP, NPN, and field effect transistors. See Figure 1-2 *C*.

The field effect transistor is different from others in that it does not need a base-emitter current in order to operate. This base-emitter current draws energy from the tuner and tends to load it, making sharp tuning almost impossible. The field effect transistor just has a voltage of the input at its gate (G), which either lets current flow from source (S) to drain (D) or blocks it. This is more like a tube, as you will see later.

Now we must consider bias in transistors.

Bias is a means of affecting an electrical device through currents or voltages so that it behaves differently from the way it would without the bias. With tubes the bias is a voltage, and it keeps too much electron flow from taking place within the tube. With transistors bias is a current, and it causes more flow of electrons (current) from emitter to collector or collector to emitter as the case may be. It was found early in the development of transistors that it was a good idea to have a small but continuous current through the emitter-base circuit whether signals are coming in or not. For one thing it keeps the barrier slightly open at all times, and greater amplification of an incoming signal takes place. Second, without bias the transistor would cut off parts of every signal. For a PNP transistor, every time the signal became less than greatly negative the barrier would hold and parts of the speech and music would be garbled. This is especially true for strong signals.

The bias for transistors is called *forward bias*, because it sends a current that increases the collector current. At first an extra battery was used in the emitter-base circuit, but later it was found that a resistor could go from base to negative battery (for PNP type) and this would make for a small, constant bias current from base to emitter. The problem is that if too large a bias current flows, the complete breakdown of the barrier will cause enough current in the collector-to-emitter circuit to burn out the transistor. Any short between the collector and base will do this. Often the transistor will heat too hot to handle for a moment, then it will be done forever. Be very careful when touching the leads with a voltmeter probe. A slip that touches the base and collector together (and they are very close together to begin with) will likely ruin the transistor. Also, a short from base to negative supply in a PNP transistor will probably ruin it. With an NPN type a short from base to positive battery is equally disastrous.

A large resistor is thus needed, or too much bias current will go through. Actually for most transistors a bias resistor of 47,000 ohms or larger will not injure the device. For the best balance of good sound, minimal noise, and smooth operation, however, a larger value is desirable. Sometimes a double resistor is used. One goes to negative battery with a large ohm value and another goes from the base to positive battery with less resistance. This is known as a *voltage divider* type bias. It works well as a battery wears down. Still, the single resistor bias is often used.

In a few circuits the resistor goes between base and collector, requiring the bias current to travel through the earphone before going to the base and emitter. Such a system causes *degenerative feedback*. Degenerative feedback bias lowers volume compared to bias directly to negative battery, but gives greater volume than no bias at all. Its advantage is that this degenerative feedback cancels some noise, and extra-loud parts of the signal are subdued. The overall effect is to give greater quality to the sound.

We will now bias our lab circuit. Refer to Figure 1-2 *F* and *G*. Find R3, which is a 330,000-ohm resistor. We will deal with color code and its meaning later. For now, find the resistor with the colors orange/orange/yellow. Start counting from the end where the color bands are nearest. First must be orange, next orange, third yellow. A silver or gold color later doesn't matter for now. Figure 1-2 *F* shows how it is put into the circuit schematically. Figure 1-3 *A* shows how it fits in physically, so you can allow space for R2 and R1. Partial drawing Figure 1-2 *G* shows just this bias resistor in place.

Connect as shown, and if you have a voltmeter measure the voltage across the transistor, Q2. It should be less than before. Why? You have established a current from (−) battery through R3 to the base, out the emitter, to (+) battery. This breaks the barrier down somewhat, so more current can go across the transistor from collector to emitter. You have, then, lowered the resistance of the transistor. With less resistance and less voltage across the transistor, but the same applied voltage,

where is the extra voltage? Try measuring the earphone voltage. It is now higher. Will the clicks be louder if you break and remake the circuit at the battery? Try it. The clicks should be noticeably louder.

Now we will try a hum test to further demonstrate that a forward bias increases the volume a transistor develops of a signal. With all parts connected, touch the base of Q2 or the junction of R3 and the base with one finger. You should hear a slight hum in the earphones (or earphone). If you hear no hum, you live in a low-signal area, and it will be necessary to add a ground to the circuit to get this hum. You will do this in the next experiment. If you do get a hum, temporarily disconnect the fahnestock end of R3. Leave the battery connected to the fahnestock clip. You are now simulating a bad bias resistor. Try the hum test again. You should hear much less hum or none at all.

This test illustrates the need for forward bias. Also if some stage of a radio or TV is weak, the problem could be a bad bias circuit. Restore the resistor to its connection to the battery.

By now you should have learned several more important facts of electronics. Disconnect the battery as you review:

1. Transistors are amplifying devices.

2. A transistor has an emitter and a collector separated by a very thin base.

3. For a PNP transistor the collector must have a negative voltage source.

4. For an NPN transistor the collector must have a positive voltage source.

5. A small current from base to emitter in a PNP transistor causes a much larger current in direct proportion to flow from collector to emitter.

6. A small current from emitter to base in an NPN transistor causes a much larger current in direct proportion to flow from emitter to collector.

7. A field effect transistor has no gate current, but a voltage affects current flow from source to drain.

A. Enlarged Cutaway View of PNP Transistor
Base
Collector
Emitter
Emitter Lead
Base Lead
Collector Lead
B. PNP Transistor, Forward Current Condition in Input
C
B
E
D. PNP Transistor, Reverse Current Condition in Input
C
B
E
C. Transistor Type Symbols
B
C
E
PNP
B
C
E
NPN
G
D
S
Field Effect

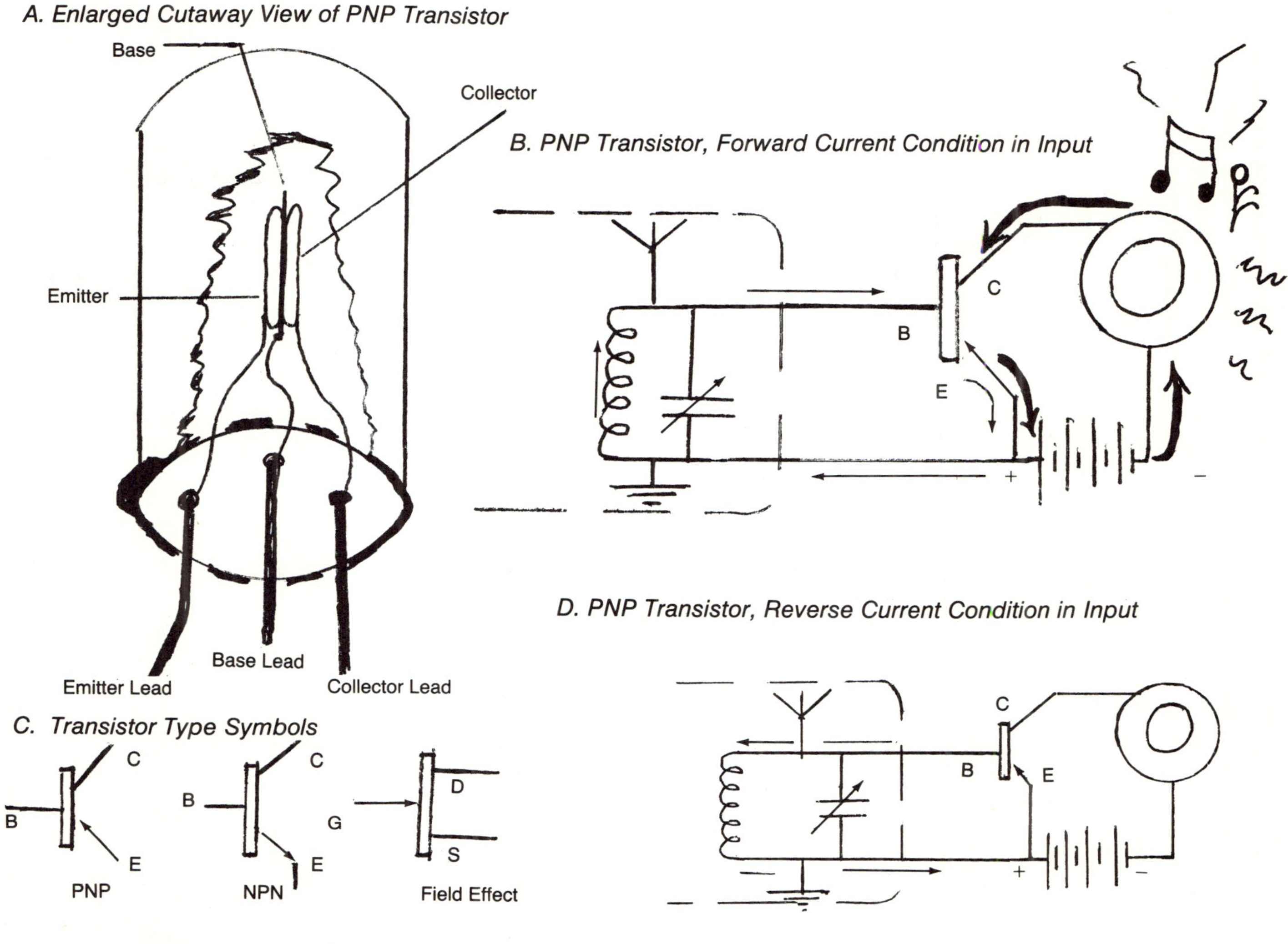

FIGURE 1-2

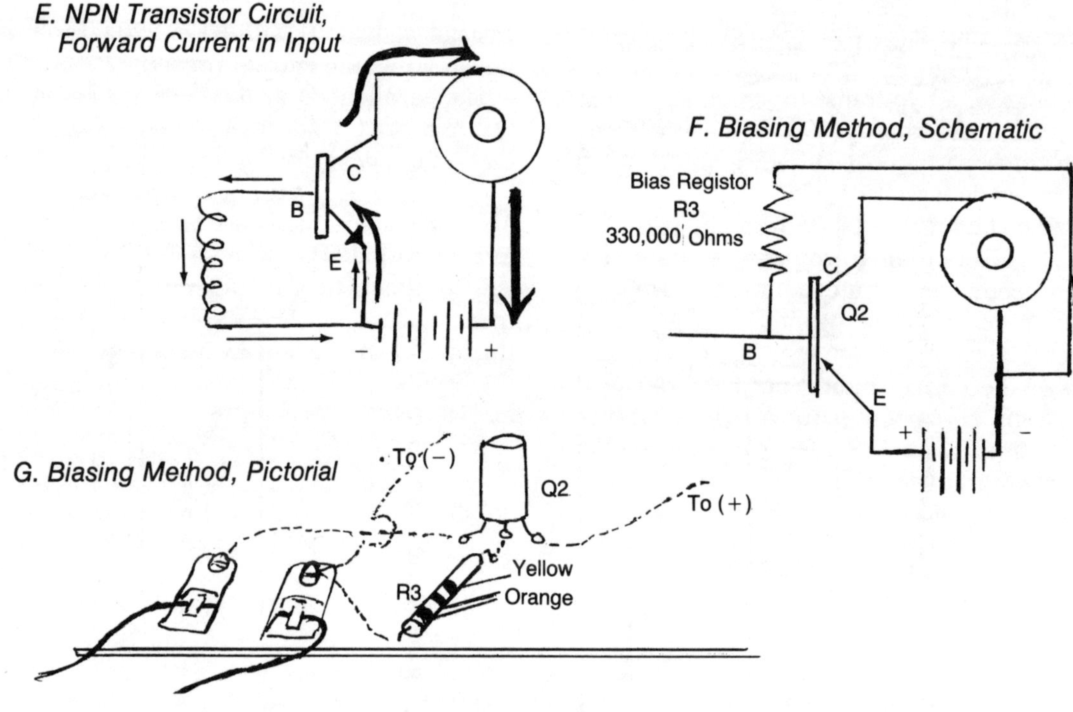

FIGURE 1-2 (*cont.*)

8. Forward bias current in a transistor (NPN or PNP) increases collector and emitter current flow, and increases amplification of the device. It also prevents garbles and other distortion.

TWO-TRANSISTOR AMPLIFIER

After you learn these principles and the symbols for three types of transistors, you are ready to construct a two-transistor amplifier.

Early in the days of radio, technicians discovered how to take the output of one tube and feed it to another tube before it went to the earphones, producing a much stronger signal for the phones. One tube or transistor can be connected to another one for the purpose of greater amplification in several ways. Though the same principle will work with either tubes or transistors, different values of parts must be used. Transformer coupling can be used in either case, but transformer coupling has certain drawbacks. First, it is rather expensive coupling, as most transformers are costly. Next, in audio amplifiers some pitches or frequencies are amplified more than others, so certain notes might not be heard well. And transformer coupling usually takes considerable space. Though transformers are necessary in certain places—usually between transistors operating the same way, such as two audio transistors—transformer coupling is no longer used.

Direct coupling is sometimes used. In this case the collector of one transistor feeds directly to the base of the next. The trouble is that some arrangement must be made to keep too heavy a voltage from the base of the last transistor, and yet have enough voltage on the collector of the first transistor, so the first transistor will operate efficiently.

Resistance-capacitance coupling, often labeled *RC coupling*, is usually employed. It is cheap and efficient; it is not frequency-sensitive; and it can fit in a small space if necessary. Figure 1-3 is an example of RC coupling for a two-transistor amplifier, which you will make shortly. You will use it along with other parts to make radios and a transmitter, so do it well.

Radio station signals, or even music from a record player or tape recorder, does not travel in a smooth, steady note, but in pulses in time with the voice or other sound. As these signals enter a base-to-emitter circuit in a transistor, they make corresponding pulses of voice in the form of current surges. These surges cause similar, much larger, surges or pulses of current in the collector circuit, as you have seen before. These pulses dropped pulsating voltages across the earphone resistance. Instead of the earphone you can put a resistor at the same point and drop the voltage of pulses across the resistor. The signal must be sent to the next transistor quickly or it will subside and be lost. Q2 now has a 1,000-ohm resistor to replace the earphone. This is R2 in both schematic and pictorial drawings in Figure 1-3. To get the signal over to Q1, the other transistor, you use a capacitor, C2.

A capacitor is a device that passes pulses of alternating current, but blocks direct current. It is often a small disc with pieces of metal inside that come very close together but do not touch. The negative charges, or *electrons*, on one side can push or repel the electrons on the other side away from the metal on that side. If you connected a capacitor to a battery, one push would result and then no more current would flow. But when many pulses occur, such as in music or voice, these all push across the capacitor and on to the next transistor or other component. Between pulses there would be a backward pulse to clear the capacitor, but the next transistor would not be sensitive to this. So the pulses can hop the capacitor forever. The smooth direct current is blocked. Thus we do not put too much current into the base of Q1. Just 1,000 ohms would allow too much current.

Instead we bias Q1 with R1 or 220,000 ohms, which limits the current properly.

The earphone is in the collector circuit of Q1. Why is C1 necessary? Some noise and high-frequency pulses build up in a double or triple amplifier. These are bypassed by the earphone so they don't set up a howl and distort the sound. The purpose of C3 is to allow you to attach this amplifier to other equipment that might have a voltage too high for the base of Q2. C3 lets the pulses of electricity representing the music or voice through, but blocks any direct current.

R3 was covered before when you built the one-transistor amplifier. Q1 takes the already amplified output of Q2 and amplifies it further. If Q2 amplifies the signal 20 times and Q1 does as well, the result should be 20 × 20 or an amplification of 400. The trouble is that there is some loss in the circuit. Not all of the pulse that develops across R2 goes through C2 to Q1. Still, two transistors hooked as shown develop a far greater signal than one alone could.

Before you build this circuit it is important to learn about the values of resistors and capacitors. You will be using them from now on, so learn them well. In Appendix E you will find a chart for reading resistor and capacitor color code. It would be well to memorize it now. Many capacitors, such as those shown in Figure 1-3 *A*, have the capacitance shown on their side. These are usually larger values of capacitance such as 0.01, 0.02, 0.05, and 0.1 microfarads. Sometimes their value is smaller, such as 0.001, 0.005, and so on. *Microfarad* means a millionth of a farad. A farad would require a capacitor so large it could not fit into a TV set, or perhaps even into a room, unless it were of the electrolytic type.

The microfarad capacitors are used in power supplies of TV sets. Often 40-microfarad and even up to 1,000-microfarad electrolytic capacitors are used in such cases. Then the capacitors that couple a signal from one transistor to the next or one tube to the next are often less than a microfarad, such as 0.01 microfarad. There is even a smaller type that allows a very small part of the pulses to go from one place to another. These are the *picafarad capacitors*. A picafarad is a millionth of a millionth of a farad. These capacitors are used in tuning sections of radios and television sets. Many capacitors have their value written on them. If they do not, use the color code to determine their value. The rectangular ceramic capacitors usually are color coded.

Resistors are usually color coded, though a few have their value in ohms stamped on them. As you gain experience with TV you will spot a brown/black/red resistor and know immediately that it offers 1,000 ohms, and that a yellow/violet/yellow offers 470,000 ohms. A brown/black/green resistor offers a million ohms and is often called a *megohm* resistor. A brown/black/yellow is a 100,000 ohm resistor. These are very common values of resistors. A red/red/red is 2,200 ohms; a yellow/violet/brown is 470 ohms; an orange/orange/black or orange/orange/(nothing) is a 33-ohm resistor. Again, look at the color code chart in Appendix E if you have not already done so and learn it. This is a must for anyone interested in radio, TV, or other electronics whether for hobby or business reasons.

Resistors of less than 1,000 ohms are often used in the cathode of tubes to ground circuit or between the emitter and ground or battery with transistors. Resistors of from 1,000 ohms to 5,000 ohms are usually used as collector loads to couple between stages. For tubes, however, often the load resistor is 100,000 ohms. The load resistor in Figure 1-3 is R2. Bias resistors will usually be from 47,000 to 500,000 ohms. For tube circuits the grid bias or return resistor will often be a megohm or more. Although you have not studied tubes yet, their requirements are noted now so you can see that transistors take lower-resistance resistors than tubes, as transistors operate on low voltages and need a reasonable current flow in both base and collector circuits to operate properly.

Resistors have not only an ohm rating, but a wattage rating as well. The larger the diameter of a resistor, the more wattage it will handle. The number of ohms of resistance it offers is determined by the material of which it is made, not its size. With the transistor circuits we are now working, wattage is hardly critical. Only a few milliamps of current flow. Wattage equals current times voltage. If 0.003 amps, also called 3 milliamps, flowed through a resistor and a full 6 volts were dropped across it, the wattage would be found by multiplying the current in amps (.003) times the voltage (6) to get 0.018 watts. Resistors are not made with that small a wattage, so any resistor of the right ohms will do. Such is not the case with power transistors or tubes.

Suppose you had a resistor in a circuit that carried 0.003 amps but had 200 volts dropped across it. The wattage is 0.003 times 200, or 0.6 watts—over one-half watt. A one-quarter-watt resistor, even if it had the right ohms, would sweat, get hot, maybe change to a higher or lower resistance value, or burn out completely. Always replace a resistor with one as large in diameter or larger, and of the same ohms, as the original.

Capacitors are rated in microfarads or picafarads, as you have learned, but they also have a voltage rating, which may be stamped on the capacitor. Often this rating is greater than the voltage to which the capacitor is subject, for heavy pulses of voltage surges might occur from time to time. In Figure 1-3, C2 should have a rating of at least 6 volts even though R2 takes part of the voltage. C3 is a different matter. Since you might use this amplifier to check out tube circuits, it might be good for C3 to have a rating of several hundred volts. If the rating is less than the voltage that can get to one side of it, the capacitor may break down and short internally, and then it is not more than a solid wire. Capacitors can also open and not make any signal contact. If a capacitor shorts you can measure the applied DC to one side on the other side. If a capacitor opens you can't get a signal across it. Resistors sel-

Earphone, 1,000-Ohm Type
Negative(−)
Positive(+)
6-Volt Battery
Q1
Q2
C
B
E
C2
.01
C
B
E
C3
.01
.01
C1
R1
R2
R3
Grounding Wire
Testing Wire
A. Pictorial
R3
330,000 Ohms
(Orange, Orange, Yellow)
R2
1,000 Ohms*
R1   220,000 Ohms
(Red, Red, Yellow)
Testing Wire
C3
.01 Mfd.
C
B
E
Q2 PNP Trx.
C2
.01 Mfd.
C
B
E
Q1 PNP
Earphone
C   1
.01
*(Brown, Black, Red)
+ (Positive) 6 Volts
− (Negative)
Ground
B. Schematic

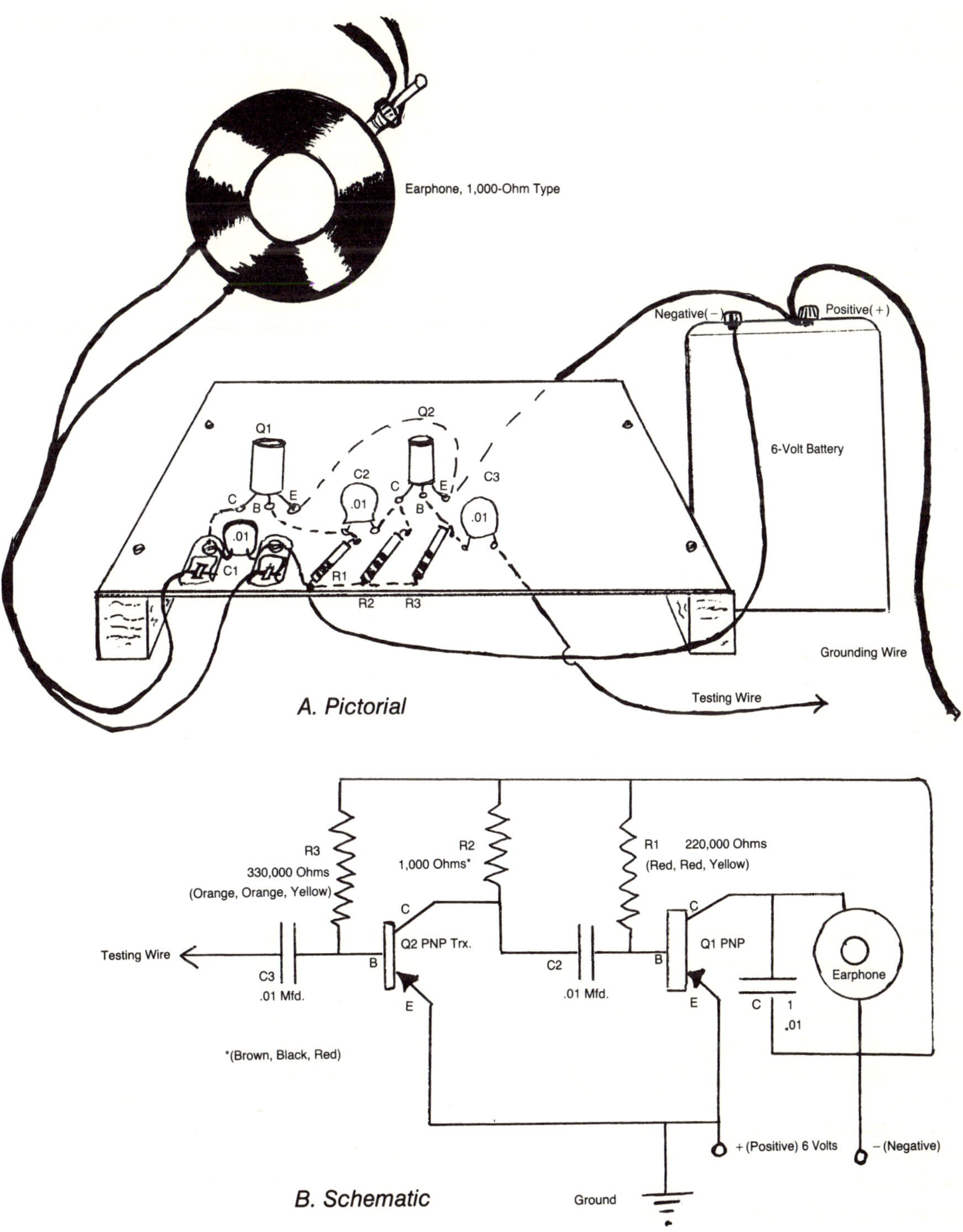

FIGURE 1-3

dom short, but they may change value or open, or one end may lose its contact.

Now we are ready to change the one-transistor amplifier to a two-transistor amplifier. Using the pictorial drawing of Figure 1-3, note that Q1 is mounted on the left part of the chassis board. You will need to drill or push holes for it, two more resistors, and three capacitors. C1 could be hung between the fahnestock clips above the chassis, or you can make two holes and hang it with the leads below the chassis. For now just twist wires together, making sure you have clean connections. Later you will solder them after you have made all changes. After you have C1 mounted, put Q1 in its place. If its lead will not reach the left fahnestock clip, make a short splice of wire. If the wire is insulated you must free each end. If the wire is enameled, scrape it all around. If it is plastic coated, just remove the insulation with a knife or pliers.

Both emitters are tied together with a lead to go to the positive battery terminal. Since you already had a lead to the emitter of Q2, just a splice across to the emitter of Q1 will complete the circuit. Do not attach the battery yet on either side. You will also need to ground the circuit at this point. Twist a long lead—long enough to reach a cold water pipe in your house—to the wire that goes to (+) battery. Scrape the cold water pipe until it is clean and wrap bare wire of the ground wire around it several times. You can also ground this circuit by taking the ground wire out the window and burying about 3 feet of it with the insulation removed for the entire 3 feet. To be effective, this should go down to damp soil.

R3 should be connected. Put R2 in its holes. R2 is a 1,000-ohm resistor and its color code is brown/black/red. Connect the side of R2 nearest you to the lead from the right fahnestock clip that you used to connect R3 before. Now connect the negative battery side of R1. Put R1 in the holes you have made for it. R1 is 220,000 ohms, or red/red/yellow. Connect it to the wire to the right fahnestock clip also. Now

all three resistors have their connections to (−) battery. Of course, you don't have the battery leads on, but you will attach them when all connections are complete. This is good construction practice, and it is necessary when you start working with high voltages. In addition, you might slip and touch a screwdriver to base and collector of a transistor and ruin it if the battery were connected.

The left side of C2 connects to base of Q1 and to the unconnected side of R1. Twist these wires together under the chassis. Be especially careful with the transistor wire. The right side of the C2 lead fastens to the collector of Q2. This is where the lead to the earphone was before. This earphone lead must be removed. R2 is also attached at the same place. About three good twists for each connection are sufficient for now. Let the remaining wires hang down, but see that they don't short to other connections. You have just put in the interstage coupling that you read about earlier. R2 develops voltage drops in accordance with the signal output of Q2. C2 allows these to go as pulses across to the base of Q1 and down through its emitter.

We have the bias for Q2 already in, but we need to attach the input coupling capacitor to the base of Q2. Connect the left side of C3 to the base and R3. The other (right) side of C3 needs a long lead, preferably insulated except at both ends. With it you will put signals into the amplifier. Check your wiring carefully. It should look like Figure 1-3 *A* and be electrically like 1-3 *B*. You now have a two-transistor, RC-coupled audio amplifier. Both transistors are forward biased. Now put it to work.

The human body is a fair antenna. It picks up all kinds of electrical energy in the air—the hum of household AC in the walls, radio, TV, shortwave, CB, and other signals. When you touched the base of Q2 in the previous experiment, you were injecting some of this radio and other energy into the amplifier. Since the human body doesn't separate stations, all the energy goes together and makes a hum. Now, with a ground wire and another stage, the hum should be much louder. Put the earphone on and connect the battery. You may hear a slight hiss of current through the transistors, or you may not, depending on the transistors used. Now hold the far end of the testing wire in your hand. You should get a much stronger hum than before. Disconnect the ground wire at the positive terminal. The hum should decrease in intensity. Without a ground the radio waves don't have so good a path. Many radios would work louder if they had grounds, but grounding would be inconvenient, especially with small portables. Electric radios do ground through the power lines.

Your amplifier can be used as an audio signal tracer. With the ground still removed, attach a lead in its place to ground to objects. An alligator clip on the end would be useful, but you can hold the wire to objects if you don't have one. Open your pocket transistor radio and attach the positive lead from your amplifier battery to positive battery of the radio. Now with the probe, or testing wire, touch the center side terminal of the volume control on the radio. If you have a station playing you should also hear it in the earphone. Now try touching several bases of transistors in the radio, remembering to touch only the base, not the collector too. Some will play through the amplifier and some will not. If you are in the audio section of the radio it will probably play, but in other sections the signal is not in the hearable or audio range and it will not work the amplifier. In some cases you may need to change the ground side to the negative battery on the radio end. No problem, as C3 on the amplifier will keep the battery currents apart.

Temporarily unsolder the tone arm leads of your record player where they come down into its amplifier. Write down or draw their arrangement so you can restore the record player to service. Now connect the (+) lead from the battery of your amplifier to one lead of the tone arm. Connect the testing wire to the other lead. You should be able to hear records in the earphone now, even though they might be too loud or badly distorted. Later you will learn about volume controls, and a control could be inserted in such a circuit to clear up the tone. If your pickup is not terribly sensitive it might play records clearly. Restore the record player to normal use.

The important thing to understand is that this amplifier could be used to see whether or not other devices are working, or to see where they work and what bad part makes them not work all their circuits. For instance, if you had a bad guitar amplifier, you could start at the input jack and listen for the guitar as you strum it. If you got music, you would then test at the collector of the first transistor. If the signal were louder there, as it should be, you would go on to the next transistor, and so on, until you found a place where there was no music. At the point where no pickup is heard, back up to the base. If you get music there, but not at the collector, possibly the transistor is defective. You would need to take several voltage tests to find out. At any rate you can see how this amplifier can be used to find bad parts of audio equipment.

Let's summarize a few more important points that you should have absorbed by now:

1. Coupling between stages of an amplifier takes the signal that one tube or transistor has amplified and in turn feeds it to another tube or transistor for more amplification.

2. Transformer coupling between stages is expensive, bulky, and doesn't respond to all frequencies of audio pitch equally, but is used in certain applications.

3. Direct coupling requires special voltages to prevent damage to the second transistor.

4. RC (resistance-capacitance) coupling is inexpensive and takes little space.

5. A capacitor is a device containing plates that are close but don't touch. It passes pulses of DC but not smooth DC. It passes AC and signals. It blocks battery DC.

6. Capacitors are rated in microfarads (the millionth part of a farad) for coupling and bypass.

7. Capacitors are also rated in picafarads (the millionth part of the millionth part of a farad) for tuning circuits.

8. Capacitors have a voltage rating that must be considered.

9. Resistors are not only rated in ohms, but have a wattage rating that must be followed. Size determines wattage.

10. Capacitors can short and allow DC to pass them easily.

11. Capacitors can open and block all signal as well as DC.

12. Resistors change value or come loose from connections.

Be sure you know all this, plus the color code for resistors and capacitors, and all symbols used in schematics so far.

Tube Amplifiers

Now that you have worked with transistors, resistors, capacitors, earphones, and batteries, let's turn our attention to tubes. Although tubes are not used in radios and TV sets as they once were, they still are used in the horizontal and vertical output of quite a few TV sets, and the actual picture tube on which pictures appear is a tube, to be sure. In addition, many older TV sets are still being used, and quite a few people will probably have these sets for many years and will want them repaired as needed. It is imperative that you know about tubes as well as solid state, transistor equipment. If you decide on a career in broadcasting you will find that certain high-power circuits in high-powered transmitters must use tubes. The transistor is cheaper and more efficient for the voltage applied, but when you need to develop 50,000 watts of power, tubes are the only answer as of now.

Use Figure 1-4 throughout this discussion. The simplest type of tube that will amplify is the *triode*. A triode is a tube with only one grid (see drawing). A two-grid tube is a *tetrode*, and a three-grid tube is a *pentode*. Other tubes have many grids so two or more signals can be put into them at once. A *diode* has no grid at all and is used to detect or to rectify.

Much of this probably makes no sense to you yet, but it will later.

All tubes, even those that have metal on the outside, have a glass envelope that keeps air out. Tubes must work in a vacuum to be effective, and often they are called *vacuum tubes*. The innermost part of the tube is called the *heater* or *heaters*. A small voltage is applied to the heater terminals, quite often 6.3 volts of AC. This is the only part of an amplifying tube that can use AC. Since the purpose is only to heat, it does not put an AC signal out into the tube. If it did, a hum would result. If the heater ever shorts to the cathode, this hum will be in the tube. In a TV set a hum might occur in the speaker, or if the short is in the picture section, it would make wavy lines in the picture. The voltage applied to the heaters is not always 6.3 volts, but it is for tubes such as 6SK7, 6V6, 6CB6, 6J5, 6AU8, and many others. (The first number of the tube gives its approximate heater requirement in volts. A 12SA7 requires 12.6 volts for its heaters, and a 50C5 needs 50 volts.)

This heating makes the cathode quite hot. The cathode, a cylinder that surrounds the heaters, has a coating of material such as thorated tungsten that will give off electrons when hot. Remember that electrons are negative charges of electricity. Old-style tubes have no cathode, and the coating is directly on the heater. Such a coated heater is called a *filament*. With early radios the filament was heated by battery so no hum was put into the radio. In later sets the filament was heated by AC, and many early radios hummed to beat the dickens. Of course they were fitted with bypass capacitors and so on to get rid of as much as possible. In modern tubes the heater and cathode are separate.

The plate of the tube is the outermost cylinder in the glass envelope. It is connected to a large positive DC voltage, often as much as 100 volts or more. Since the electrons given off by the cathode are negative charges of electricity, and since they are attracted to a positive charge, they travel to the plate by the

Glass Envelope
Plate
Grid
Cathode
Heaters
5
6
3
7 8
A. Pictorial and Schematic of a Tube
5
6
7
8
4
5
3
6
2
7
1
8
9
Widest gap
4
5
3
6
2
7
1
8
Key
B. Pin Identifications
+
+
(+)
C.
+
+
D.

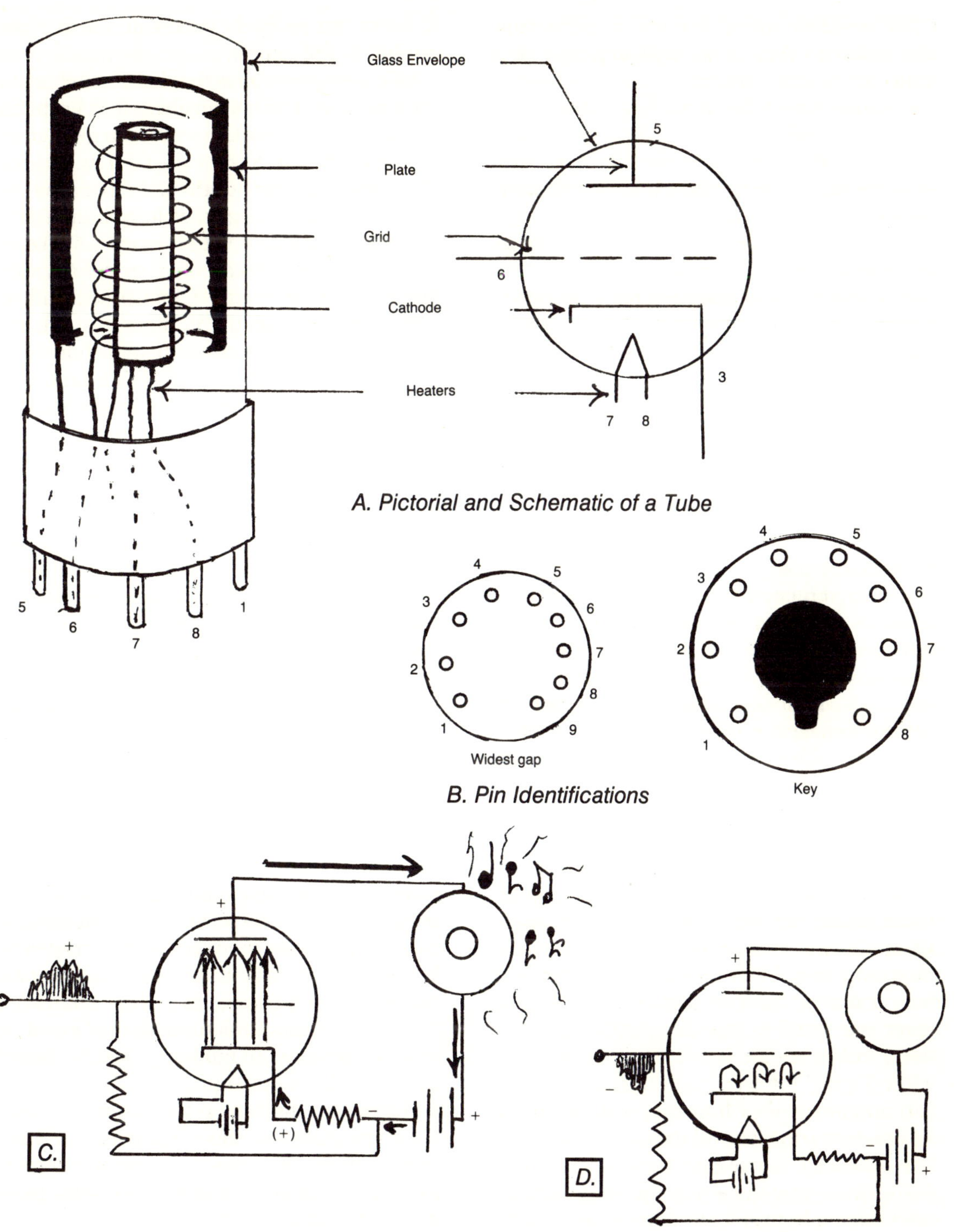

FIGURE 1-4

millions. They would not travel if the tube contained air, but in a vacuum they travel quite easily from cathode to plate as long as the plate is highly positive.

Between the cathode and plate can be one or more grids, which are either coils or mesh or screenlike material that is stiff and quite stationary. A grid cannot be solid. The electrons must travel between the wires or holes in the grid to get to the plate. Suppose a positive voltage is applied to the grid as shown in Figure 1-4 C. In this case the voltage represents a positive part of a signal from a radio station. Electrons travel from the cathode toward the grid, because negative charges go to a positive charge. They would stop at the grid, but this positive voltage is very small compared to the positive charge on the plate, so they go on through and to the greater positive charge before they enter the plate and go out to the earphone.

Now in Figure 1-4 D a negative part of a signal is placed on the grid. Because that is negative to negative, the electrons are repelled back to the cathode. Why don't the electrons on the grid go to the plate? They are not heated and their surface on the grid is not a material that will emit them, or kick them off. Sometimes in an old tube the grid will pick up the cathode coating and give off its own electrons. A tube in which this occurs is a gassy tube and must be replaced. Enough heat gets to the grid to cause this unfortunate occurrence. Normally only the cathode emits or gives off electrons. As you can see, then the incoming signal can let lots of electrons, or few, or none flow to the plate. In this way the incoming signal controls the output current to the earphone. As the electrons strike the plate and go down the wire and out the pin to the earphone wire, they become a current again. Within the cathode until they get to the plate they are called an *electron flow* or a *beam*. They are a current in a way, but since they are not jumping from atom to atom of wire or metal, they are not a normal current.

Tubes generally do not operate with a positive grid. They are more or less negative, depending on what is being transmitted at the radio station. The reason they are kept slightly negative is that with great voltages on the plate too many electrons would flow and ruin the tube. Bias for tubes must cut down on flow rather than encourage it. In Figure 1-4 C and D the cathode has a resistor that develops the negative bias for the grid. It does not make the grid negative, but it makes the cathode more positive than the grid, so this affects it the same way. Wouldn't this resistor in the grid-to-battery circuit make the grid more positive too? No; very little current is in the grid-to-battery circuit so no voltage is developed. On the other hand, a good-sized current is in the cathode-plate-earphone circuit so a large current is developed. This makes a fair voltage across that resistor between battery and cathode.

Thus, with a positive cathode in relation to the grid, the grid is more negative than the cathode and holds back some electrons. A typical bias for tubes might be minus 3 volts from grid to cathode. This would be enough. The cathode is still negative in relation to the plate. The resistor in the cathode-to-battery circuit is known as the *cathode bias resistor,* and the other as the *grid return resistor.* The cathode bias resistor often is 500 ohms or less. The grid return resistor is often a megohm or more. Because tubes have large voltages on the plate and because a very small signal such as a thousandth of a volt can affect the number of electrons that get from cathode to plate, tubes can amplify as much as 100 times. Since no current flows from grid to cathode as it does from base to emitter, the tube doesn't consume the input signal, so tubes will tune and separate stations better.

Tubes are even more inconsistent about their connections than transistors. With the demonstration tube you will see that pin 5 goes to the plate, pin 6 to the grid, 7 and 8 to the heaters. Every tube is different. In one type pin 1 might be the plate, and in another pin 6. If you have a good schematic of the circuit it usually will give each tube number and the pin numbers will show at the edge of the circle around the elements. In addition, tube manuals are available listing all United States-made tubes and giving their pin connections, information about the best use of each tube, voltages to use, and other valuable information. Many tubes are coated internally so you can't see their connections, and it is best to learn from a schematic or manual and also apply what you learn in this chapter.

Look at Figure 1-4 *B*. Most modern tubes fall into two general categories—those that have a key in the center of the base so that they can only be inserted into the socket one way (right drawing) and those that have a wider gap between pins at one point. The drawings show the bottoms of the tubes looking up toward the tube. Looking at the underside of the chassis at the tube socket would give the same view. Now, starting at the widest gap or at the key notch, go clockwise around the tube with pin 1, pin 2, pin 3, and so on. Most tubes today have either 7, 9, or 12 pins. A few octal (8-pin) tubes with the keyway still show up in older TV sets. Always be sure to get a tube in its socket properly. If you don't, you can injure pins, or possibly put a high voltage on the grid or heater and ruin the tube. Tubes are mechanically fragile and should be handled with care to avoid cracking the glass.

Tubes drop a voltage across themselves just as transistors do. If the voltage from plate to cathode is much too high, the tube may be dead. It would be the same as a resistor of millions of ohms. If the voltage is much too low, probably the bias is defective somehow and the tube is running "wide open." If you can't measure any voltage at the plate, the output must be defective. It could be an earphone or a load resistor, or the power supply could be defective. You can put the black lead of a DC voltmeter on chassis ground and back

up with the red probe from the plate back to where you find voltage. That way you would find the defective part or connection.

By reversing the leads and placing the positive lead on ground and the black lead to the grid you can measure the bias. In some cases you need to measure from grid to cathode. If you have a good schematic of the set you will be working on, it will give pin numbers and normal voltages and may have footnotes that tell you where to connect the negative lead, such as to chassis, or to a certain terminal. There are several capacitors in a tube circuit, and you will learn about them later. One is a bypass capacitor across the cathode resistor. It lets the signal get to ground without fighting through the cathode resistor. When there is a screen grid (the next grid up from the main grid) a capacitor often goes to ground. These capacitors can give a lot of trouble in tube circuits.

EARPHONES

We have not yet considered how an earphone works. Figure 1-5 *B* shows a standard set of earphones. Part *A* shows a cutaway side view of an earphone. Your high school science teacher probably demonstrated how a permanent magnet would pick up certain metal objects and how an electromagnet would attract or pick up metal objects only when the current flowed through its coil.

Both kinds of magnets occur in an earphone. The diaphragm is a thin piece of circular metal that the permanent magnets are always pulling inward. Though the cap and frame hold the diaphragm in place the attraction of the permanent magnets is there. When pulses of current go through the coils of the electromagnets the current either adds to the magnetism of the permanent magnets if it is the same polarity or subtracts some of the magnetism if it is opposite. This adding or subtracting affects the pull on the diaphragm. Only the outer edge of the diaphragm is secure. The rest can move in and out a tiny amount. This vibrates the air and produces sound in your ear. The pulses of electricity that the transistor or tube sends to the earphone, then, are changed to vibrations in accordance with the music, voice, and so on from the radio station.

Carefully remove the cap from your earphone. Note that the permanent magnets are trying to keep the diaphragm in place. Remove the diaphragm too and observe the coils. It is best not to touch anything inside, as the wire is very thin and, once broken, is very hard to repair. There are literally thousands and thousands of turns of this very thin wire in each electromagnet. They are what make up the 1,000 or more ohms of resistance. Heavier wire, if it would fit in the earphone, would not offer so much resistance. For best sound the diaphragm should be screwed tightly down by the cap. Replace both.

Loudspeakers work just a little differently from earphones. They have only a few turns of wire on a moving coil form. See Figure 1-5 *D*, *E*, and *F*. Since a speaker does not offer 1,000 ohms of resistance but usually only 8 ohms, it takes a stepdown transformer or other arrangement to match a speaker to a tube or transistor. Note that the permanent magnet is inside the moving coil to an extent. When current is in one direction the poles of magnetism between the permanent magnet and the moving coil, an electromagnet, are the same and the coil pushes outward. This moves the cone, which is of heavy paper, and the cone vibrates the air and makes sound. If the current reverses, the magnetism is reversed and the cone is pulled inward. For a speaker to work well the moving coil must move freely and not be bound. Neither the coil nor the cone can be torn. There must be a complete circuit for the coil. The permanent magnet must have a good magnetic charge. A spider (not the 8-legged creature) holds the cone-and-voice-coil mechanism in place.

Other kinds of earphones include crystal phones that depend on current vibrating a crystal to make sound, and 8-ohm phones that have very few turns in their electromagnets. An old style of speaker uses a large electromagnet in place of the permanent magnet. The current for this large electromagnet is usually DC developed by the power supply. It then has a small-turn voice coil also. Such a speaker will have four or five wires going to it instead of two.

Let's review the important points again:

1. Tubes must have an internal vacuum in order to work.

1. A heater heats the cathode, causing it to give off tiny negative charges called *electrons*.

3. Electrons are attracted to the plate by a large positive voltage on the plate.

4. The grid has a signal on it that controls the flow of electrons that goes to the plate, thus a small signal controls and makes a big one.

5. A bias protects the tube from overwork. A negative charge is kept on the grid at all times.

6. Pin numbers on tubes are identified by looking at the tube upside down, finding the wide gap in pins, and counting clockwise.

This first chapter has covered a lot of material. Some of it will be repeated in part later in the book, but it is best to study this chapter again before continuing. Leave the 2-transistor amplifier as is; you will be making it into a simple radio in the next chapter. Be sure you have learned the codes for resistors and capacitors.

QUIZ FOR CHAPTER 1

▶ 1. If you have a series circuit composed of a transistor that offers 1,500 ohms resistance between collector and emitter, and a set of earphones of 2,000 ohms total resistance, and you attach a battery and connect up the circuit, which item other than the battery will have the largest voltage drop across it?

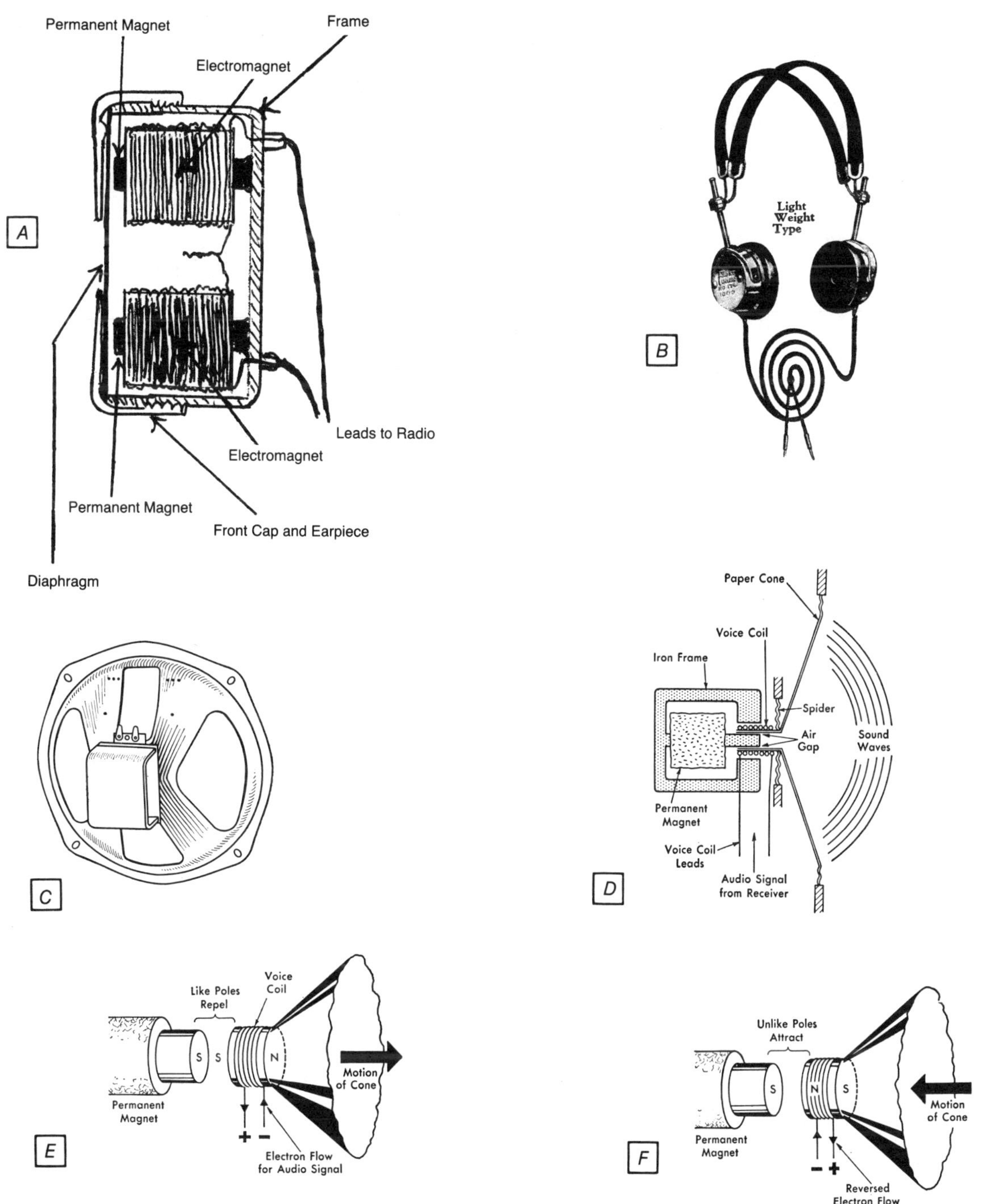

FIGURE 1-5
EARPHONES AND SPEAKERS

▶ 2. A resistor color coded red/red/orange (orange toward the middle of resistor) would have what value in ohms?

▶ 3. Which resistor would ordinarily have the larger value in ohms—a collector load resistor or a base bias resistor? Why would it need to be larger?

▶ 4. How has the base of a PNP transistor been doped in manufacture? How has the emitter been doped?

▶ 5. What physical characteristic of the base is important?

▶ 6. If a collector compares roughly to a plate on a tube, and an emitter to a cathode, what compares roughly to a base? They both have to be ______.

▶ 7. An earphone has both ______ and ______ type magnets. The part that is attracted is the ______.

▶ 8. In RC coupling between stages, what is the function of the resistor? Of the capacitor?

▶ 9. If neither voice nor DC would go through a capacitor, what would be wrong with it? If both voice and DC would go through it, what would be its defect?

▶ 10. Describe how a simple audio signal tracer could be made and how it would be used to troubleshoot audio equipment such as a guitar amplifier or record player.

2. Detection and Tuning

ANY SIGNAL-RECEIVING device—radio, TV, CB receiver, or other receiver—must accomplish the following in order to achieve acceptable reception:

1. It must capture the radio (or TV) waves from the air in electromagnetic charges and convert them to electric currents. The antenna and ground, if any, do this.

2. It must accept only one frequency of signal and reject all others. Coils, capacitors, and sometimes crystals select the desired frequency.

3. It must detect the desired signal; that is, it must change the high-frequency AC signal to pulses of DC.

4. It must convert these pulses of current to sound waves by means of an earphone or speaker.

5. If the signal is weak it must amplify it, either before or after detection or both. Tubes or transistors produce this amplification.

In the case of a TV set, the receiver must convert the detected pulses into degrees of brightness and darkness, and other pulses into magnetic fields that move this brightness or darkness across a screen. The picture tube and deflection yoke do this. We will explore TV reception and conversion in depth later, but an understanding of radio tuning and detection goes a long way toward the same understanding of TV.

Point 5 above raises the question of amplification if the signal is weak. Although nearly all receivers manufactured today have amplifiers, many receivers made in the early days of radio consisted only of a tuning device, detector, and earphones. These were called *crystal radios,* as the detector was a type of mineral such as galena, a crystalline substance. Crystal radios work only within a few miles from a station, as all the radio energy that works the earphones must be captured from the radio station. Usually crystal sets require a very good antenna and ground. The nice thing about such a radio, however, is that it takes no purchased energy source to make it work. It is literally "free" radio. Study Figure 2-1 throughout this chapter. If you could leave off the amplifier and attach the earphone in place of C3 to the diode detector, D1, and connect the other end of the earphone to ground, you would have a crystal radio. Since many readers who will study this book are miles from stations, it would be best to use the audio amplifier with the radio and get reasonable results.

ALTERNATING CURRENT

Direct current flowing from a battery through a circuit always goes in one direction. It does not produce radio waves. Alternating current (AC), because it is constantly changing polarity back and forth through a circuit, develops radio waves. Next time you drive under high-power electric lines note the effect on your car radio. The power lines are acting like a radio station to a degree. True, they don't play music, but they do broadcast a hum for a few hundred feet.

AC as supplied to customers for home use in the United States is 60 hertz, meaning that the current goes in one direction through the wires, reverses itself, and goes back the other way 60 times a second. Each back-and-forth flow of current is termed a *cycle,* and there are 60 of these cycles each second. This kind of current used to be called "60-cycle AC," but recently the term has been changed to *hertz,* or *Hz* for short.

As you listen to the buzz of the disturbance caused by high-potential power lines on your car radio, you are hearing a pitch of 60 Hz. Such a pitch or frequency is quite low. Only cellos and bass violins play down in this region. The term *frequency* refers to how often the cycles occur, so the higher the frequency, the more cycles of change from $(-)$ to $(+)$ in the wire of AC.

This same frequency is used to measure the speed of vibrations of sound. It would be correct to say that a piano was playing at a frequency of 256 Hz, which is middle C. It would be just as correct to say that a radio station was producing a carrier frequency of 650,000 Hz (WSM, Nashville, Tennessee). The alternations of sound vibration produce no radio waves, but alternations of electric current do produce radio waves. Low-frequency AC such as the 60 Hz from the power line produces radio waves that travel only a short distance. Fortunately, it would take tremendous current and voltage to make 60 Hz AC broadcast for many miles, if it didn't, radio and TV sets would hum beyond belief unless we could tune out such signals. As a general rule the higher the frequency of AC the farther the radio waves will travel for a given amount of current. Very, very high-frequency radio waves traveled all the way to the moon and back, not only with sound, but also with pictures.

RADIO FREQUENCY

Any alternating current whose frequency is high enough to broadcast voice or pictures satisfactorily is called *radio frequency,* or *RF*. Though some broadcasts are below 500,000 Hz, most are above this frequency. The RF signals that operate on 500,000 Hz are for navigational purposes. Just above this is 540,000 Hz, where commercial broadcasting begins. To avoid using all the zeros we use the prefix *kilo-* to mean thousands, so normally this frequency would be called 540 kiloHertz, or 540 kHz. Radio stations operate up to 1600 kHz. Perhaps you have heard a statement

such as the following as a radio station signs on or off the air: "KWRE operates on a frequency of 730 kHz with an effective radiated power of 1,000 watts as assigned by the FCC. . . ."

TV stations operate on much higher frequencies consisting of millions of AC alternations. A million hertz are called *megahertz* or *MHz*. Channel 2 of TV commercial broadcasting, no matter where it is in the country, operates on 54 to 60 MHz. The other channels operate on even higher frequencies. See Figure 12-1 *A* for a list of all VHF channels and their frequencies. Perhaps you have seen microwave stations with their towers, big dishes, and small buildings. These radio devices operate on super-high frequency. Some are in the 4,000-MHz to 6,000-MHz range. It is hard to believe that electricity could change its direction that many times—as many as 6 billion times a second!

The frequency on which a radio or TV station operates is called its *carrier frequency*. (The carrier is the composite of all positive and negative alternations or cycles combined and radiated into space at the transmitting station.) This frequency is not heard at a receiver. It is well above audio range and our ears can't perceive it. Sometimes you can "hear" the carrier go on the air at a radio station at your receiver before anyone talks or the national anthem is played. Actually what you hear as a hissing noise is not the very-high-frequency carrier; you hear its effect on your radio. It causes the tubes on transistors to conduct more—thus the hiss. The actual carrier is too high to hear.

The name of the carrier suggests what it does. It carries the messages to your radio miles away. The carrier is generated by a device called an *oscillator*. With a tube or transistor set up a certain way, and a crystal that vibrates at only one frequency, plus some coils and capacitors to tune it to that one frequency, the carrier is made of very fast alternating current. Through a series of amplifiers this carrier is made very strong. It is ready to travel for miles by being applied to an antenna and ground. (It carries no music yet, however.)

See Figure 6-1. *D* represents a higher-frequency carrier. *E* represents a lower-frequency carrier waveform. Note the difference. They could be the same in power, and they are in this case (represented by height) but one occurs faster or more times per given time period than the other. If you wanted to send out code messages, all you would have to do would be attach a telegraph key into the oscillator circuit so you could start and stop the carrier at will. At *F* you can see how that would affect the carrier to send "Di Di daah," or the letter *U* in international code. Sending voice or music is much more complex, requiring speech amplifiers, microphones, turntables, and an all-important device called a *modulator*. The modulator either enlarges or cuts down the height or power of the carrier, depending on what is being said, sung, or played. *G* shows how this is done.

Later you will build an oscillator and a simple modulator, but for now you only need to learn that the carrier is constant in frequency but varies in size or *amplitude,* to represent the transmitted voice, music, or whatever. (Engineers reading this are going to argue the point, so let's say that music played does very, very slightly change the carrier frequency. OK; but a change of 256 Hz played by a piano on a carrier of 540,000 Hz or larger isn't going to bend the carrier enough even to show on the drawing.)

FM stations are different. They depend on considerable change in the carrier to carry the sound. The change consists of either increasing or decreasing the frequency of the carrier. For instance, the carrier without modulation, with no sound being broadcast, might look like Figure 6-1 *D*. The sound might spread it as at *E* or might bunch it even closer together than at *D*. These are exaggerations, but the bunching and spreading of the carrier frequency is what FM transmission is all about.

THE RECEIVER

Now let's examine AM or regular broadcast programs. AM stands for *amplitude modulation*. The modulation, or music, voice, or whatever, affects the amplitude or height of the carrier. Every station operates with a carrier at a different frequency so stations can be separated from each other. Many stations may operate on, say, 1050 kHz, but they are miles from each other and in any given area of the country, only one will be receivable.

The first thing a radio receiver must do is change the carrier with its modulation from electromagnetic waves to electric currents. At the transmitter the carrier with modulation is fed to an antenna. The actual radio waves that leave the antenna are magnetic waves. These travel through space, still with the modulation on each of them. The receiver must have some kind of metallic antenna. It could be a long wire, or it could be the ferrite metal of the core of a loopstick coil in the back of a pocket portable radio. This antenna is stationary. As the magnetic waves break across the antenna, they develop tiny currents of electricity. These currents have the frequency of the transmitter and the modulation in the form of the size in amperes of each current. The currents that originated at the radio station changed to magnetism in the air and became currents again at the radio receiver miles away.

With a simple radio such as the one shown in Figure 2-1, a long antenna picks up the waves of magnetic pulses, which are constantly traveling between antenna and ground as high-frequency AC or radio frequency, RF. To go up and down the antenna-ground path, some of the radio station waves are going through the coil, L1. Others go through the capacitor, C4. Signals from many stations are traveling this route and must be sorted out.

A coil, sometimes called an *inductance,* has a natural tendency to resist high frequencies; the higher the frequency, the more the coil offers resistance. A capacitor has a natural tendency to reject low frequencies. It blocks DC altogether, and the lower the frequency the more a capacitor reacts to it as if it were DC. The lower-frequency station carriers with their modulation are going up and down the coil to antenna and back down to ground. The higher-frequency station carriers are going up and down C4, the capacitor, to antenna and ground. Right in the middle will be one radio station carrier that finds it equally hard to go through either L1 or C4. It goes through both with difficulty, and they act as a resistance to the one middle frequency. This makes a load of voltage somewhat like a collector or plate load, which is fed on into the rest of the radio, as it has no other place to go.

Wonderful, but suppose the station you want is not operating at the exact center frequency for which the coil and capacitor are fixed? If it is not, you can change the number of turns in the coil or the number of plates in the capacitor or both to send a new station into the set. Such alterations would take physical labor and changes in the set for every station selection. You could make a slider that would short out some of the turns of the coil to change stations. Fewer turns result in a higher frequency received. Such a system was used in crystal radios of 50 years ago. As an alternative, a slug of ferrite metal can be used to go into and out of a coil to change the frequency it receives. The frequency that a coil and capacitor tune for the set to receive is called the *resonant frequency* of that coil and capacitor. In the case of Figure 2-1 *A* or *B* the coil and capacitor are in parallel with each other, making a parallel-tuned circuit. Figure 2-1 *C* is a series-tuned circuit. You will see how it works later.

One other possibility remains. You can vary the amount of capacitance that C4 offers to change the resonant frequency of the tuned circuit, and thus change stations. We just suggested that plates could be removed from the capacitor, C4, but that would be a clumsy solution. Look at C4, the actual capacitor that you will use in building the radio. If you turn

it on its side and back and slowly rotate the shaft, you will see that many plates mesh with many other ones, but that they are separated by thin plastic. When you turn the shaft one way the plates are fully meshed between the other ones to give maximum capacitance, or about 365 picafarads. Turning the shaft of the capacitor fully the opposite direction will lift the movable plates completely away from the stationary plates and make the capacitor offer about 10 picafarads of capacitance.

The greater the capacitance offered, the lower the frequency of the station received. Because this capacitor can be varied or changed at will, it is called a *variable capacitor*. The arrow in the schematic represents this variability. If you have enough turns wound for L1 and turn C4 with the plates fully meshed, you might get 550 kHz, or KSD in St. Louis, if you live in the Midwest. If you open the plates of the variable capacitor apart more you will get higher-frequency stations such as 650 kHz or 990 kHz.

The length and type of antenna you use will greatly affect the resonant frequency of your tuned circuit and therefore the adjustment of the variable capacitor to get the station you wish. After you make the radio, tune in a station. Then go out and add fifty feet of antenna to the radio. You will find that you will have to unmesh the plates of the capacitor (cutting down on picafarads) to get the signal as well as you did before. You have effectively added turns to the coil and must remove capacitance to compensate.

Signal Detection

Now that you understand how a radio receiver "captures" station signals and selects only one, let's deal with the detection of the signal. Figure 2-1 *B* has some little waveforms that show how the signal is affected throughout the receiver. At point *J* is a small selected signal from one station. The tuned circuit, L1 and C4, selected it, as just explained. Note that the carrier has identical positive (above

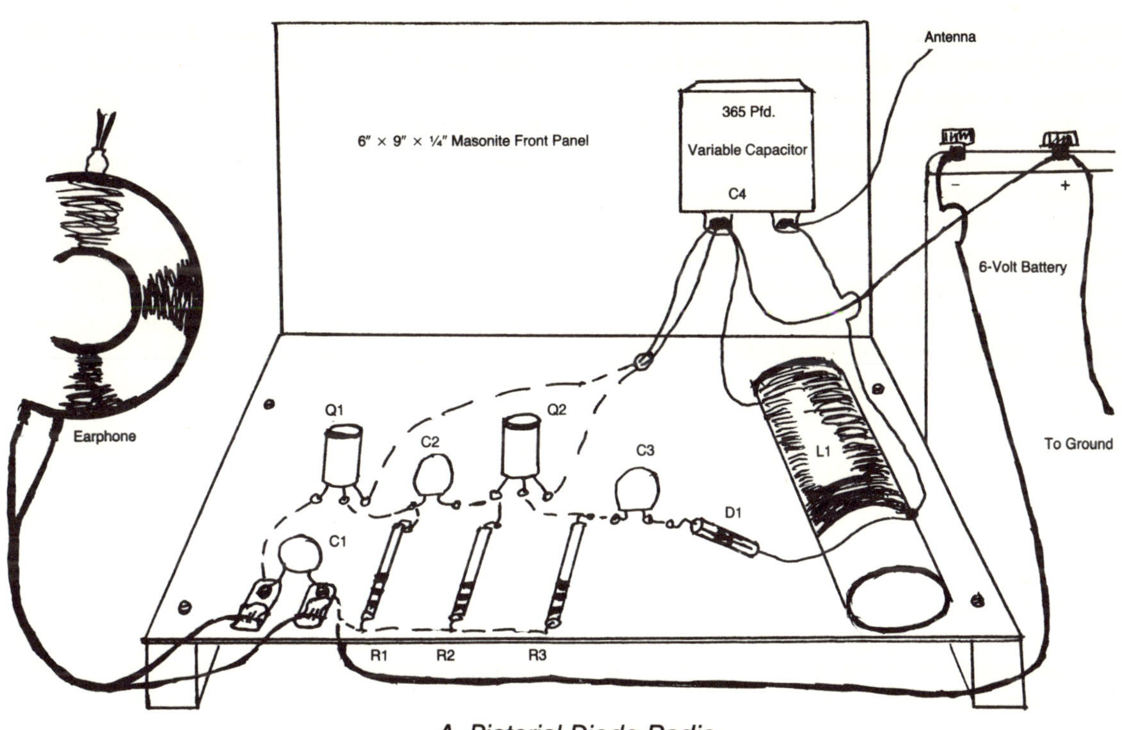

A. Pictorial Diode Radio

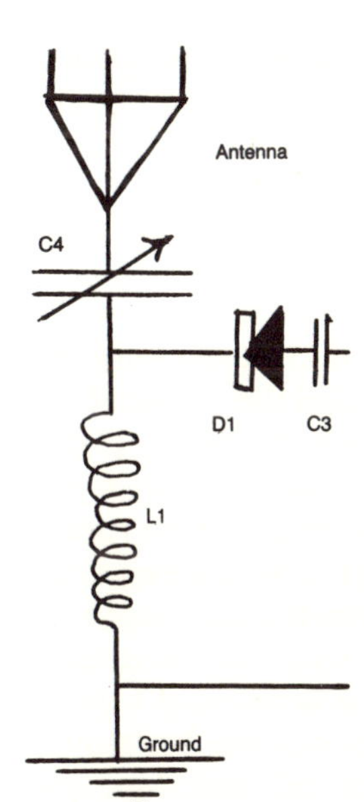

B. Schematic Parallel-Tuned Circuit Diode Radio

C. Series-Tuned, Circuit Diode Radio (Partial Circuit)

Figure 2-1

the line) and negative (below the line) modulations. If this signal were connected directly to Q2 through C3 it is doubtful that any signal could be heard. The negative part of the signal would affect the transistor one way, but the positive counterpart would affect it just the opposite. They would cancel each other out, so it would be the same as receiving no signal at all. Now the bias of a transistor can be adjusted so that it will respond to only the negative or positive half of the signal, but this changes the function of the transistor from amplifier to detector. (This will be discussed in Chapter 3.)

All standard broadcast station signals are double as shown. Some amateur radio broadcasts clip off part of the carrier before they transmit it. Regular radio stations have the double-sided signal, however, and we must cut off half of it through a process known as *detecting* the signal. D1 is a diode, the modern counterpart to the crystal of long ago. You might also think of it as two-thirds of a transistor. The left-hand vertical line is known as the *cathode* and could be compared to the base of a PNP transistor. The arrow part is known as the *anode* and is comparable to the emitter. Current can easily travel from cathode to anode (base to emitter in a PNP transistor), but it is all but completely shut off if it is trying to flow from anode to cathode (emitter to base in a PNP transistor).

Try measuring the resistance of D1 with an ohmmeter. One way it will be very low and the other way it should show thousands of ohms. As the pulses of station carrier come in, with the modulation making them larger or smaller to carry the music or voice, they can go through the diode, D1, across C3 (they are AC, not DC), from base to emitter (it is a PNP transistor), and back to ground at left. On the return swing D1 blocks all current flow so it is lost. We have now cut off all positive going pulses as shown at *K*. Note that we still have the carrier frequency.

The current through base to emitter of Q2 caused current from collector of Q2 to go to emitter of Q2. This builds up pulses across

R2. Note at *L* the signal is all positive. That is right; as current travels from battery through R2 and down through the transistor, the top of the resistor, R2, would take on a negative polarity and the bottom positive. The signal is now positive, but it keeps its shape and some of its carrier. It has grown considerably in size, as Q2 amplified it. Some of the carrier disappears, as Q2 is an audio transistor and doesn't amplify RF (radio frequency) as well as it does audio frequency.

At point *M* the signal has gone through another transistor that has inverted it again and made it negative. From base to collector a signal always inverts. The same is true with tubes from grid to plate. With TV this inversion of signal is very important; with radio it is worth noting, but either half could work the earphone. The signal is very much amplified by now and would be drawn larger if space permitted. The carrier is gone. Going through the second transistor helped get rid of it, as it too is an audio type. The capacitor, C1, bypasses what is left of the carrier away from the earphone. We have no high frequency now, just the rising and falling pulses of current that represent music and voice. The earphone changes these to sound.

Some tubes will detect. A tube with a cathode, heater, and plate, but no grid, is also known as a diode. Such a plate could be in the circuit, if there is a heater voltage to heat the cathode. There would be no high voltage on the plate. Instead, the plate would be at the C3 end. As the signal tried to go inward there would be more electrons on the cathode than the plate so electrons would flow. On the opposite cycle current pulses going across C3 would make the plate negative and repel electrons back to the cathode. Such a tube was used in earlier radio and TV.

MAKING THE RECEIVER

Now you will actually make the radio. You will need to make a masonite panel 6 by 9 inches by a quarter inch thick. This will be the front of the radio. You may find it easier to make the hole for C4, the variable capacitor, before you mount the panel in place. If you use a Radio Shack variable capacitor, handle it with care. It is rather fragile at best. The first nut is loosened to attach the capacitor to the panel, but you must not bother the inside nut. It will allow the plates to turn and it is hard to synchronize them properly again. Beware of moving the inner nut! The outer or mounting shaft may not penetrate the masonite panel completely. You can either cut away some of the inside of the panel material or mount the capacitor on a piece of tin can metal, with a small shaft hole, and fasten the ends of the metal to the back of the panel with screws in the form of a bracket.

If you use a heavier variable capacitor you may need to mount it to the chassis board and make a hole in the front panel for the shaft. At any rate mount the capacitor and shaft so the shaft turns freely. A knob on the front is a very good idea. Without it you will change frequency slightly as you touch the metal to change stations. When you let go the station will not be as loud. The knob isolates the hand capacity, or human body capacity, from the radio circuit. The panel is mounted to the rest of the radio with wood screws that go into the lower supporting pieces of wood in the front. If you wish, you can fasten angle brackets between the panel and chassis masonite boards for more support. However, this is not absolutely necessary.

Connections to the variable capacitor must be soldered. You might wrap wires to the terminals, but chances are you'd get scratchy, intermittent reception. Before soldering, you must make L1. From polyvinylchloride (PVC) pipe 2 inches in diameter, cut off a piece 5 inches long. This plastic pipe makes an excellent coil form. Starting about a quarter inch from one end, wind No. 20 enameled insulated copper wire either for about 40 turns if most of the stations in your area are in the 1000 to 1600 kHz region, or for about 65 turns if most of the stations are in the 540 to 1000 kHz region. This radio will not track the entire broadcast band, so you must make some com-

promise. All turns must be closely spaced and neat. Secure both ends of the coil with electricians' tape to hold it in place. Leave ends long enough to reach to the capacitor and to D1. The ends must be scraped clean all around.

(If you want to try for all stations, you could twist a loop at the middle of the coil up above it and scrape insulation clean for this loop. Be sure both halves of the coil are attached nicely to this loop. Make another loop at the bottom end, nearest the variable capacitor. Mount a wire from the variable capacitor ground side to an alligator clip. Now you can connect all 65 turns or just 32 if you prefer, and then the variable capacitor should be able to get about half the dial with each setting of the coil. This is quite optional, but it might be fun to try.)

The leads from the emitters are now fastened to the left terminal of the variable capacitor, and so is the coil, and a lead to battery (+), to ground transistors, variable capacitor, and coil. Solder the antenna lead and upper coil lead to the right terminal of the variable capacitor. Now carefully twist one lead of D1 to C3. This is where you did have the test wire. Connect the other end of D1 to the top side of L1. You will need to scrape insulation from the copper wire to connect here. You also will probably need a short splice of wire to make this connection. To be consistent with the schematic you should connect the cathode of D1 to the antenna, variable capacitor end, and the anode to the C3 end. This is the way we traced it in our study. Some diodes have the arrow and mark on them. Others have a single band around them to mark the cathode end.

Will the radio work if you fasten the diode in the wrong way? Yes, it will still detect, but it is a little hard to understand it with the current having to go in the opposite direction to detect. Here is roughly what happens. Current can't flow up from emitter to base of Q2. Instead, it must go from ground actually through the battery, around, up, and through R3. Pulses go across C3 to the diode, now turned with cathode in and arrow out. Current pulses go out to the antenna. In reverse the diode blocks them. In this case when pulses go across C3 it actually takes away from the bias current for Q2 and makes it less than normal. This closes down on current flow to the emitter, and in turn cuts collector-to-emitter current below normal. Electrons pile up at the collector of Q2 and the signal is negative. The signal has just been turned over from the way it was at first. At the earphone you would now have a positive signal. As you can see, a signal can either add to the bias or take away from it. Either way the pulses carry the music, voice, and so on. The highest part of the signal would be a very loud word or note.

Some thoughts about the antenna are in order. Bare wire picks up signals a little better than insulated wire, but if you are in a congested area where electric lines could touch your antenna in a storm, insulated wire is best. An inverted L antenna is quite good. Connect one end either with an insulator like that used for electrical wiring or by wrapping the end to a piece of plastic clothesline. Secure this to an appropriate tree. The other end should be strong, from 50 to 100 feet long, and as high as practical. Loop it, with the insulation on, around a second tree and come down with the rest as a lead-in to the radio. Though you could make a lightning arrestor, it is best to disconnect the antenna during thunderstorms. For the best antenna possible, stranded bare wire such as TV antenna guy wire is excellent. Use insulators at both ends. Solder a lead-in wire near one end. The lead-in wire should be insulated.

If you are near stations, an alternate can be a loop antenna around the ceiling of your room at the corners with the walls. This gives an omnidirectional antenna; all four sides receive equal currents from the electromagnetic waves. Do not connect the ends together, but bring one down for the antenna and leave the other unconnected. You can sometimes use the dial clip of your telephone for an antenna. The ground we discussed before is very important with this radio. Be sure you have good connections.

Now connect the battery. As you move the knob of the variable capacitor, you should be able to get stations. With this diode radio signals will come in rather broadly and may be hard to separate.

After listening a bit try the series-tuned circuit. Figure 2-1 C shows the part that is different. To try it you will need to unsolder the wires from the right terminal of the variable capacitor. Resolder the emitters, coil the lower end and lead to (+) battery together but not to the variable capacitor. Just fasten them to each other and tape them. Leave the top of L1 to the right terminal of C4, but move the antenna lead-in to the left side of C4. Now you have a series-tuned circuit diode radio. Try tuning stations with the battery connected and the earphone to your ear. You will find that they come in a different place from before. Have a good listen, restore the radio to parallel-tuned circuit as before, and listen some more. Leave the radio as a parallel-tuned circuit set, as you will use it that way when you make a transistor detector radio in the next chapter.

The series-tuned circuit functions just a little differently from the parallel. The coil blocks all higher frequencies from traveling from antenna to ground. The capacitor blocks all lower frequencies from the same circuit. The one middle frequency can go all the way in but finds it easier to go through the diode circuit than across the coil, so the frequency signal enters the diode and works the rest of the radio. You could say that a series-tuned circuit offers least resistance to the resonant frequency. A parallel-tuned circuit offers greatest resistance to the resonant frequency. Usually radios use parallel-tuned circuits, but some TV tuners have series-tuned circuits.

Important points to remember are:

1. Radio waves travel as electromagnetic waves through space.

2. The antenna or loopstick of the receiver picks up electromagnetic waves as current pulses.

3. These pulses are oscillating at a certain frequency for any station. Their size is

changed (modulated) to represent the transmitted message, such as music or voice.

4. A combination of a coil and capacitor make a tuned circuit that will resonate at only one frequency. This frequency signal is fed on through the radio.

5. Signals can be detected by a diode, which only allows current to go through it one way.

6. A tube or transistor always inverts the signal between input and output.

7. The carrier frequency is limited by audio stages and bypassed so as not to interfere with speech.

Quiz for Chapter 2

▶ 1. Why must a broadcast signal be detected in a receiver?

▶ 2. If a carrier is amplitude modulated the voice or music would be affecting the ______ of the carrier.

▶ 3. If a carrier is frequency modulated the voice or music would be affecting the ______ of the carrier.

▶ 4. If a variable capacitor is tuned to 365 picafarads and the coil is properly made for the broadcast band, would you receive low- or high-frequency signals?

▶ 5. Does a diode amplify a signal? Explain your answer.

▶ 6. If a radio had two diodes connected back to back—that is, with cathode to cathode and the anodes going to antenna and coupling capacitor to a transistor— would the radio work? Could it detect better than a single diode or not at all?

▶ 7. Explain how a resonant frequency is developed in a parallel-tuned circuit.

3. Amplifying Detectors

Project: Constructing a Three Transistor
TRF Receiver.

Now that you know how a diode detector circuit removes half the incoming signal from the radio or television station, you will learn how a transistor or tube can be used to detect radio or TV signals too. You will experiment with high-frequency transistors and learn how they are different from audio transistors. You will learn how potentiometers work and how they are used, and you will learn about devices to sharpen the tuning of a radio, such as antenna coupling capacitors and RF transformers. Then we will simulate problems that could develop in this radio and other similar equipment and you will learn how to find these problems and repair them.

You will use the same audio amplifier you build in Chapter 1, but the diode must be removed. Leave the coil (L1) and the variable capacitor (C4) connected, but change them to a parallel-tuned circuit arrangement. Now you can begin with the new circuit. Use Figures 3-1, the schematic, and 3-2, the pictorial drawing as you build this set. Try to rely more and more on the schematics and less on the pictorials, as you will need to use schematics almost entirely in TV work.

The radio frequency transistor, Q3, is specially designed to operate above the audio range. This is a part of its internal construction. There is less capacitance between the emitter, base, and collector. Q1 or Q2 might work in this particular radio if used as the detector, but if you were trying to get short-wave they would not. Always remember to re-place transistors with one of the same frequency range. Also be sure to use PNP transistors instead of NPN if the circuit has been developed for PNP types. Be very sure your transistor to be used as Q3 is a PNP type. If you use a Radio Shack high-frequency type it may be a half cylinder with a flat side. Usually, but not always, these will have the leads arranged as shown in Figure 3-2. The best bet, however, is to look at the lead identification paper that comes with the transistor to see which is the emitter, which is the base, and which is the collector. If the transistor comes without lead identification, don't buy it.

As shown in the parts list in Appendix A, an SK3007 RCA transistor may be used in this case for Q3. Its leads are the same as those of Q2 and Q1. As before, drill or push three small holes in the masonite base and insert the leads of Q3. C3 is connected to the collector along with a load resistor, R4. For R4, 1,300 ohms is good, but anything from 1,000 to 2,000 ohms will work well. Remember to attach the opposite end of R4 to the negative battery supply by wiring across to R3. As before, unless instructed differently, just twist the wires together with about three tight turns.

Some of these connections have to be changed from project to project, so soldering only slows the work down. Certain parts, however, such as the variable capacitor (C4), and the potentiometer (R5), which we will take up soon, must be soldered to get a good connection. The emitter of Q3 is connected to ground by connecting to the ground lug on the variable capacitor, C4. This connection must be soldered at the capacitor, but don't solder it yet, as two other leads must go there too. The ground side of L1 should already be there from the diode radio. The two new leads are from the left side of the potentionmeter and from the L2 winding, which we have not made yet.

Examine R5. A potentiometer is a resistor with a sliding shorting device that can touch any part of the resistor. This device has three

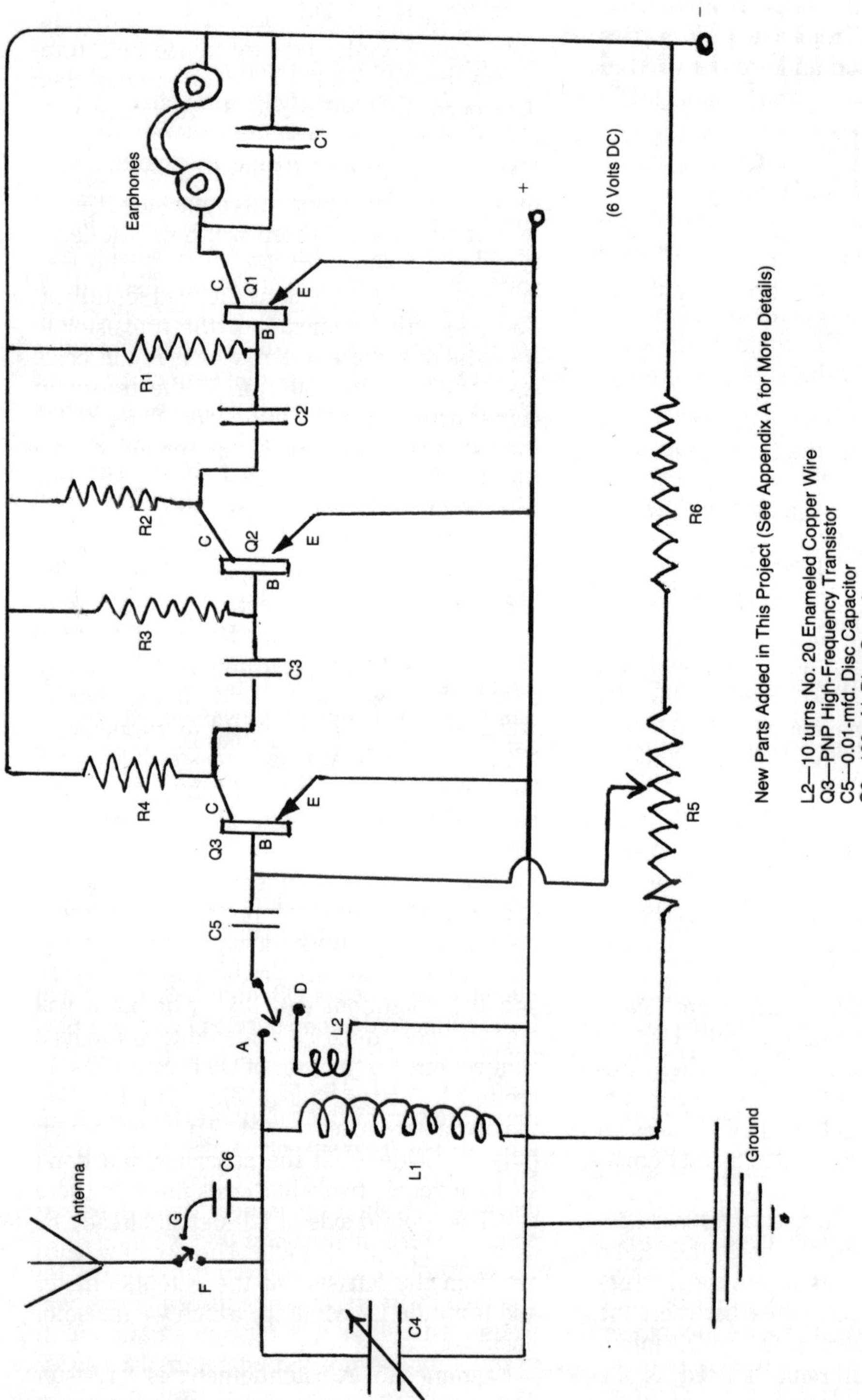

FIGURE 3-1
SOLID STATE AMPLIFYING DETECTOR RADIO

lugs on the side near the back. It might also have two lugs on the extreme back. These two lugs are an on-off switch and are not used in this construction. Before you connect the potentiometer to the chassis panel, test it with an ohmmeter if you have one. Measure across the two outside terminals and you should get 100,000 ohms regardless of how you rotate the shaft. Figure 3-4 (top) shows the parts of a potentiometer.

Now connect the ohmmeter between the middle terminal and one of the outside ones. Note that as you rotate the shaft, the resistance varies equally. You will also find if you change the leads that as the resistance becomes less between the middle and the outside terminal, it becomes greater between the middle and the other outside terminal.

Remember, the more resistance a circuit component offers, the more voltage will be dropped across that component. You can control voltage across parts of this potentiometer and feed that voltage anywhere you desire. In many radio and TV sets such a device acts as a volume control. In such a case the outside terminals would be connected as a load from a transistor or tube. The middle terminal would allow some of the voltage developed across the entire load—as much as you set by the shaft—to go to another transistor or tube. This voltage would be in the form of pulses of music or voice or whatever is being broadcast. Such a device also could control the brightness, contrast, or color of a TV set. Another name for a potentiometer is *variable resistor,* because it varies some of the resistance.

In your work in this chapter, you will use the potentiometer in a different way. You will control the forward bias to Q3 for its base current. This will affect the volume of the radio, but there is a more immediate need for adjusting the base bias continuously. You could have put in a bias resistor such as R1 or R3 if Q3 were only to amplify. Since it must detect also, its bias current is very critical. You must be able to set it while the radio is on. The variable resistor, R5, makes this possible. You will see later why the bias must be set carefully.

Make a hole in the panel with a drill or the reaming blade of a knife. Place a file with the handle removed, point end in, and ream until the shaft and threaded outer shaft will fit snugly. Insert the potentiometer into the hole with a lock washer inside the panel and a smooth washer outside the panel. Attach a nut of suitable size and secure the potentiometer, obviously with the shaft outside and the body of the device with terminals inside. It is good practice to attach a knob on the shaft so your hand capacity will not affect the radio. If you have none, wrap a layer of electricians' plastic tape over the shaft where your fingers will touch it.

R6 is a protective resistor. It prevents the full 6 volts from being fed to the base of Q3 and possibly ruining Q3. Solder R6 to the right-hand terminal of R5 and connect the opposite end to the B-, right side, fahnestock clip. You will need a length of wire to complete this connection.

The emitter of Q1 can be tied to the left-hand lug of R5, and the previously mentioned connection from C4 to the left lug or terminal of R5 is now connected. A long lead from left terminal of C4 to ground and (+) battery is attached. The battery should not be connected to the battery leads during construction, however. The middle terminal or lug of R5 has a lead to the base of Q3. Solder the R5 end but twist to the transistor. C5 is also connected to the base of Q3, and the other end of C5 is connected to one of two places. For now connect it to point *A*, which is the "top" or high side of L1. The terms *top* and *high side* mean the part of the coil that is farthest from ground. The high side of L1 also goes to point *F*, but this should already be connected. Solder in C6 as shown. Connect the antenna at point *F* for the first part of the experiment.

Now for coil L2. Wind ten turns of No. 20 enameled wire in the same direction that you wound L1. Place it about one-quarter inch from L1 for starters, but tape it in place so L2 can be moved. The lower end of L2 is grounded. This is the end nearest L1. Now you can solder the left terminal of C4, the

Antenna
G
365 pfd.
C4
C6
100 PF
F
R5
100,000
On-Off Terminal
R6
100,000
Radio Shack High-Frequency
PNP Transistor
Leads,
Bottom View
C
B
E
PNP Audio Transistors,
Bottom View
(Also SK3007 High
Frequency)
C
B
E
Q1
Q2
C3
.01
C2
.01
C
B
E
C
B
E
220,000
R1
1000
R2
330,000
R3
1300
R4
C1
.01
Q3
C
B
E
C5
.01
L1
L2
A
D
To Earphones
– 6 Volts DC
+
Ground

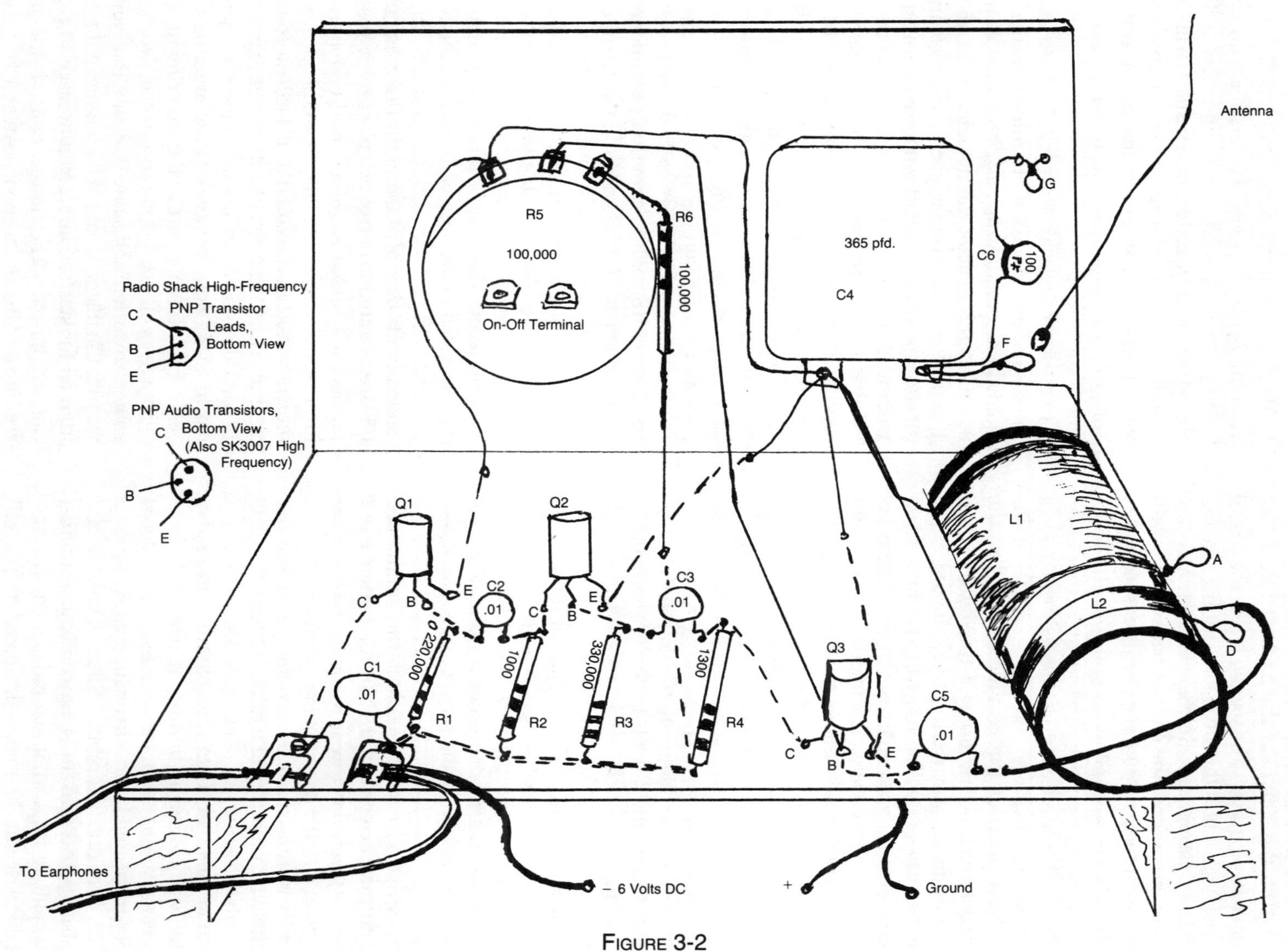

FIGURE 3-2

variable capacitor, and you have all four connections in place. The top of L2 is point *D*, and later you will connect Q3 through capacitor C5 to that point rather than to point *A* as it is now.

Check your wiring carefully. Remember that all wire at connecting points must be free of enamel or coverings of any type. At other points the covering is needed to prevent shorts. Be sure your ground is good and the earphones are on. Now connect the battery, carefully observing polarity. Start with R5 all the way down—that is, counterclockwise looking from the front of the panel, not from where you see all the parts. Advance R5 slowly while trying to find a station on C4. You will find a place where stations are loudest on the R5 setting, but if you advance the shaft farther they will quit.

Transistor Detection

Now we are getting to the detection principle of transistors. A diode detects, remember, because it cuts off either the negative or the positive part of any radio transmission, depending on which way you set it in the radio. No voltage is applied to it except that which comes from the radio station. Transistors are different in that they need a current to flow from base to emitter to make a large battery current flow from collector to emitter. If the current is too small in the base-to-emitter circuit, no current will flow in the collector circuit, so you don't hear music or voice. (That would be with R5 set in lowest position.) With certain transistors, if you are near a large station you might still hear something, but normally nothing will happen.

On the other hand, if you advance R5 up all the way, so large a current goes from base to emitter that a very large current flows from collector to emitter. This will amplify not only the negative half of the radio signal but the positive half too. Since both halves are being amplified, the signal is not detected and cancels itself out. The only way to make a transis-

tor work as a detector is to set its bias so that only the negative pulses of signal allow current to flow in the collector circuit. That way the positive pulses are lost, and the transistor detects. In addition, since large currents in the collector circuit flow in proportion to the negative currents in the base, the signal is amplified.

Remember that the advantage of a transistor or tube detector is that you make the signal much louder in addition to getting detection. But there is a distinct disadvantage; can you figure it out? You'd probably have to set the bias on the detector transistor every time you turned on the set. People won't put up with complicated equipment. Also think how a radio or TV owner would complain if he or she turned up the volume and the set got louder, then turned it up more and it quit altogether. You have by now found the critical point on the control, R5, where Q3 will detect. It will be between a low nothing and a high nothing. At the critical point there is just enough current flowing in the base circuit of Q3 to amplify negative parts of signal but to reject positive ones.

Now that you have a working radio of the transistor detecting type, let's find ways to improve its performance. One big disadvantage of any radio that detects at the point where the antenna enters is that such detection absorbs a lot of the radio energy that is placed in the tuned circuit. For a tuned circuit to tune in only one station sharply, the resonant frequency must build into the circuit. If the transistor or diode drinks it in fast, adjacent frequencies also are detected and amplified. If you live where stations are close together on the dial, the radios you have made thus far will probably do a very poor job of separating them. A diode lets current pass one way with very little resistance. This is the case of the radio you made in Chapter 2. The radio you have just made is as bad, for negative pulses travel from base to emitter very easily. Since such a radio will not separate or select stations well, we say that it has poor selectivity.

IMPROVING SELECTIVITY

Selectivity in a radio set can be improved in several ways. In superheterodyne radios (most pocket sets are this type) the signal is not detected until after other processes occur. Regenerative radios, which we will deal with in the next chapter, improve selectivity because of their construction. With transistor or diode detector types, though, you have two choices: (1) you can separate the antenna from the tuned circuit, or (2) you can separate the tuned circuit from the detecting transistor or diode.

Let's separate the antenna from the tuned circuit. Change the connection of the antenna lead-in from point *F* to point *G*. Now the antenna is connected by a 100-Pfd capacitor to the tuned circuit of C4 and L1. Actually you are doing two things here. First you are allowing less energy from the radio stations to get to the tuned circuit. With less energy the stations will separate better. Think of four big football linemen pushing against each other. If they were four little linemen they would not fill up the space so well and would not push so hard. Similarly, this capacitor limits some strength of signals and makes separation of stations a little easier. Try it.

But you should notice something else too. The stations will not come in the same place on the dial setting as they did before. That is because the capacitor in the antenna circuit effectively shortens the length of the antenna and thus changes the input to the tuned circuit. Without the capacitor, C6, if you removed about four turns of the coil of L1 you would notice the same retuning, but the set would not tune more sharply. Remember, a capacitor in the antenna circuit shortens the antenna for purposes of the station signals. Placing a coil in the antenna lead-in will effectively lengthen the antenna. CB radios use this principle in the whip antenna for mobile use. It would take too long an antenna for a car or truck for CB to work at its best, so a coil is built into the base of the antenna.

The capacitor in the antenna gives some station separation. It shifts the tuning a bit, and it probably lowers the volume of the stations received. As you conquer one problem you create another. Selectivity can also be improved by separating the detector from the tuned circuit. Return the antenna to point *F*. Now move the lead from C5 from point *A* to point *D*. You now have L2 in the circuit. Certainly the tuned circuit is not so greatly attached to Q3. How will radio signals get to the base of Q3 at all?

TRANSFORMERS

You have made your first transformer. A transformer is a group of two or more coils so arranged that the electrical pulses from one will produce similar pulses in one or more others. Remember that a coil (also called an inductor) produces magnetism when current goes through it. This magnetism travels through the air to nearby objects. If another coil is nearby, as the magnetism travels through the second coil it produces a voltage in it, and a current travels in the second coil. Transformers don't pass DC from one coil to another, but they do pass AC, and since radio waves are high-frequency AC, also called RF, they pass. Every time a pulse of voice or music goes through L1 a tiny magnetic charge builds up a voltage in L2. This in turn makes a current that travels with the same music or voice pulse into Q3 through C5, which also passes pulses of AC.

The rule of transformers is that the more turns there are in the secondary (L2 is the secondary in this case) in proportion to the primary (L1) the more voltage is passed across. You have lots of turns in the primary (L1) but very few (10) in the secondary, so very little voltage is passed. The detector gets samplings of the signal without taking too much. This is why it will separate stations better. Of course some loss in volume will be noted.

Try the radio this way. Try removing the tape and move L2 closer to and farther from L1 to see how much variance you can put into the selectivity and volume of such a set. Also try changing the antenna to point *G* for even more selectivity. Finally, try winding L2 directly on top of L1 with a thin paper between. As you can see, the more you couple them together the more signal you get and the less selectivity. This transformer type coupling to a tuned circuit is used in a lot of pocket radios. Even though they are superheterodyne, you need all the station separation you can get. Usually both coils are wound on a ferrite rod. There is no external antenna, so the rod acts as an antenna.

Radio frequency (RF) transformers never need to pass great amounts of electricity, so they have only a ferrite rod core or nothing at all as with our set. See Figure 3-4 (middle and bottom). A very different kind of transformer that raises and lowers voltages for power supplies is used in television sets. Because it sometimes passes hundreds of volts from coil to coil, it has a heavy iron core of sheets of iron. The iron helps transfer the magnetism from the primary to the secondary. We will study this matter in detail in Chapter 11. For now remember that the more turns there are in the secondary in relation to the primary the more voltage is transferred to the secondary.

Troubleshooting the Radio

We are now going to fake a few troubles with this radio so that we will learn to spot problems with any electronic device. If your radio works loudly as it is now, you could do these exercises without changing it, but if not, change it back by placing the antenna at *F* and the C5 lead to *A*. Temporarily remove the ground lead from L1. Keep everything else connected to the left lug of C5. You may need to move the C5 connection to *A* for this test if you haven't done so. What do you hear? Why the hum instead of radio programs? Because you have broken the tuned circuit. It

will not resonate to any particular frequency, so the set is picking up random electrical noise.

Any time you get a loud hum in a set with no power supply problems—and this set is battery-operated so a power problem is not likely—the trouble is probably a coil off the ground. It could be a bad solder connection, a break in the coil wire itself, or something similar. Return the ground connection of L1 to its proper mode. Be sure the radio is working normally. Nothing is as bad as having two faults at once, so always be sure to repair the first fault before continuing.

Take a jumper wire and temporarily short across the variable capacitor lugs, or do this with a screwdriver blade while listening. This trouble simulates a shorted variable capacitor or a short from the top of the coil to ground. Your set should be nearly dead. The capacitor, C5, prevents the bias from being grounded, but all the signal energy is grounded. You may hear some noise or hiss from the transistors, but no radio programs. When you suspect a bad variable capacitor, try rocking the shaft back and forth. Chances are some place on the dial it won't short, so you'll get intermittent reception and know that the capacitor needs replacing. Return the radio to normal function.

Lift one end of R1 out of its connection, thus removing the forward bias on Q1. It is good practice to disconnect one battery lead every time you work on the equipment, and reconnect only when ready to listen. You save your battery, but far more important, if you slip and touch the base and collector of a transistor together with a metal object you won't ruin the transistor. Now listen to the set without the bias to Q1 working. The radio will probably play, but with low volume and a great deal of distortion. Get used to this sound; when you hear it, you have either a defective transistor, improper voltages supplied to the set (perhaps a shot battery) or, as in this case, an ineffective bias. The resistor could change value or it could come loose from a circuit board. It will usually look OK, but by moving it slightly with a pencil eraser with the set on you can catch the trouble.

Replace R1 to service and try the same thing with R3, the bias resistor for Q2. Again, there probably will be reception, but it will be poor in quality and usually lower in volume. Replace this resistor. Now disconnect one end of C1. Chances are you will get a howl or squeal. In any multistage radio (one having more than one transistor or tube) the set has a natural tendency to squeal. Each device is drawing current from the battery in pulses, and these feed through from one stage and affect another. The howls can be minimized by bypass capacitors such as C1. They allow this high-pitched noise to pass by and not go through the earphones. Bypass capacitors can be at other places in a radio or TV. Sometimes they go from a collector to ground. Why doesn't all the signal go to ground and get lost? Bypass capacitors are designed to be of values too small for this to happen, but high enough to get rid of howls. Note the difference in this sound from that caused by a defective bias resistor. Reconnect C1.

Now we are going to simulate a completely inoperable set. Remove the battery side of R4 so that it gets no voltage. If you have very strong signals you might still get reception by Q3 acting like a diode between base and collector. Normally the set would not play, but if it does, change the antenna to *G* and C5 to *D*. This should take care of even a good signal. Now you have a dead radio. Remove the antenna and use it for hum tests.

Though you know where the trouble is, pretend you don't. Start back at the last transistor, Q1. You should hear a little hum or at least a click as you touch the antenna to the base of Q1. If you do, go across C2 and test at the collector of Q2. If you didn't hear the hum or click, C2 would be *open* and need replacing. A capacitor is said to be "open" when it won't pass signals. Suppose C2 had been shorted. You would still have heard the click. The problem then would be that a heavy negative voltage would be at the base of Q1,

and it could burn it out. Shorted capacitors usually cause other problems too. Open capacitors usually do not.

Anyway, everything is OK at the collector of Q2. Now hum test at the base of Q2. Again you will hear the hum and it should be louder, as Q2 would amplify it. If you hear nothing or no increase in volume, either Q2 is bad or some item connected to Q2 is making it do poorly. As we found out earlier, if R3 were not connected, Q2 could not amplify properly. Now continue and go to the collector of Q3. Here you would get a hum, as C3 is good and the fact that Q3 is not operating would make little difference. You can easily get fooled.

But now test at the base of Q3 and you will get nothing. It should be a loud hum by now, so you know that something is wrong with the circuit of Q3. You could push it a little and think that it has poor connections. Sometimes this will reveal the trouble. Today, however, you must take voltage readings. Use a DC voltmeter set on the lowest scale above 6 volts. Attach the positive lead of the voltmeter to (+) battery or to the lead to ground. Try at the collector of Q3, using the black or negative probe of the meter. No voltage! Try both sides of R4. Again nothing. In this case the transistor Q3 is not shorted, but the following test will tell if it were: When you get no voltage at a collector try removing the collector lead from the circuit. If you now get voltage across the load resistor the transistor is shorted. When the transistor was in the circuit it shorted the voltage to ground. This can be a troublesome fault. Remember it.

In this case though, you find voltage back at R3 but not across to R4, and you know you have a bad connection. Repair it, and you have a working set again. This shows how you can backtrack through a radio or TV and find the defective stage. Then with a voltmeter you can test and find the bad part and replace it.

An Equivalent Tube Circuit

We will now consider a tube equivalent to the set you have just built with transistors.

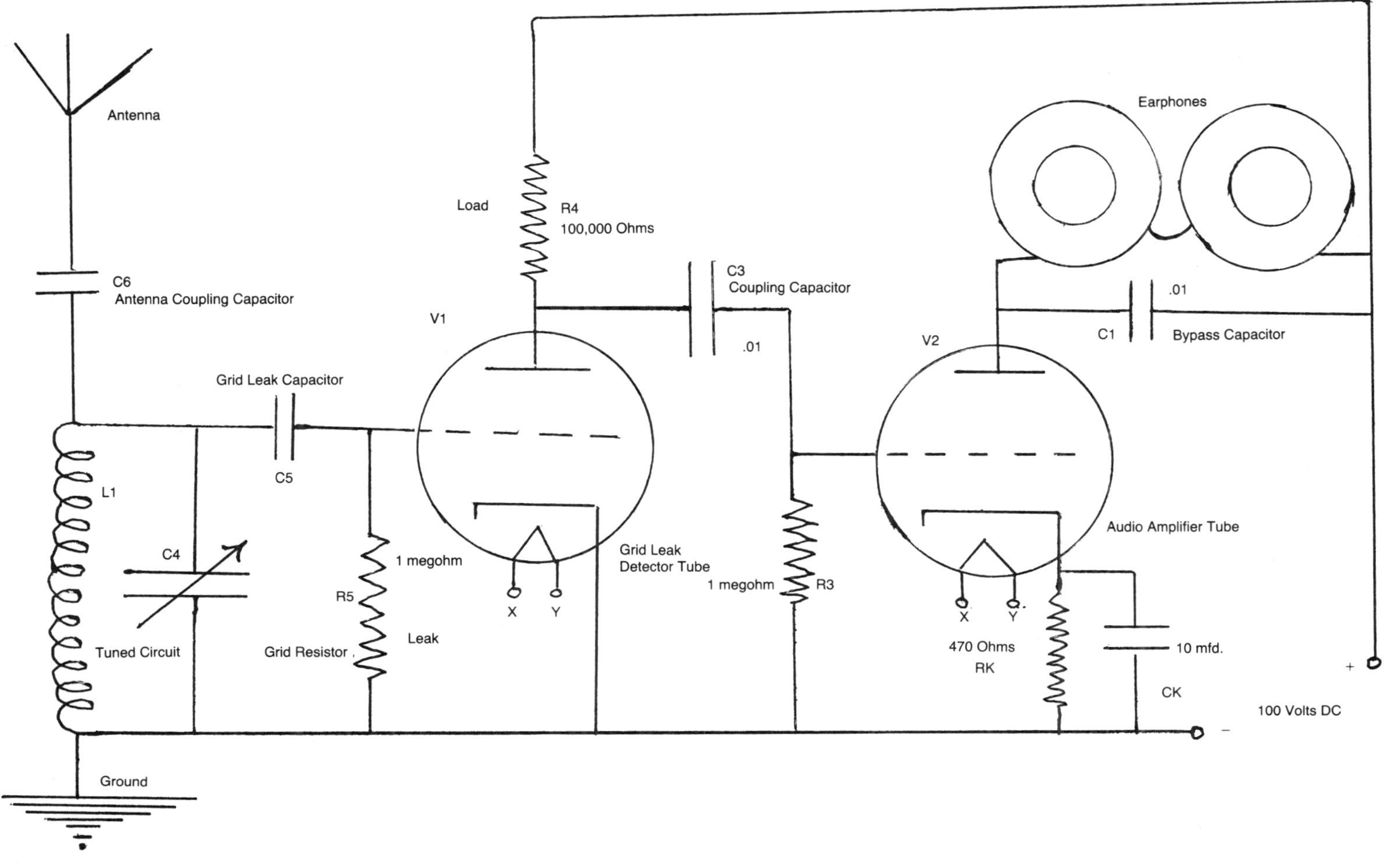

FIGURE 3-3
TUBE AMPLIFYING DETECTOR RADIO

See Figure 3-3. It is not necessary to build this set in order to learn about tubes. This circuit will illustrate the difference between tubes and transistors for similar functions. First, two tubes would likely do the work of three transistors. The higher voltages used account for some of this greater gain. Also, since tubes don't draw much current from a tuned circuit it is not necessary to have a secondary winding such as L2. The tube itself will separate stations better than a transistor. C6 is a good idea though, and does aid some.

I have used the same number to identify parts where practical, but the equivalent parts don't always perform the exact same function. C6 would be the same in Figures 3-1 and 3-3. So would L1 and C4, although L1 might need more turns to get the stations at the same place on the dial. C5 serves the same function but must be much smaller in capacity. It is in the picafarad range (and a picafarad is a millionth of a microfarad). If C5 were measured in microfarads, it would not make up part of the grid leak bias (this will be explained shortly).

Tubes are usually labeled V1, V2, and so on, whereas transistors are given a designation such as Q2. C3 would be the coupling in either circuit, but with tubes it must be heavy enough in voltage rating so that 100 volts won't break its dielectric, the material that separates the metal plates, and short it out. C3 in the transistor radio need not have more than a 6-volt rating. Usually this rating is printed on the capacitor along with its value in microfarads or picafarads. Note that R4 has a high resistance in ohms for the tube model. Since 100 volts is applied to this circuit, great resistances must be used. C1 is the same in both sets, but must have a 100-volt rating here.

Now let's consider tubes. In Chapter 1 you learned that the bias is negative, but the tube has a big positive voltage on the plate. In tubes the bias holds back some of the electron flow to the plate. You learned how RK, as it is labeled in Figure 3-3, develops a bias for the tube, V2. For V1 a self-biasing method that

also detects the signal before the tube amplifies it is needed. Such a system is known as *grid leak bias*. It is still used in tube amateur radio equipment. C5 and R5 work together to develop it.

First, there is no incoming signal so there is no bias. Tubes need bias or too many electrons flow to the plate and could damage the tube. With grid leak bias the set must be designed so that without a signal coming in, the tube is not in danger. In this case the large value of R4 cuts the plate voltage down enough. Some sets also have a cathode bias that works until a signal is picked up. Now, a signal comes in and is tuned by L1 and C4. Since it is high-frequency AC in pulses it goes across C5 without difficulty. Every positive pulse puts a (+) charge on the grid. Electrons travel from cathode to grid, and many go on to the more positive plate. On negative signals they don't. Thus the signal has been detected. Some of the electrons stay on the grid and travel a return path on R5 toward ground. Some pile up on the inside plates of C5 and leak off to go down R5.

Remember that when a current enters a resistor, the side it enters on is the negative side. This develops a weak but steady voltage across R5 with the negative side attached to the grid of the tube. Thus, the tube is held negative most of the time, except when the large positive pulses from the radio station come in. The tube is protected by keeping the grid negative through much of each cycle of carrier. For a small part of each pulse of music or voice, current travels from cathode to grid and out to C5. Then it drains off C5 through R5 and biases the tube. Most of the electron flow goes to the plate, so detection, amplification, and bias all occur in one operation.

Such a system of detection has a little distortion, so it is seldom used in high-quality radios. TV sets tend to use diode detection almost exclusively. The heater leads x and y are not drawn all the way to battery, but it is understood that they go there.

In this chapter you have experimented with transistor detection.

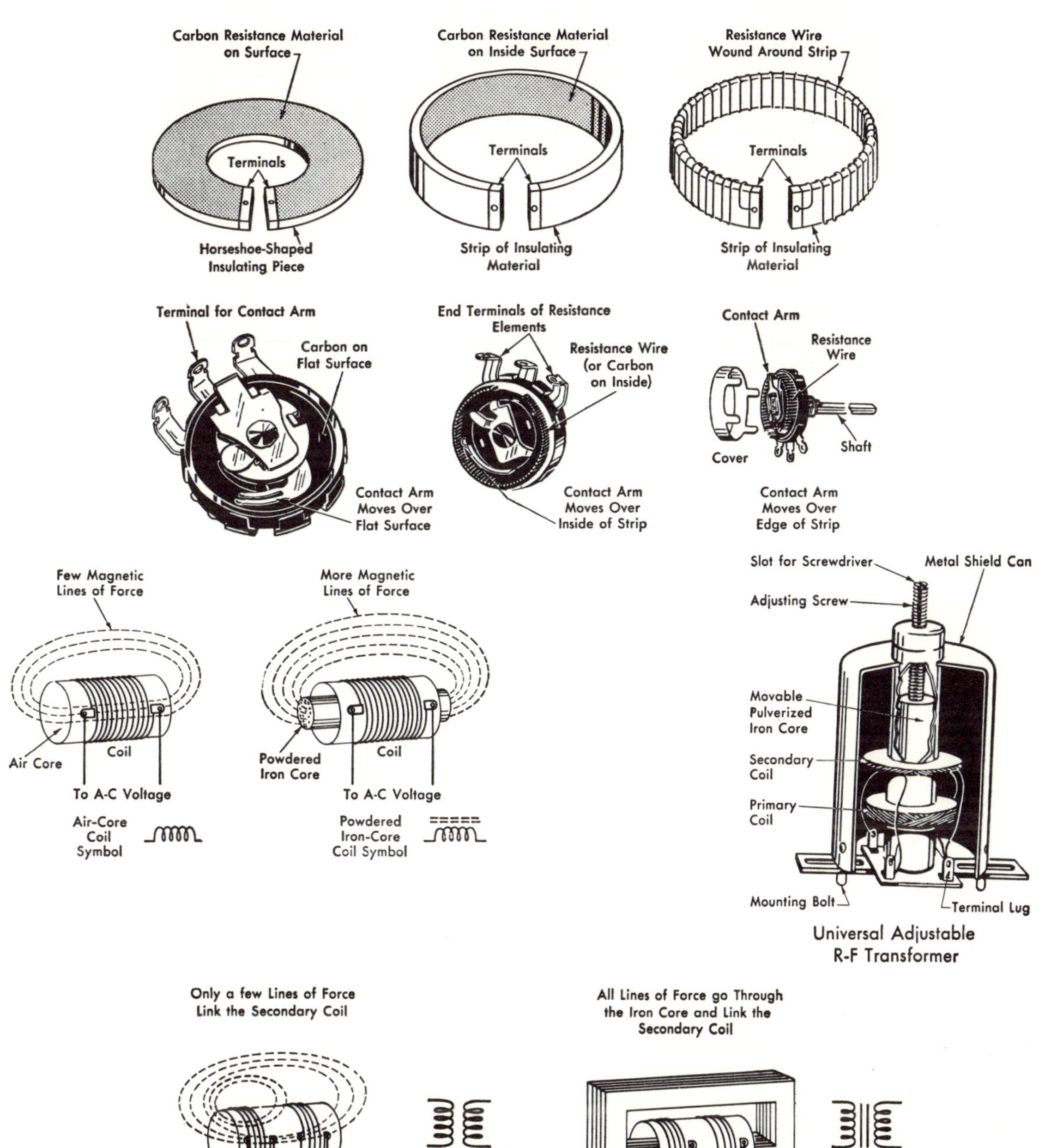

FIGURE 3-4
POTENTIOMETER CONSTRUCTION, RF
TRANSFORMERS

1. Transistor detection gives greater gain than diode detection but is more tricky to obtain.

2. Radio frequency transformers separate the detector from the tuned circuit for greater selectivity.

3. Trace backward through a set collector, base, new collector, base, and so on to find a defective stage.

4. Voltmeters and ohmmeters reveal shorted and open capacitors. Bypass capacitors must be connected and in good condition to prevent hum.

5. Improperly biased transistors cause distortion and low performance.

6. Tubes don't draw current as badly as transistors and give better selectivity. Tube capacitor circuits must have a voltage rating at least equal to the applied voltage.

7. Grid leak detection is an effective way, at the expense of quality sound, of detecting signals.

QUIZ FOR CHAPTER 3

▶ 1. Would a capacitor be open or shorted if no signal would travel across it? Why?

▶ 2. A coil in series with an antenna would have what effect on it electrically? Would an antenna coil make the set more selective?

▶ 3. What type sound gives evidence that a coil might not be properly connected to ground? How could you find the defective coil or connection?

▶ 4. A device that transfers voltages from one coil to another one is known as a ______ . If such a device had a primary of 100 turns and a secondary of 10 turns, would the voltage be greater or smaller in the secondary than in the primary?

▶ 5. Why will a transistor not detect if its forward bias is too great?

4. Regenerative Receivers

PROJECT: Constructing a Broadcast Band Regenerative Radio

WHEN MAJOR Armstrong got his earphone close to the tuning coil of the radio he was experimenting with, heard the message get louder, then heard a big whistle, he took radio to a new level. He had discovered regeneration—not the kind that makes man a new creature, to be sure, but the kind that takes an output signal from an amplifier and feeds it back to the input of the same amplifier to boost the signal.

The Armstrong patents were numerous and were said to cover everything except the paper on which they were written. Regenerative radios were the ultimate in getting the most for the least during the late 1920s and into the 1930s. A three-tube regenerative receiver could pull in stations about as well as a six-tube superheterodyne receiver, and it would separate interfering stations almost as well. True, the "superhet" did surpass it slightly on both counts, but the better performance was far outweighed by the fact that superhets cost a great deal more, used batteries much faster, and, in those days, required considerable skill to operate and align.

Nevertheless, the superhet won out, and today most radios with four or more tubes and six or more transistors are superheterodyne circuits. All TV sets, to my knowledge, are superheterodyne, as are the receiver parts of larger CB transceivers. Why did the regenerative radio bite the dust? First, superheter-odyne sets have been improved a great deal since that time and now can be operated relatively simply. Even a TV set can often be operated with just a turn-on and channel change. In the old days a superhet radio had two dials that had to be carefully tuned at the same time, and the readings on both would not be the same for a station. Maybe you'd set one dial at 90 and the other at 67 to get 550 kHz. You'd have to remember this or log it on paper. Second, superhets take fewer tubes or transistors than they once did, and they are made a lot less expensively. You will learn about superhets in the next chapter.

Even with improvements in superhets, stage for stage, transistor for transistor, a regenerative radio will outperform a superhet in pickup and will tune and separate stations nearly as well. Regenerative radios are used in a few instances today. Handheld CB receivers are often a type of regenerative radio when in receive mode. Two- and three-transistor radios are often regenerative or reflex, which is similar to regeneration. Amateurs still use regenerative equipment, and so do radio hobbyists. Though it is worthwhile learning about regenerative radios for these reasons, it is also a logical step before studying transmitters, as many transmitters use a circuit very similar to a regenerative receiver to produce the carrier.

THEORY OF REGENERATION

Consult Figure 4-1. You have been using L2, the secondary coil, as an input to the transistor, Q3, to limit loading of the tuned circuit. Note that in a regenerative radio, L2 is in the collector circuit instead of the base, or input point. Here is what happens now. Tune C4 to a station. All higher-frequency stations go up and down C4 with very little resistance. All lower-frequency stations go up and down L1 between antenna and ground. The resonant frequency finds an easier path through the transistor, Q3. Now you could set the potentiometer, R5, so this transistor would detect the signal as you did in the last chapter. You

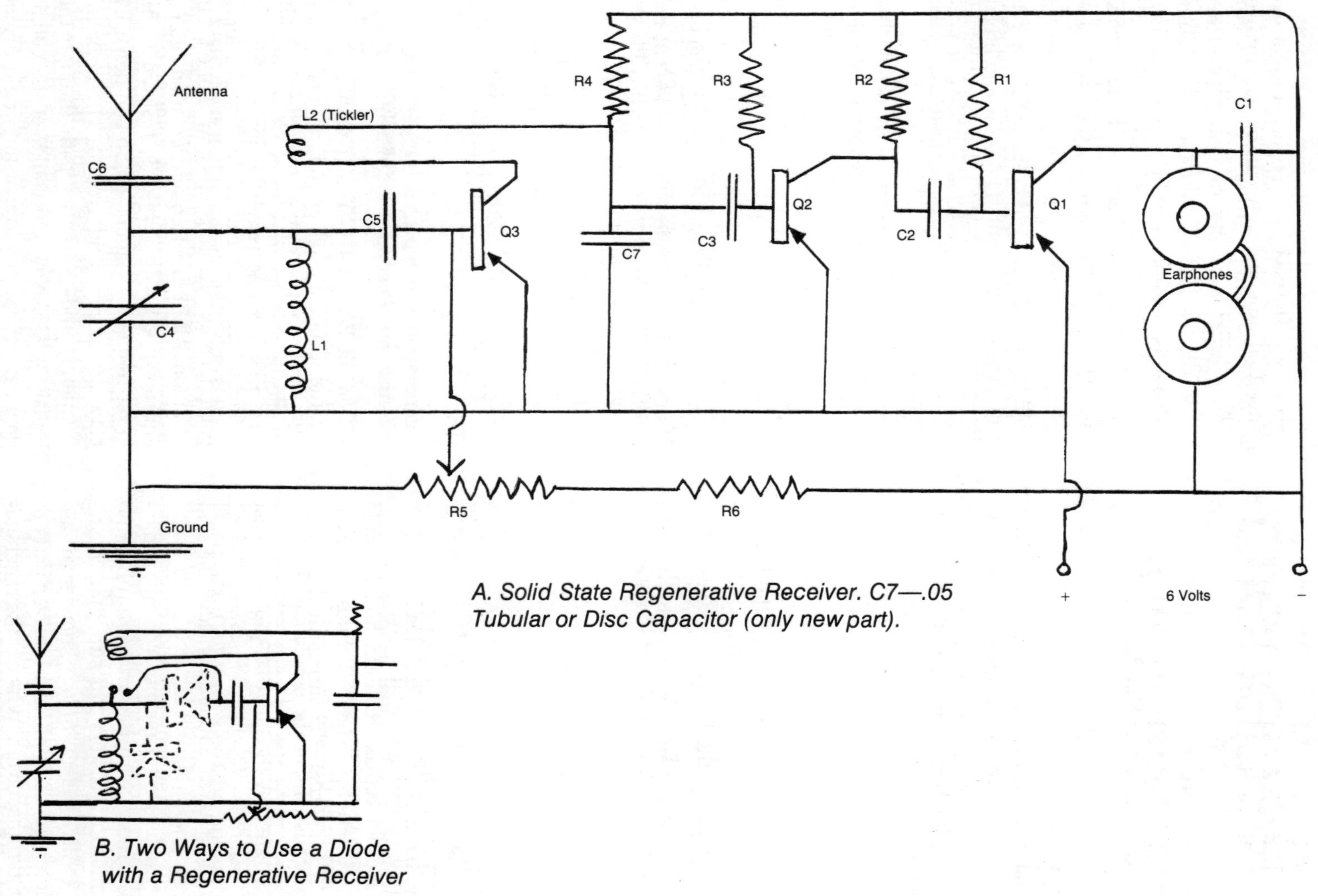

A. Solid State Regenerative Receiver. C7—.05
Tubular or Disc Capacitor (only new part).

B. Two Ways to Use a Diode
with a Regenerative Receiver

FIGURE 4-1

don't need to worry about this, though, as the regeneration will detect the signal itself.

An incoming pulse of current will go down from base to emitter and to ground of Q3. This will open the barrier wider, and current will travel through from collector to emitter of Q3. To get there it must start at negative battery, go down through R4, through the tickler coil, L2, and then on through the transistor and back to positive battery. This is all a series circuit, and when the barrier opens up from collector to base of Q3, a pulse in time with the bit of information from the broadcast station goes through all these places. As the current pulses through R4 a stronger signal is built up that goes through C3 to Q2 for more amplification. The rest of the amplification to earphones you should know by now.

L2 is connected rather close to L1, although you wouldn't know it by the schematic. The pulse that goes through the transistor must first come through L2. Because it goes through a coil at L2 it develops magnetism that cuts across L1 and adds current to L1. In theory it can raise the effective amplification of the transistor to the second power—that is, if a transistor has an amplification of 50, with regenerative feedback it has a potential amplification of 50×50 or 2,500. One problem exists, however. If it is allowed to feed back this much signal the transistor and circuit will set up oscillations and become a miniature radio transmitter or oscillator. The trick is to send back as much signal as possible just short of this oscillation point.

One way to do this is to separate L2 from L1 physically just shy of the oscillation point. In many early tube radios L2 was on a shaft that could be rotated just above or even inside of L1. Adjusting a knob on the front panel moved L2 to the critical place. Or one could make R4 a potentiometer and control the amount of collector current the transistor would have. If less current were supplied, the pulses across L2 would be less.

Another way is shown in Figure 4-1, controlling the forward bias of the transistor by adjusting R5. In this way you can limit the

general opening of the barrier between collector and base, and thus control the amount of collector-to-emitter current. This limits currents across L2 equally. R5 would be set just below the oscillation point for most volume and station separation. When the set starts to oscillate it will whine in the earphone or earphones. Cranking up R5 even higher would stop the whine or howl, but the set would be oscillating above the audio range and you just wouldn't hear it. It would oscillate at the rate set by the tuned circuit. The set would broadcast its own carrier, and if tuned to a station, it might rebroadcast the signals from the station to other radios nearby.

In the early days when many people had regenerative radios some would turn theirs too high and would rebroadcast signals from stations far away. Neighbors with crystal radios would pick them up and think that they were receiving these stations directly.

Remember, a signal coming into a tube or transistor as a negative half of a cycle will be a positive half of a cycle when it goes out the plate or collector. Either device reverses or inverts the polarity of the signal. For this reason turning the coil L2 one way would feed back signals just wrong, and they would tend to cancel out the input signal. Turned the other way they increase it, as the polarity is right for the magnetism to cut across to L1, the primary coil, and add to its current. As shown, for regeneration the top of L1 must go to the antenna end. The top of L2 must go to R4, or nearest battery $(-)$. The collector side of L2 must be wound closest to the high or antenna side of L1. For greatest effectiveness L2 should be slid far away from L1 so that R5 must be turned nearly all the way up for the set to go into oscillation. It might seem that L2 should be a tuned circuit the same as L1 with many turns of wire and a capacitor across it.

Such circuits were used in early days, but they were not terribly good. The reason is that this circuit detects the signal through the feedback path. Once a circuit is detected, further tuning is useless or nearly so. A detected signal is at audio frequency and not at carrier frequency anymore. A few turns of wire for L2 are sufficient.

The detection principle is rather interesting. With a weak negative signal going from base to emitter of Q3, a large pulse in time with it feeds back from L2 to L1. This in turn is very quickly amplified even more and produces very strong signal amplification. The negative pulse is much stronger. The other half of the signal that is positive is not stopped by a diode, but though it might travel the R5 route, it won't cancel the negative half, as the upper resistance of R5 would limit it. Also the strength of the feedback signal would overcome the opposite charge. With tubes or NPN transistors in this circuit (and appropriate batteries, of course), the positive half of each cycle of the signal would be amplified and fed back. As is, however, only the negative half of each cycle of the signal is amplified, fed back, and detected. The positive half is lost in the circuitry of R5 and overcome by the feedback, which is negative in this case.

A point of confusion might arise here. Some books refer to regenerative feedback as "positive feedback." This does not mean that it is necessarily positive in polarity, but that it strengthens the signal. The term "positive feedback" originated when only tubes were used, before the PNP transistor, so it literally was positive in nature. These same books refer to degenerative feedback as "negative feedback." Degenerative feedback was explained in Chapter 1. In the case illustrated there, the signal would have been of negative polarity, but it would not always be so. A positive voltage to buck against another voltage and reduce the signal strength would also be degenerative or "negative" feedback.

Let's summarize what we have covered so far:

1. Regeneration is a process of taking the output of a transistor or tube and feeding it magnetically back to the input in order to strengthen the signal.

2. The regeneration must be controlled, for it can result in oscillation if too much feedback is achieved.

3. It is possible through regeneration to raise the amplification almost but not quite to the second power (multiplied times itself).

4. Regeneration not only increases the amplification but increases selectivity of the radio receiver.

The last point has not really been explained. It is quite simple, however. Since only the tuned-in station is being fed back, this makes it stronger in the tuned circuit. Remember, one reason a transistor doesn't tune sharply is that it takes away currents from the tuned circuit. It not only takes current from the resonant frequency, but it takes a little from those on either side of it. With regeneration, current is being put back into the tuned circuit, so this drain is minimized.

Some regenerative circuits don't use magnetic or inductive feedback directly. They have a capacitor that takes some of the collector current back to the tuned circuit. This would make the signal out of phase with the base. It takes a circuit such as that in Figure 3-1 using L2 as a secondary, plus this feedback capacitor, to accomplish regeneration. Since between L1 and L2 there would be magnetic coupling, magnetism still plays a role here. Such a system is harder to control.

BUILDING THE RADIO

Now you are ready to construct a regenerative radio for the broadcast band. The circuit of Figure 3-1 will be used and changed somewhat, and C7, a bypass capacitor, will be added. Though the radio would work without C7 in theory, C7 helps make the control of regeneration easier. It also gives a cleaner signal. Though the regeneration circuit detects the signal, some RF does continue on. C7 bypasses it to ground.

Since C7 is 0.05 microfarads and C3 is 0.01 microfarads, it would seem that much of the usable signal would travel to ground instead of going through the smaller capacitance and transistor. Actually, plenty of signal will go through C3 and all will work properly. C7 puts the top of L2 at RF ground potential and this aids in feedback, as L2 is a little more like a tuned circuit if that side is grounded for RF. True, not much RF is left, but any amount could be a nuisance. A convenient place to add C7 to the circuit would be between R4 and the left lug of C4, the variable capacitor. Connect one end of C7 to the junction of R4 and C3. Solder the other end to the left lug of C4. This end grounds C7.

Remove the connection of the collector of Q3 to the junction of R4 and C3. Remove the lower end of L2 from ground (left lug of C4), and instead fasten the lower end of L2 to the collector of Q3. Connect the lead from the right side of C5 (as shown on Figure 3-2) to point *A* if it is not already there. Connect a lead from the high side of L2 (point *D* as labeled on Figure 3-2) to the junction of R4, C3, and C7 (which you just installed). Recheck your wiring and compare it with Figure 4-1.

The antenna should be at point *G*. Connect the battery and the ground if you have it off, and put the phones on. Since this radio will squeal unpleasantly, it is a good idea to put the phones near but not on your ears (maybe on your cheeks) until you get the hang of operating the set. Turn R5 up enough to hear stations. Tune in a station strongly and then run R5 up until it starts to howl. Back off just a bit. You now have that station at its best tuning.

You will find that slight retuning and touch-up of regeneration will help bring in hard-to-get stations. Chances are you will need to readjust the regeneration control, R5, as you move the tuning capacitor. You may wish to make a dial and paste it on the front of the radio showing where different stations may be found. If you want to further separate stations and are willing to settle for less volume, you can make a tap about two-thirds of the way down L1. Leave C4 and the antenna capacitor, C6, still attached to the top of L1. Change the lead to C5 down to this tap.

Stations are much easier to separate now, but you will have to adjust the regeneration control carefully to get them in without oscillation. This set will usually work without a ground, but as you bring your hands near the dial knob to change stations you will change frequency just by being near. In addition, if you touch the earphones or change your sitting position and lengthen or shorten the stretch of the earphone wire you'll change frequency. It's best to have a ground.

As a further experiment, you might wish to try the diode in two different ways with this regenerative radio. See Figure 4-1, lower left. With the parallel diode you must have a wire to connect the top of the coil, antenna, and variable capacitor to C5. When you use the diode in series mode, no wire is used. It may be necessary to move L2 rather near L1 to get regeneration with a diode in the circuit. For best control of regeneration in any regenerative radio, the tickler and tuning coils should be separated as much as possible, so that regeneration can only be achieved with a high setting of the control resistor or other device. If the coils lap each other the set will jump from very low reception immediately to squeal with very little room for reception between. Keep this circuit intact (original without diodes) for the next chapter.

QUIZ FOR CHAPTER 4

▶ 1. Another name for the regeneration coil, L2 in your set, is ______ ______ .

▶ 2. What type of radio is superior in sensitivity and selectivity to a regenerative radio, although it might need more stages to operate?

▶ 3. If a regenerative detector circuit had a transistor with a gain of 20, which of the following is a reasonable gain for the entire circuit? 40, 360, 450

▶ 4. List and explain three practical ways of controlling regeneration in a regenerative detector system.

▶ 5. Regeneration beyond the point of howl produces ______ .

5. Shortwave Radio and the Heterodyne Principle

The term *shortwave* needs explaining. All radio waves have a length, related to their frequency. The lower the frequency that is generated in an oscillator, the longer the waves that will radiate from the antenna. The lowest frequency allowed for commercial broadcast in the United States, 540 kHz, would produce waves that would be approximately 1,822 feet or 555.5 meters long. These waves would occur at the rate of 540,000 a second.

Remember that the modulation affects their height or amplitude. An antenna long enough to give off a full wave at a time would have to be 1,822 feet high or would have to be a long wire of that length strung between towers. Either is impractical, so a quarter length antenna (often called a *Marconi antenna*) is used at such a radio station. A quarter of 1,822 is 455.5, a length that is not impossible for an antenna. The other three-fourths of the signal is fed into the ground. Sometimes with a very long antenna even a full quarter wave is not made. Perhaps the antenna would be 200 feet high, and a big coil would be loaded up in series with it. Such an arrangement works fairly well.

The top of the broadcast band, 1600 kHz, is at considerably higher frequency, so the wavelength is less. In this case approximately 613 feet make a full wavelength. A quarter length antenna would be only 153.25 feet. Even so, this is a much longer antenna than is needed for shortwave broadcasts. In general, shortwave is considered as anything above 1600 kHz. The longer the wave and the lower the frequency, the more power needed to transmit any given distance. Actually, even with standard broadcast it would help to have an antenna hundreds of feet long for a pocket radio. Since this is impractical, radio stations bombard the air with thousands and thousands of watts of power to get the message out.

The higher the shortwave frequency, the farther a message will go for a given amount of power. The trouble is that the higher the frequency, the more the message, carrier and all, travels in a straight line. We could bounce a signal off the moon with very little power. A few transistors or tubes would do it at times when the moon appears on the same side of the earth as the transmitter, because the waves could go in a straight line. Trying to get shortwave to follow the curvature of the earth is the problem. Two amateur broadcast bands just a little above the standard broadcast band of frequencies are low enough to follow the earth's curve reasonably well. These are the 160-meter band (1.8 to 2 MHz) and the 80-meter band (3.5 to 4 MHz).

You can adapt your regenerative radio to get the interesting broadcasts on these frequencies. You will also find foreign broadcasts—many in English intended for listeners in the United States. At higher frequencies the straight-line problem increases, of course, but amateurs, foreign stations, and so on use what a CB operator calls "skip." The signal goes upward, bounces off the ionosphere (the electrically charged layer of thin air just above the atmosphere), returns to earth, maybe bounces off the earth, goes up again, down, and so on. This is not terribly reliable broadcasting, but it does work, and many persons have communicated with radio operators on the other side of the world with bounced signals. Such broadcast doesn't call for a lot of power, just good luck and favorable atmospheric conditions.

TV stations operate from 54 MHz upward, using line-of-sight or straight-line broadcast. After about 30 miles, depending on the terrain, the signal goes out into space because the curve of the earth begins to fall away. Up to about 50 miles a high receiving antenna can usually bring in good pictures and sound. Beyond 50 miles it usually takes a very tall antenna and a good amplifier system like those used for cable TV. Even so, reception is terrible at times. TV stations operate with tremendous transmitting power. This power helps because the small part of the signal that doesn't go out into space is stronger.

Figure 12-1 *A* shows measurements of wavelength in inches for each half of the antenna. For instance, a full wavelength for Channel 2 would be about eight feet. For Channel 13 it would be only a little over two feet—quite a difference from the 1,822 feet for the lowest-frequency radio station. The waves really are short in shortwave. UHF TV wavelength is very low in feet, and microwaves are literally inches long. The big dishes you see at microwave stations are catchers or reflectors. The actual antenna is a little component in the middle. Microwave stations are usually about 25 to 35 miles apart and are constructed in high hills with high antennas so that nothing will get between them and bounce the signals away.

Since shortwave signals are of higher frequency than standard broadcast, a shorter antenna and tuned circuit of less coil and capacitor is necessary for the reception of the signals. Leave your long antenna alone, however. For the signals you will try to hear, an antenna of 200 feet would be none too long. You needed one of over 1,000 feet for the radio stations. So unless you are striving to pick up CB (and even then a straight wire of over 30 feet would be needed), keep your antenna up.

The radio now needs some attention. Fortunately you have a regenerative detector transistor that will amplify and detect higher frequencies. An audio type transistor will fail altogether at about 1500 to 2000 kHz. Some won't even detect the broadcast band, at least the upper part of it. A high-frequency transistor cannot usually be identified by its appearance; you have to know the number of the unit and its characteristics. It is very poor policy to replace a bad transistor in a radio or especially in a TV with any one that matches emitter, base, and collector and PNP or NPN. You should also know whether it will handle the required frequencies, whether it needs an insulation shell, and so on.

Remember that the more turns there are in a coil the lower the frequency it will tune in, so to speak. Actually, more turns make it harder and harder for low frequencies to get through. It was already rejecting the high ones. Since low frequencies can't get through the capacitor, adding turns to the coil keeps pulling the resonant frequency, the one that is equally repelled by the coil and variable capacitor, down lower and lower. Now you want to tune in shortwave, which is high-frequency. It seems logical that removing turns from the coil would tune in higher frequencies. Or you could use a variable capacitor of less capacitance such as one that would swing from absolutely no picafarads (an impossibility) to about 50 pfd. The trouble is that you would not be able to tune in a very wide range of signals. It is best to keep the capacitor you have and cut the coil.

Let's explore how a coil of fewer turns will make the radio tune to higher frequencies. Removing turns makes the coil less able to produce magnetism when a current is run through it. The higher the frequency of this high-frequency or RF current, the more the magnetism builds and bucks against the incoming RF. That is why a coil resists higher frequencies and passes lower ones. With a coil or fewer turns than before, the coil has less ability to build magnetism and reject RF signals. This allows many more medium-frequency signals than before to go up and down through the coil to antenna and ground. They are shorted away from the radio detector.

Regular radio stations such as KMOX in St. Louis, WLW in Cincinnati, XERF in DelRio all now go through the coil, whereas before they could be made to travel to the diode or other detector by adjusting the variable capacitor. With fewer turns in the coil they are lost. Higher frequency hams, foreign broadcasts, barge navigation traffic, code, and so on, all of which the coil had been rejecting, are now not rejected by it so strongly. They were rejected so hard before that they were forced across the capacitor to ground and antenna. Now they find an equal resistance between coil and capacitor, and depending on the setting of the variable capacitor, one of them will now be the resonant frequency and go to the detector. Only the resonant frequency that finds equal resistance between coil and capacitor in the tuned circuit will travel through the detector.

There is a formula for number of turns to tune various frequencies, but because transistors load a circuit, and because coil forms also change its inductance, a few practical suggestions will be considered instead, as the formulated turns ratio usually doesn't work very well. If you want to get the area just above the broadcast band and hear the 1600-kHz radio stations with the plates of the variable capacitor fully meshed, 17 turns for L1 will be about right. This assumes a fairly long antenna and a good ground. If, on the other hand, you want a compromise coil that should tune in code, hams, foreign broadcasts, and the like, about 9 turns wound close together will probably do

the trick. To go to the highest level at which this transistor is capable of producing regeneration, 6 turns should be used.

In these experiments with shortwave you must have the least feedback that will barely go into oscillation at the highest setting of the potentiometer. For that reason the turns of the tickler coil, L2, also should be cut down. Depending on the strength of your transistor, you might get by with four turns, five turns, or at most six. You will have to experiment.

Before doing the actual cutting think how you want to achieve the new coil arrangement. Later you will need to restore the broadcast band regenerative radio. You could use another piece of PVC pipe and wind entirely new coils, or you could remove most of the wire from L1, save it, and splice it back on later. Either way, don't just junk all the enameled wire. You will need it for the transmitters in the next chapter. If you splice later, be sure you scrape clean the point where you rejoin the wire.

Adapt the coil, L1, for shortwave and experiment with L2 to find the minimal number of turns that will work. With L2 you can lift turns off, retape the wire, and so on; you need not cut the wire. Later you will need all ten turns again. When you have finished, try tuning in. Although shortwave reception is normally better at night, you have a fair chance of getting a few things in the daytime. At night the ionosphere bounces signals much better.

You should be able to pick up code rather easily. Some of it, modulated code, already has a tone. At the transmitter a tone modulation is added to the carrier so that it will produce this tone in the receivers. Other code is not modulated and will come in as hisses of "di daah's." Now here is one good feature of a regenerative radio. You turn the regeneration just up to the point of squeal, and it will squeal in sync with the di daah's. A superheterodyne radio for shortwave has a beat frequency oscillator (BFO) to beat against this unmodulated code to produce a note that can be heard. This requires an extra circuit. This

BFO is not the regular oscillator that tunes the set, as you will soon learn. With the regenerative radio, however, you can hear unmodulated code by just adjusting the regeneration.

You probably will be able to get WWV for time signals without too much difficulty. Each tick you hear is a standard second, exactly produced. Then the voice will give the time for 0 degrees longitude, which is the time at Greenwich, England. If you live in the central time zone in the United States, subtract six hours from the time given, but leave the minutes the same. For instance, if the announcement over WWV indicates that the time at the tone is ten hours, thirty minutes, Greenwich mean time, in the central time zone it would be 4:30 A.M. During daylight saving time, it would be 5:30 A.M. Times are given on the basis of a twenty-four hour day, so 2:00 P.M. would be 1400 hours. If the announced time is fourteen hours and ten minutes, central standard time would be eight hours, ten minutes, or 8:10 A.M. and daylight saving time would be 9:10 A.M. Try to see if you can figure out the time in your zone with this message. For eastern standard time, subtract only five hours; for mountain standard time, subtract seven hours; and for Pacific standard time, subtract eight hours.

Foreign broadcasts will come and go. They will fade, then get strong, all because of the effects of the ionosphere. Sometimes, if you are lucky, you can hear an entire broadcast from across the ocean. Such broadcasts generally have a propaganda purpose, but it is interesting to see how much you can pick up. You can hear much mobile shortwave—delivery and dispatch orders, call-in from crews of utility workers, barge traffic on large rivers and lakes, perhaps even mobile phones on ships at sea. Chances are you will hear amateurs—hams—communicating with each other. In some cases your set might get CB, depending on conditions. If your area is fairly saturated with CB operators you may hear them.

After you have had some fun with this radio, review the following summary of what you have learned so far in this chapter. Turn the radio off; you will go back to it for more work later.

1. The frequency of a carrier signal and the length of the waves it produces are inversely proportional. High frequencies have short waves. Low frequencies have long waves.

2. Wavelength can be expressed in meters or feet. A meter is 3.28 feet.

3. Decreasing inductance (the number of turns in a coil) or capacitance increases frequency in a tuned circuit.

4. Lower-frequency signals need longer antennas than do higher-frequency signals.

5. Most radio stations use a quarter-wavelength antenna and depend on the ground to carry three-quarters of the wavelength.

6. High frequency requires less power to broadcast, but the signal travels in a straight line.

7. TV stations use great power to help get a signal across land better, even though most travels out into space.

8. Superheterodyne, shortwave receivers often have a beat frequency oscillator (BFO) so that unmodulated code can be given a tone.

9. The ionosphere bounces signals back to earth, especially at night, to permit skip broadcast reception at high frequencies.

THE HETERODYNE PRINCIPLE

The heterodyne principle of radio dates back to the early years, and some people think of an old radio when they hear the term "superheterodyne." The truth is that most of the radios in use today are superheterodyne. TV uses the same superheterodyne circuit designed for higher frequencies and for a band of several adjacent frequencies, as you will learn later. Because it is also used for TV, you should understand the heterodyne principle for radios. (You also may wish to repair radios occasionally.)

The idea of heterodyning is to beat two frequencies against each other in some kind of circuit. This mixing can produce several things. You can tune in either one of these carriers or their mathematical sum or difference. If you had one signal of 1000 kHz and another of 1500 kHz and you heterodyned them, you could get either 500 kHz or 2500 kHz with the right circuit. The amazing thing is that if one of them had modulation it would still occur on the frequency produced by adding or subtracting. The idea, then, is to change the frequency of the carrier but to leave the intelligible music or talking alone—that is, to keep it as is for detection later.

Two good things are accomplished by heterodyning:

1. Changing the carrier frequency gets rid of frequencies on either side almost completely, so selectivity of the radio is much improved.

2. Since most tubes and transistors amplify best at low frequencies, a lower carrier is produced to be amplified with its modulation.

For radio, the new frequency is usually 455 kHz. This new frequency usually is amplified by one tube, or two in the case of transistors. Since only the one frequency is amplified, this also helps get rid of stations on either side of the tuned-in station. The new frequency is called *intermediate frequency* or *IF*. If a radio has one tube for this amplifier, it has one IF stage. Of course, the IF has to be detected as all RF signals do. Any radio waves are both positive and negative. Half the carrier must be removed or the waves will cancel each other out in an audio amplifier. In IF amplifiers they work all right, as the bias is different.

See Figure 5-1 *A*. The tube, V1, is known as a *pentagrid converter* when used this way. The cathode and first grid, pin 5, along with other items, form a miniature radio station oscillator. It makes a carrier without any modulation. At the same time the signal from the

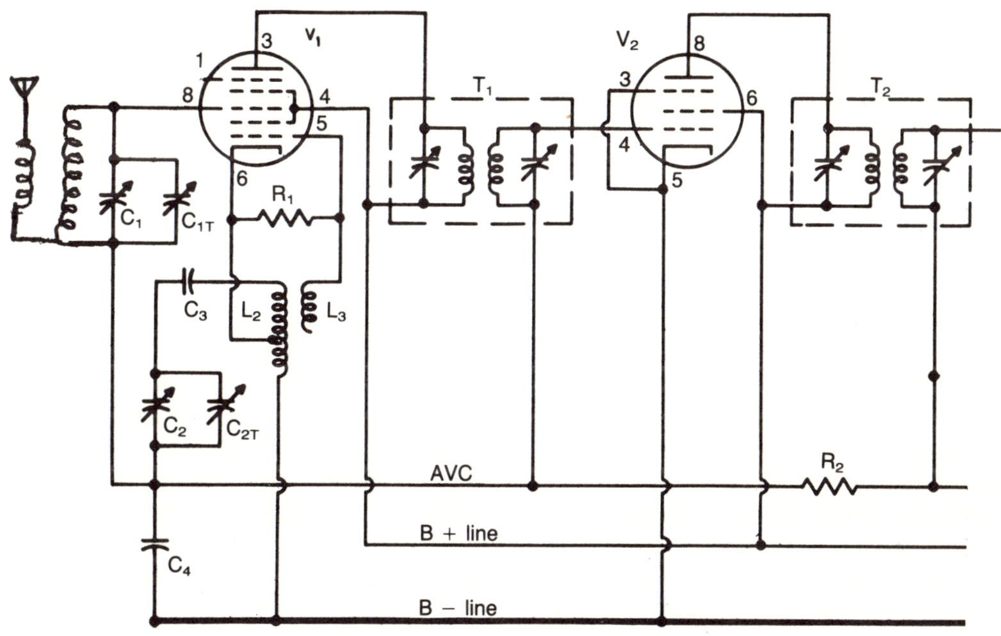

Figure 5-1 A. Oscillator-Mixer and Intermediate Frequency Stages of a Tube Radio

radio station is being tuned by a coil, the secondary of the antenna transformer, and capacitor, C1. This signal is applied to grid pin 8. The tube will mix these carriers together along with the modulation coming with the carrier from the radio station. There are two tuned circuits in T1. Each is tuned to the difference between the incoming signal from the radio station and the unmodulated signal made by the tube and other parts. This is set at 455 kHz. C1 and C2 are on a shaft, and C2 is enough smaller than C1 so that the difference between these signals will always be 455 kHz.

Let's see how the oscillator works. Current always travels from cathode to plate in the form of electrons. First the current must come from the power supply. To get from B − to the tube the current must go through L2. Most of the electrons leave the cathode (the heaters are understood but not shown) and go toward the plate. Some strike the grid with pin 5 designation and send a signal down to L3. L3 is a type of feedback or tickler coil with only one attachment. The current, feeble though it is, travels across R1, the grid leak resistor. Signal pulses are fed back, setting up oscillations, which are tuned by L2 and C2 so that the oscillations are 455 kHz above the incoming signal at all times.

As the shaft is turned to tune in stations, C2 is set as C1 is set so that this oscillation in the cathode, first grid circuit, is always 455 kHz higher than whatever signal is being tuned to by C1. After oscillating, the carrier (with no modulation) goes on up through the cathode by other grids, and since the signal controls electron flow at pin 8, the signals mix. The oscillator is similar to the regenerative receiver, except that the regeneration is uncontrolled. You want it to go beyond the point where it howls. You want it to make radio waves.

The tuned circuits in T1 tune in only the difference frequency. You are finished with the incoming carrier from the radio station and the carrier you made in the lower circuit. You now have only the IF. V2 amplifies this intermediate frequency greatly with the modulation. T2 is also tuned to 455 kHz in both sides. The double-sided signal is now taken to some kind of diode to cut off one half and make the signal capable of audio frequency amplification.

The automatic volume control (AVC) line might need explaining. A sample of the detected signal at the diode is developed across a resistor. A small voltage is fed back to a grid of both V1 and V2. If the signal is very strong a strong negative voltage is fed back, biasing these tubes so they amplify less and cut down the signal. This is automatic volume control. Even when you set a manual volume control to the desired volume for a program, the AVC helps hold the signal down if it starts to get louder. If it starts to fade, the voltage at the diode will be less, and less bias is applied at pin 8 of V1 and pin 4 of V2, so the tubes work harder and try to build up the volume. Sometimes a program fades anyway, of course, because the tubes can only do so much. Remember AVC. A TV has a similar mechanism.

In a television set the frequency is not dropped down as far as 455 kHz. Since sound, picture, and color signals are being received, all on different frequencies, there would not be room for all of them. Instead, the frequency is set to an IF in the 40 MHz range. (This is lower than any station, however, as they start at 54 MHz. Special tubes and transistors have been developed that can amplify well at that frequency level.)

Figure 5-1 *B* shows a transistorized version of a superheterodyne receiver. These are NPN transistors and have positive voltage at the collectors and negative at the emitters. The first transistor at the left has the oscillator in its emitter circuit tied in by C3. The feedback of signal path to produce oscillations is a tickler coil out of the collector. Then the line goes on to the first IF transformer. The coil with C1A is a loopstick and replaces an external antenna. The input from the tuned signal is at the base of the first transistor. Mixing takes place within the transistor. There are two IF

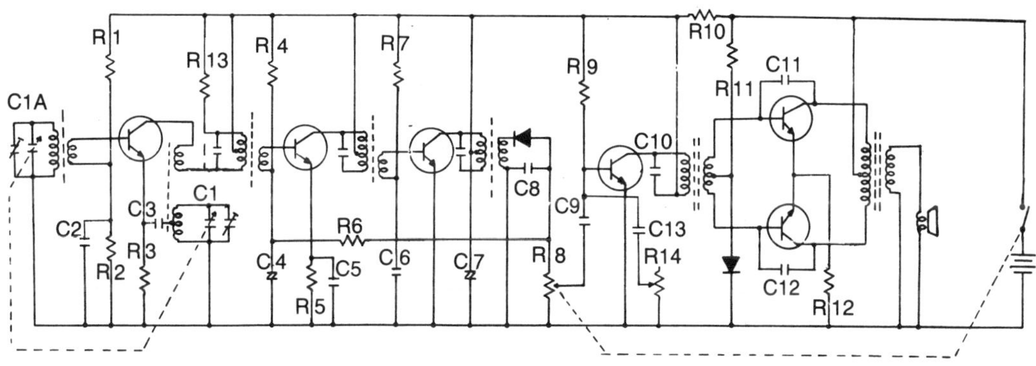

FIGURE 5-1 *B. Transistor Superheterodyne
Pocket Radio with NPN Transistors*

transistors and three IF transformers, one before the first transistor, one between the two, and one following the last that goes to the diode. Note that only the primaries have a tuned circuit of capacitor and coil. The secondaries have only the coil, thus matching transistor base inputs better. With three transformers in the IF, it is not necessary to tune both primary and secondary.

C8 bypasses some of the carrier as the diode detects. R8 is the set volume control. The AVC is developed here and sent back through R6 to the base of the first IF transistor amplifier. In this drawing wires are connected only if a dot is shown. Wires drawn across each other without the dot are not connected. You will have to get used to many variations in electronics. The negative AVC current is correct. These are NPN transistors. A negative base would cut down on amplification. With PNP transistors it would increase it.

R14 is interesting. C13 will allow some signal to escape to ground, and the setting of R14 determines how much. Since higher frequencies go through a capacitor better than low, some of the higher audio frequencies can be

sent to ground. What will this do? It will emphasize the bass. R14 is a tone control. The double transistors before the output transformer and speaker are called a "push-pull" stage. The first audio transistor, just before C10, gives its output to a transformer that has two wires in its primary and three in its secondary. The push-pull transistors, between C11 and C12, are so arranged that if a pulse of current went down through the primary, it would go up through the secondary.

The lower transistor by C12 would operate and feed a big pulse through the output transformer to the speaker. An opposite pulse would operate the transistor near C11. The one that is not operating develops a big positive voltage, as it is drawing no current and acts like a large resistor and drops a big voltage across itself. This voltage is also at that end of the output transformer. The one that is conducting will have a much lower collector voltage and added up a big potential difference is across the primary of the output transformer, since one collector is very positive and the other less positive, current flows between them and sends one healthy surge out to the speaker. Push-pull systems often develop more than twice the power of either one of the transistors. They also give quality sound.

To review:

1. The superheterodyne radio beats the incoming signal with a manufactured signal.

2. The manufactured signal is created by an oscillator that often is part of the same stage that mixes the two signals together.

3. The manufactured signal usually is arranged to be 455 kHz above the input signal for standard broadcast receivers.

4. The output of the mixer is fed to one or more intermediate frequency (IF) stages for amplification.

5. This amplified signal must then be detected.

6. Superheterodyning gives great sensitivity, because signals can be greatly amplified at the IF, intermediate frequency.

7. Superheterodyning gives great selectivity, because only the difference signal is amplified in frequency and several IF tuners help eliminate close frequencies to the input frequency.

8. Automatic volume control (AVC) takes a small sample of the detected signal and biases the stages before it to cut down on very loud parts of signal and to allow weak signals to be amplified.

Heterodyning Signals

Some TV schools insist on having the student build a superheterodyne radio. If the radio is built properly and if the student comprehends how and why every part goes where it does, building it can be a valuable experience. If, on the other hand, the student is told to "connect part 24 between terminals 1 and 14" without explanation or interruption to teach the principles, very little practical learning will occur. Building a color TV is also worthless unless one carefully analyzes it as he builds it. It is far better to build simple circuits with good direction and plenty of study.

We will see the effect of heterodyning signals without building a superhet radio. With your shortwave regenerative radio you will have the input tuner, detector, and amplifiers. Now take a pocket radio of the transistor type (with four or more transistors to be sure it is superhet) and wrap the antenna leads just where it leaves your regenerative radio, about three turns around the pocket set. The rest of the lead should go on to the real antenna as before. This wrapping will couple some of the oscillator energy from the pocket radio to your regenerative set. Do not connect the radios electrically in any other way. Now turn on the regenerative radio and adjust it for midrange on the tuning knob. Turn the regeneration a little below squeal. Now turn the pocket radio on. Keep the volume control on the pocket radio very low so you will only hear what comes in on your regenerative shortwave set. The pocket set can be standard broadcast.

Move the dial setting on the pocket superhet. You should start hearing different things in the earphones. By changing the oscillator setting of the pocket radio you are sending an unmodulated signal into the regenerative set. As certain frequencies from the oscillator beat against the input of your shortwave set, a signal is produced that is in the audio range, and you can hear it. You don't have any IF. When you have a station selected by the oscillator in the pocket radio, try changing the tuning of the shortwave set a little to bring the station in more clearly. Remember that with superheterodyne reception not only the oscillator but also the input must be tuned. You should be able to get quite a few stations this way, and the tuning will be quite sharp.

Perhaps you are wondering how an oscillator for broadcast stations can get anything in shortwave. First, the oscillator tunes above the broadcast band for about half of its swing in order to drop the IF to 455 kHz. When the radio is set at 1600 kHz. The oscillator is set at 2055 kHz. In addition, you may be operating on a second harmonic. Oscillators send out not only their own frequency but they send out somewhat weaker harmonics or multiples of their fundamental frequency. The second harmonic of 2055 would be 4110 kHz.

Perhaps you have tuned in a ham or amateur who sounds as if he were holding his nose. This "whank whank" unintelligible talk is single-sideband broadcast. In order to crowd more signals into the amateur bands of frequencies, some, in fact many, amateurs broadcast single-sideband. They prepare a carrier and modulate it normally, then cut much of it off. The carrier then can be amplified more or sent out as is. What is left of the carrier still keeps the transmission on frequency, but when received, parts of the modulation fall into the place where the other sideband usually is. The detector tries to cut off half of the half signal. As a result some of the modulation is lost.

The only way to receive single sideband (SSB) well is to reinsert the carrier in proper amount at the receiver. Expensive shortwave radio receivers have a built-in oscillator that produces this carrier to put back in the signal. By just the right amount of regeneration, and input of oscillation from the pocket radio, you may be able to clear up SSB. Don't expect a miracle. It will still sound tinny, and perhaps the man will sound quite nasal or perhaps even feminine, but this is about the best any equipment can do it. Such transmission is not intended to sound beautiful. It is just intended to be understood. Try adjusting until you clear up the SSB to understandable quality. Both the oscillator and the regenerator are adding to the carrier that has been mutilated at the transmitter. This is tricky business, and very touchy.

Another interesting thing you can do is take off the earphones, turn up the volume on the pocket superhet radio, and tune the shortwave radio. You should be able to inject shortwave into the pocket superhet. The superhet must be between stations, possibly near the top of the dial. Since the shortwave regenerative radio has an antenna, it picks up the signals well. The coupling of antenna loop feeds into the pocket superhet. Since the regenerative radio is set at very close to oscillation it is injecting or rebroadcasting signals it picks up. After you have enjoyed the shortwave radio for a while, return it to standard broadcast mode by putting L1 and L2 back as they were for Chapter 4. You will be making the set into a transmitter, and it will need to be for standard broadcast. We *could* make a shortwave transmitter, but it might be on someone else's frequency and cause problems.

QUIZ FOR CHAPTER 5

▶ 1. In most radios is the local oscillator tuned to a higher or lower frequency than the incoming signal? Usually by what amount?

▶ 2. Identify with words: IF; AVC; RF; SSB.

▶ 3. Which coil would get the shorter-wave broadcasts, all other things being equal: one of 30 turns or one of 42 turns?

▶ 4. With a shortwave receiver containing a variable capacitor that tunes 10 to 365 picafarads, which end would give the highest-frequency tuning?

▶ 5. Why are the oscillator and tuner capacitors connected by a single shaft?

▶ 6. Do all superheterodyne radios have one IF stage and two IF transformers? If not, where are you likely to find a different arrangement in a manufactured radio?

▶ 7. Explain how push-pull audio works and its main advantage.

▶ 8. What type of information does WWV give?

▶ 9. What upper layer of charged particles bounces signals back to earth and aids in skip reception?

6. Oscillators and Transmitters

PROJECTS: Adapting the Regenerative Transistor Stage to Produce a Carrier; Adjusting It to Produce Modulated CW; Connecting an Amplifier and Modulator System to Broadcast Voice

WHEN THE Federal Communications Commission (FCC) approved the new class of broadcasting, citizen's band, CB was intended for the serious business of allowing people in cars and trucks to send necessary messages that they would communicate by telephone if they could. Nobody anticipated that everyone and his grandmother would get on the air and have a ball just gabbing for the fun of it. Finally the FCC saw that it was hopeless to try to enforce the original purpose of CB and allowed everyone to use it for fun or other ends they desired. In the tiny Midwest community where I live, some people who live two houses away from each other persist in CB conversations. They could talk in their yards and hear each other. People in trucks and cars call their friends as they drive by their houses: "Just wanted to give you a shout, old buddy, as I drive through."

What these examples show is that just about everyone would like to put some kind of a broadcast transmitter on the air and say something over it. You will construct several transmitters in this chapter, and you will be on the air. Your transmissions will be low power and quite legal; they will be on the broadcast band, and will be receivable by any radio within a short distance—a few hundred feet.

You will remember that a radio carrier is high-frequency alternating current. Once you have a carrier you can send information on it in several ways. You can interrupt it and make dots and dashes as shown in Figure 6-1 *F*. You can modulate it with voice as in Figure 6-1 *G*. You can modulate it with picture information. Once you have a carrier you can adapt it as needed. First, however, you must make the carrier.

Any alternating current produces radio waves to some degree. As you know, the higher the frequency the farther the waves will travel for a given amount of alternating current. At the point at which the AC travels a reasonable distance in frequency, the AC is called radio frequency or RF.

Before tubes were used to create radio carrier waves, the high-frequency AC was created either by a very rapidly turning alternator or a spark transformer. The alternator was something like the alternator in your auto. It was turned at a speed as high as possible so that higher-frequency RF was generated. Usually this RF was fed to an arc light, and the RF going up and down the arc radiated the waves. While this transmitter was often a code type and merely needed to be stopped and started in accordance with the desired dots and dashes, it also was sometimes voice modulated. For code use, the alternator remained on and a telegraph key was usually connected in the antenna or ground circuit to feed the RF to the antenna in short bursts. For voice, an audio frequency transformer was connected in series with the antenna circuit. The secondary was in the antenna and the primary went to a carbon microphone with a battery in series with it. As a person talked into the mike and made the battery work, the pulses of current produced currents in the secondary that either held back or increased the RF in its travel up to the antenna. Such modulation was poor at best.

The second type of transmitter, the spark

transformer, was just a large version of an auto ignition coil. The primary had an interruptor that kept making and breaking the circuit, causing the current supplied by the battery to take on an AC characteristic. The secondary had thousands of wire coils, so the voltage was stepped up a great deal. The speed with which the interruptor (a type of buzzer) could affect the primary determined the frequency of the waves produced. The secondary, of high voltage, went to a spot at the top where sparks jumped between terminals. A wire went from one side to the antenna, and from the other side to ground. A telegraph key was in the primary circuit with the battery.

Despite what you might think, these old-time devices are not being discussed for their own sake. You should be able to see that RF carriers can be created by any device that produces very fast alternations of current. A tube or transistor alone normally cannot produce radio alternations. Two other things are needed: a tuned circuit and a means of feeding the output back to the input. This description sounds a lot like the regenerative radio receiver, and if you turn the regeneration control up high on the regenerative receiver, it becomes a transmitter.

The part of a transmitter that produces the RF carrier (or carrier waves or radio waves) is the oscillator. In simple transmitters the oscillator is just about the whole transmitter. In complex transmitters at radio and TV stations, the oscillator feeds the carrier to a buffer stage that isolates the oscillator from the rest to avoid loading problems. The buffer stage, a tube and tuned circuit, then feeds to drivers that build up the strength of the carrier far more than the oscillator could. Finally, RF power amplifiers feed the antenna. In some systems these tubes are called "finals." Throughout the transmitter, tuned circuits keep the oscillations on frequency. These final tubes may be over a foot tall and several inches in diameter. They are often connected by clamps, not just plugged in. Some are wa-

ter-cooled; others have air-cooling fins and a fan to circulate. Other big tubes, voice amplifiers and modulator tubes, modulate the transmitter.

In TV transmitters the oscillator often operates on a lower frequency. The first set of drivers produces a harmonic of the oscillator frequency (double or triple the frequency) by circuits tuned to that harmonic, and later drivers double or triple the frequency again. This system allows the oscillator to operate at much lower frequency, and it tends to be more stable at that frequency. TV station oscillators also have a crystal in their grid circuit to keep the device on frequency. Crystals can be cut to oscillate at lower frequencies better than at very high ones.

Look at Figure 6-1 *A*, which represents a simple oscillator and transmitter. If the key is closed so that battery is supplied to the collector and base bias of the transistor, and if the potentiometer is set low, this is a receiver and will pick up, detect, and amplify signals. It would be a more effective receiver if the antenna were attached to the tuned circuit rather than to the feedback coil. Now assume that the potentiometer is turned up high and you have enough feedback to set up a howl in the earphone. The set is oscillating and producing an RF carrier.

Let's examine the way an oscillator oscillates. First you have a tuned circuit tuned to the frequency on which you wish to transmit. At this point you are not yet feeding anything back, and you are not receiving anything. The tuned circuit would develop no pulses of AC. Now you have a bias on the transistor that makes the transistor open its barrier well. If you close the key, current travels through the earphone, the tickler coil, and the transistor from collector to emitter and back to positive battery. The bias current goes from key to base through the resistors and to emitter and positive battery, thus keeping the barrier open. As the current went from earphone to collector, it went through the tickler coil, which sent a pulse of current to the tuning

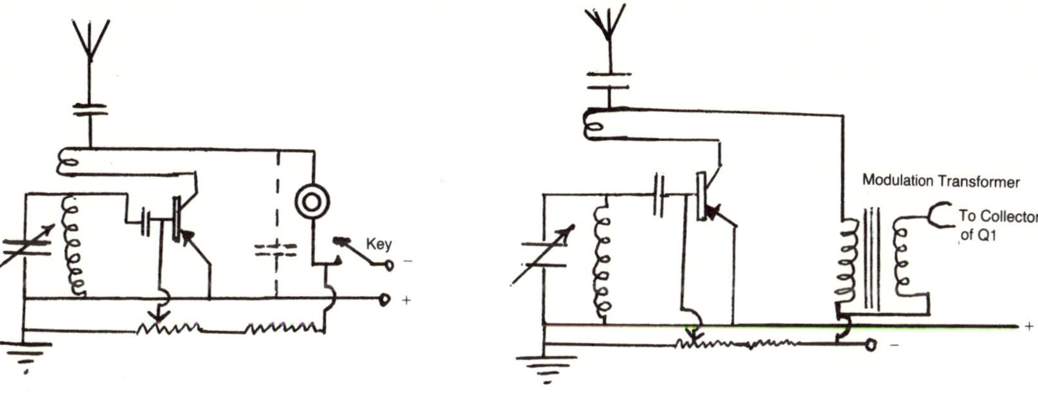

A. Code or Nonamplified Voice Modulated Transmitter

B. Voice Transmitter Using a Modulation Transformer

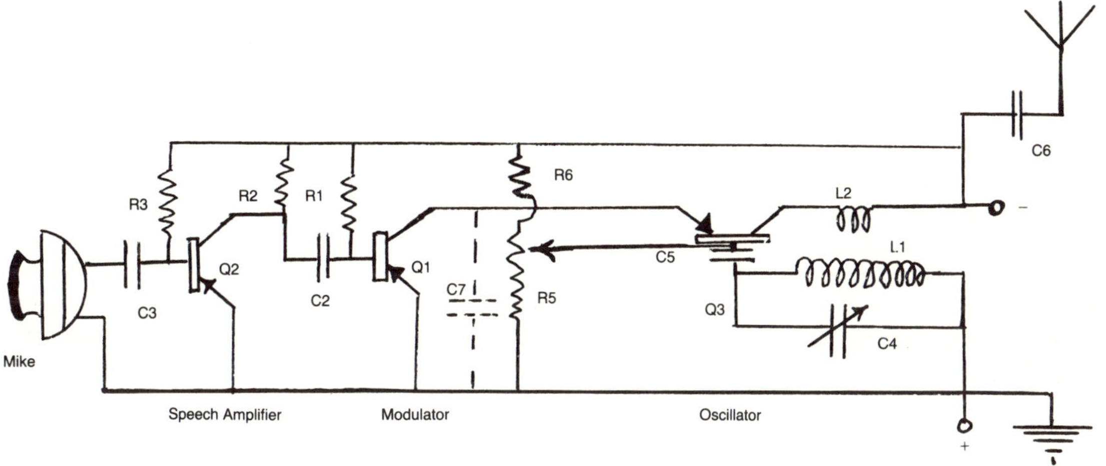

C. Low-Power Broadcast Band Transmitter

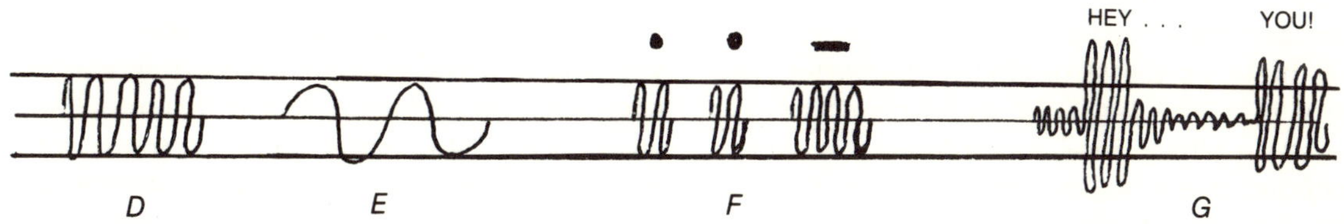

FIGURE 6-1

coil by magnetic force. This pulse is now in the tuned circuit. The frequency to which the circuit is tuned determines how long the tuned circuit resists the current—a very small fraction of a second.

Since the tuned circuit is resisting the current momentarily, the pulse pushes across the capacitor to the base and opens the transistor very wide. Now it really conducts, but by now the tuned circuit stops resisting the current, and it travels the opposite way. This cuts down the current from base to emitter, and the current from collector to emitter in the transistor. However, the very large current just before has put a new charge into the tuned circuit by the tickler coil. This new pulse is resisted by the tuned circuit momentarily, so the transistor is triggered for another big conduction.

All this opens up and closes down the transistor so that it alternately conducts heavily and lightly in phase with the frequency the tuned circuit is tuned to. You have now created a carrier. This keeps happening as long as the key is closed. If it is opened, the circuit comes to rest immediately.

Some of the high-frequency alternating current pulses are taken through the capacitor at the top of the tickler or feedback coil and fed to an antenna. The ground is connected at the bottom of the tuned circuit and transistor. Thus some of the high-frequency AC is sent out as radio waves to broadcast to a receiver some distance away. The current goes up and down and produces magnetism between ground and antenna, and this magnetic force leaves the wire and earth and travels through both to a receiver.

By alternately closing and opening the key, you can send out a series of code messages, each with several alternations of current. The earphone in the circuit could just as well have been a resistor so far. But now let's put it to more practical use. An earphone converts electrical pulses to sound by electromagnets pulling and pushing a diaphragm. It will work just the opposite way too. If you talk into an earphone rather directly, your voice will vibrate the diaphragm, cutting the magnetic lines of force the permanent magnets are producing in the earphone all the time. The result is a very small current in the electromagnets, and this current will be either in phase or out of phase with any currents traveling within the earphone from the outside. You could lock the key to the on position now, and the carrier would be continuously transmitted.

Talking into the earphone builds up currents that either strengthen the battery currents and enlarge the RF or buck against and cut down on the battery currents and decrease the size of the RF. That is what voice modulation is—it affects the size or amplitude of the carrier. The modulation system just described is about as simple as they come, and it has certain limitations. Let's make such a radio transmitter now.

PRODUCING A CARRIER

Disconnect the high side of L2 from R4, C3, and C7. Disconnect the collector of Q1 from the left fahnestock clip. Connect the high side of L2 to that left fahnestock clip. Connect the antenna through a capacitor of 50 to 300 Pfd to the same fahnestock clip. You have disabled the audio amplifier as far as operations are concerned, and although some of it will have currents, they will not affect your work. To simulate the telegraph key, you will touch the negative terminal of the battery with the lead to the right fahnestock clip in dots and dashes. At first you will need it on continuously, so connect the battery and proceed.

Use your portable radio just a few feet from the transmitter at first. Tune to a quiet place on the dial. Now turn the RF transistor, Q3, bias control until you hear a howl or tone in the earphones. Next adjust C4, the tuning capacitor for the transmitter, until you hear the tone in the receiver. Chances are you can hear it with several settings of the tuning capacitor on the transmitter. One will be loudest on the receiver. This is the fundamental frequency and not a harmonic. Set the capacitor there. You should be able to carry the portable radio around the room and even outside and still get the tone. Going close to the antenna, you should be able to receive the tone as far as your antenna runs, and then some.

Now, with the radio at the far side of the room, try sending some dots and dashes of international code. You will find a copy of the international code in Appendix D. The pitch you hear is modulation developed by setting the feedback just into oscillation. Try increasing the bias on Q3 by raising the adjustment on R5. Note that the tone changes to a hiss. This is clean RF, the kind a radio station generates. It will be harder to hear the code now as you send it.

As you learned in the last chapter, many amateurs send such unmodulated code, and it takes a beat frequency oscillator (BFO) in a shortwave receiver to add a tone to it. If you are interested in becoming an amateur you should practice the international code until you know it well. Think of the dots as "di" and the dashes as "daah." Learn to think of letters in code groups; think of *a*, for example, as "di daah."

When you can send five words per minute, start listening to hams, or amateurs, on your shortwave set and copy down what they say. It will take patience to find those that send slowly enough for a beginner. When you can read them quite well, find an amateur with a general class license or better and he will tell you how to go about getting your novice license. You will need to drill on some theory and laws before taking the test. Some of the material you will have learned in this book, and some you will need to learn elsewhere. Of course, not every person learning electronics cares to become an amateur, but in almost every class of technicians at a trade school one or more becomes a ham, so this much information has been thrown in just in case you are interested. In order to be a TV technician,

you certainly do not need to be a ham operator.

Voice Modulating a Transmitter

Now let's try to voice modulate the simple transmitter of only an oscillator. Connect the negative terminal of the battery and talk into the earphone rather directly. You will find that if the radio is quite close you will hear yourself. In fact, if you are too close you will set up an acoustic feedback. Keep the radio volume control low to avoid this. The bias of Q3 must be high enough so that you are not self-modulating with whistle, but if it is set too high, very little voice will be heard. Note that farther away you will still hear the hiss of the carrier, but the voice modulation will be absent. You will need someone to talk into the mike as you back the radio receiver farther away. Why does the carrier go farther than the voice?

Perhaps you have experimented with CB a bit and have learned that percentage of modulation is important. Some CB operators use a power microphone to get the modulation as close to 100 percent as possible. With the transmitter you have just made the percentage of modulation is very low. The 6-volt battery supplies far more energy to the transistor and other parts of the oscillator than can be built up in the process of yelling into the earphone used as a microphone. With this setup you can't increase the modulation even if you yell yourself hoarse. You can, however, cut the power of the transmitter and make the modulation a larger percentage of the oscillation.

Use a single flashlight battery instead of the 6-volt one. You will need to solder a terminal to the negative terminal (the flat back) and to the positive terminal (the tip on the top). With this battery of only 1½ volts you may need to move L2 right on top of L1 to get enough feedback for oscillation. You might even have to increase the turns of L2. This all depends on your RF transistor. After you have connected the small supply voltage, adjusted L2 if necessary, and have a smooth hiss coming from the receiver, talk into the earphone. Though you will find that the carrier will not travel as far as before, the modulation will travel farther. You have increased the percentage of modulation, not to the coveted 100 percent, but at least a little bit. Of course, the better way to increase modulation is to have voice amplifiers and full power for the oscillator.

A 1,000-watt radio station, also referred to as a "kilowatt station," would have a carrier frequency with enough power to radiate the 1,000 watts. This could be done with several tubes feeding a final amplifier or set of push-pull amplifiers that have 500 volts plate voltage and a current of 2 amps. The station would also need tubes that would amplify the voice so that it would have 500 watts of power. That is quite a lot. Stereo systems of 100 watts fill an auditorium with sound, and we are talking about 5 times that amount.

With 1,000 watts of carrier and 500 watts of audio to modulate it, you would have 100 percent modulation. See Figure 6-1 G. The two outer lines above and below the middle represent full carrier power—in the case above, 1,000 watts. When the modulation is at its loudest, an extra 500 watts of power is put in by the modulator. On the loudest voice the transmitter is putting out 1,500 watts. During periods of silence nearly no wattage of carrier is put out. In the control room an engineer watches a volume unit (VU) meter. This meter indicates modulator strength in volume units and in decibels. The decibel is a measure of sound intensity. Also, this meter will have a percentage of modulation on a scale. He must set the volume of the voice or music so that it never overmodulates. If the voice produced more than 1,000 watts it would go beyond the 1,500 watts one way and fall below no carrier at all the other. It would produce garbles and perhaps take the transmitter off its assigned frequency at times.

Overmodulation is not only undesirable, but also illegal.

MODULATING THE CARRIER WITH AMPLIFIED AUDIO SIGNALS

Now let's try to construct a modulator and audio system that gives a better percentage of modulation to the carrier than the previous system did. Figure 6-1 C is a means of doing so. Talking into the microphone develops tiny currents. Use the earphone as a mike. At recording studios and radio stations a mike similar to the earphone is called a *dynamic mike*. Other types are crystal mikes and capacitor mikes. The crystal mike contains a thin crystal material. When you talk into the mike, a diaphragm vibrates and flexes the crystal and makes it give off very tiny electrical charges. (Similarly, in a crystal pickup unit on a phonograph, the needle vibrates the crystal.) The capacitor mike has a current fed to it. As you talk into it, the talking moves two plates of thin metal closer together, thus changing the capacitance between them and changing the current slightly. Such a mike is used less than it once was.

The old-fashioned carbon mike, just like the transmitter section of your telephone (the part you talk into), has a current furnished to it. As you talk into it, the diaphragm compresses carbon grains together in a button, and the current increases with each sound when the grains are compressed. Carbon mikes are seldom used at radio stations, as their sound quality is not excellent. At one time, however, they were the only type that existed.

Talking into the mike makes tiny pulses of current that go across C3 to the base of Q2. Q2 amplifies them in the usual manner and couples them to Q1. By now the audio signals are amplified so well that you might be able to talk a few inches from the earphone used as a microphone. Q2 is the audio or speech amplifier. Q1 is the modulator. (We deal with transistors in these experiments. At radio and TV stations many of the circuits are tubes, yet in some modern equipment the oscillator might be a transistor. The first few audio amplifiers could be transistors too; then when great amplification is necessary, tubes would be used.)

Though Q3 is turned on its side, its hookup is much as before with one great exception. The emitter does not go to ground, but instead goes to the collector of Q1. This means that all currents from battery through feedback coil L2, through collector to emitter of Q3 must also go through collector to emitter of Q1 before returning to the battery. This should give a fairly good percentage of modulation. If the transistors offer equal resistance, equal voltages will form across them. Since no current can flow through Q3 unless Q1 allows it, and since the voice coming into Q1 will determine it, Q3 will develop oscillations only in proportion to the sounds put into the mike. This arrangement probably will have close to 100 percent modulation.

Now we will change the circuit to construct the transmitter. Consult both Figures 4-1 and 6-1 *C*. The microphone/earphone is removed from the collector circuit of Q3. To remove it, disconnect the battery, then connect the high side of L2 to the right-hand fahnestock clip. Move the antenna lead there also. Disconnect one end of C1, as it is not used in this experiment. Remove the right phone tip from the right fahnestock clip and connect it to ground. You will need to wrap a wire around the tip carefully and connect to positive battery. Take a lead, insulated except at the ends, and connect it to the left fahnestock clip, leaving the left phone tip in place. Disconnect R4 and C7 from C3, but leave their other ends alone. Connect the lead from left fahnestock clip to C3, connecting the earphone serving as a mike to Q2. Connect a lead to the collector of Q1 (you have already disconnected this collector from the phones).

Disconnect the emitter of Q3 from the lug

on C4. Instead connect the emitter of Q3 to the collector of Q1 with the wire lead you just connected to Q1. The free end of C7 can be connected to this wire between the emitter of Q3 and the collector of Q1. Try it, and if the transmitter works better with it leave it on. Otherwise remove it. After trying the set and deciding whether C7 helps or hinders, you should find that you can broadcast the voice much better. It will take some manipulation of R5 and C4 to get things just right. You may find that changing the circuit affected frequency somewhat.

One limitation of the transmitter you have just made and tested is that since Q3 and Q1 both use part of the 6 volts supplied, you do not have as much power of carrier as you did with the simple transmitter when you had it on 6 volts. The series modulation you have used can be changed to parallel modulation if you have an interstage transformer like those used between stages of a transistor radio in their audio coupling. Such a device as Radio Shack part No. 273-1378 or a similar driver transformer will work. It must have impedances of several thousand ohms in both primary and secondary. A speaker transformer of 500 to 3.2 ohms or 8 ohms will not do it. This impedance value is not DC resistance and will not measure as high a value of ohms on an ohmmeter. Impedance is a resistance that the circuit offers to AC.

By wiring the driver transformer as shown in Figure 6-1 *B* you can have the full 6 volts for oscillator and for modulator. This arrangement allows a very large current to travel through these circuits, so it might be well to put a 1,000-ohm resistor between each transistor emitter and ground. Note that in this case both Q1 and Q3 have emitters to ground. The 1,000 ohms protect from too much current flow. This exercise is optional. If you have the transformer you can try it. The important thing to observe is that a transformer that connects the outputs of oscillator and modulator modulates with full power to both transistors. Such modulation is often used with tube transmitters between plates of modulator and final of RF stage. The plates would be connected where collectors are connected in Figure 6-1 *B*. This is called *plate modulation*. The power from the modulator is magnetically placed on the oscillator by the modulation transformer. If the matching of the impedance is right it is possible to get 100 percent modulation this way.

Try this circuit if you have the parts with high impedance in the oscillator circuit, and then reverse the transformer and try in the modulator circuit. Which works best? Soon you will be broadcasting as if you were a real station. Use whichever transmitter you liked best for broadcasting. If you don't have the driver transformer, then the transmitter shown in Figure 6-1 *C* is your obvious choice.

Let's review the main things you need to know from this chapter so far:

1. An oscillator feeds energy to the tuned circuit by a feedback or tickler coil.

2. The setting of the tuned circuit determines the time interval for which the feedback opens the transistor or tube wide for current flow.

3. The setting of the tuned circuit also shuts down this tube or transistor flow a half cycle later. The pulse that just resulted reopens the tube or transistor by magnetic feedback from tickler to tuning coil.

4. The rapid opening and closing makes the alternations of RF. The tuned circuit times them for the proper frequency.

5. At a commercial radio or TV station an oscillator does not produce powerful enough carriers. Several driver amplifiers and a final amplifier must amplify the carrier.

6. A modulator stage controls collector or plate current of the RF final amplifier and puts on the modulation this way.

7. A 100 percent modulated signal is 1½ times the unmodulated size.

8. Speech amplifiers build up the speech or audio to a proper level to drive the modulator so it can affect the carrier sufficiently.

OPTIONAL: SIMULATING A RADIO
STATION CONTROL ROOM

As an optional exercise you might like to build a replica of a radio station control room. This is considerable work, and its value is greater if you have aspirations to study for a first-class radiotelephone license and work at a radio station as an engineer and disc jockey. Several TV servicemen have made the jump and are happy with the change. Actually, TV servicing now pays about as well as work at a radio station, unless you work up to a large independent or network station. But some of us like to be heard, to have a job that puts us before the public. If this is your goal, radio announcing or DJ work might be your destiny. If your interest is less, perhaps you will just rig up a turntable (use your record player without its amplifier) and mike at a table.

At any rate, a complete simulation of a radio control room is shown in Figure 6-2. At a real station there would be amplifiers for each input, but here just control potentiometers R1-R5 will suffice. These should all be high-resistance units such as 1 million ohms or one megohm each. You might pull them from a junked TV. They all go in parallel to C3, the input of the audio amplifier of the transmitter. The mike you had at C3 can be used in the control room or studio. Each input can be connected to the proper potentiometer if its switch is in the play position. In the drawing and schematic, turntable 2 is playing and the switch, S5, is set for the pickup unit of turntable 2 to feed to R5. All other switches are in the audition position. There is no S4. Since the control room mike is the one the operator is going to use it is unlikely he would want to audition himself. It is important that R4 be kept completely down until the operator wants to talk on the air. Note that R5 is about halfway up. Its setting would depend on the type pickup in the turntable and the loudness of the record being played.

While the record is on the air, the operator or DJ can put a record on turntable 1 and cue

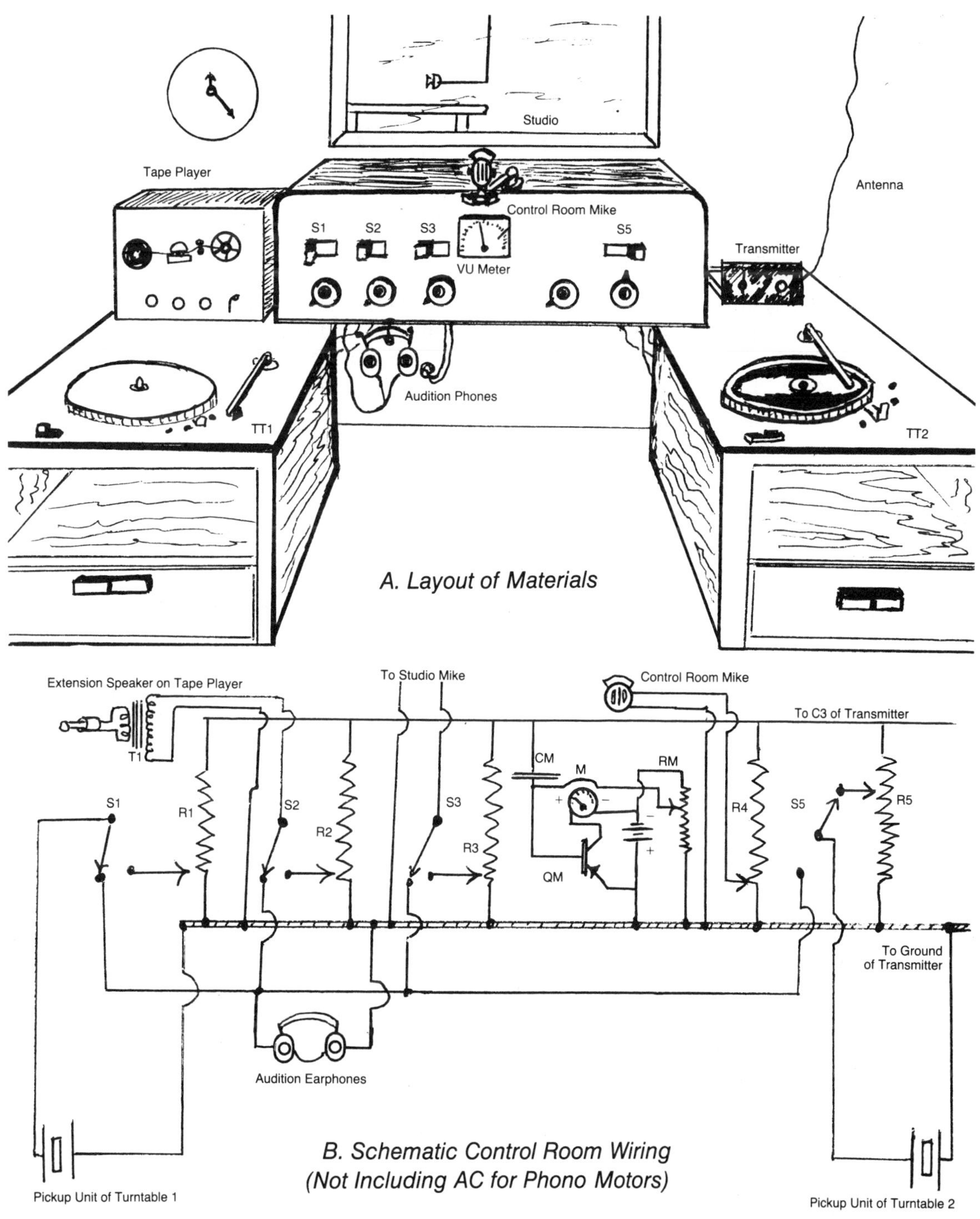

FIGURE 6-2
RADIO STATION CONTROL ROOM

it by listening in the earphones. Some stations have an amplifier on the audition system and a speaker in the room. Often there is a connection through R4 so that when the control room mike is opened, the audition speaker is killed. Otherwise an acoustic feedback from speaker to mike would howl and roar, and all this would go out on the air. With our setup of phones this is not a problem. You would put a record on turntable 1 (TT1) and put the tone arm in the proper groove ahead of the recording you desire to play. With the motor off and the turntable set in neutral you would spin the record back and forth until you find the point at which the recording is just ready to begin. Then back it off a half turn. When the record on TT2 finishes you'd quickly turn R5 to zero and R4 up and announce the next record and possibly read a commercial. Just before you were ready to spin the recording on TT1 you would hold the record lightly, turn the motor on below it, flip S1 to the on-air position (right in this case) turn R1 up some, and let go of the record. It should be up to speed in half a turn, and you would learn how much to talk as you let go of it so it would come on just as you finished. Immediately you would cut R4 off.

Cueing a record and putting it on the air is very awkward at first. It takes practice. The idea is to avoid dead air time between the talking and the beginning of the record. If you try to put the turntable tone arm down while you are on the air with the turntable and the mike, chances are you'll hit within the recording or hit the tail end of the previous selection in the case of LP albums.

Do watch the speed of the record. After you cue the record be sure to flip to the proper speed from neutral. Usually LPs are 33 rpm. Those with large holes are 45's. Some smaller recordings with small hole are 33's. Any older record is a 78. A few large transcriptions cut some years ago ran for 15 minutes each side at 16 rpm. You'd play one side, make an announcement as you turned it over, then start the second side. Thus the record produced a 30-minute program. If you had an announcer to talk while you cued the second side you were fortunate.

A modern tape player probably has an external speaker output of 8 ohms. The only way to match this to the console and transmitter is to put in a transformer like T1. The 8-ohm part goes to the tape machine, and the higher ohm side goes to the console. An audio output transformer will work in this case. You would have to cue the tape with the earphones while a recording was on the air or while someone was talking from the studio. You may not wish to invest in a window and such for a studio. Instead, you can have someone on the other side of the console with a mike to represent the studio.

Use signs to tell the studio person when to begin and, if the time is running short, when to stop. He won't be able to see your operation of the switches and pots (potentiometers) so you must cue him. Pointing a finger at him is the sign to begin. Moving the hand to the mouth means for him to speak up louder or get closer to the mike. Moving the hand outward means to get farther from the mike. A circular motion means to keep going, add more, or do something to fill the time. A motion across the throat means cease talking.

Some DJs prefer to set the level of their pots for turntables in advance and just flip from audition to on air and back. You must be alert, however. It's quite noticeable if the control room mike is on instead of the record. The radio audience hears nothing for a while and then, "Oh, gosh dang, that thing's not playing . . ." followed by a record in the middle of a song.

The VU meter circuit lets the DJ or control room operator know if he is putting out close to 100 percent modulation. This device shown is only partly useful, but it does give some practice to riding gain for 100 percent modulation on loudest sounds. To make it you will need a 1-milliamp DC meter. (See Figure 6-2

B.) The battery should be 3 volts. The PNP transistor is an audio type. RM should be 100,000-ohm potentiometer. You will have to experiment with CM to find a value that will work with the particular transistor. Start with a 0.1 mfd. and try smaller values such as 0.02, and so on. The circuit is wired to be an on-air meter system and not record audition material.

With this whole system working, listen to an on-air radio some distance from the transmitter. Have a recording playing, preferably one with a steady tone. Open up the pot on the turntable until the record distorts slightly over the radio (garbles or loses quality). Now you have slightly over 100 percent modulation. Back the pot to a just acceptable level. Run the RM pot up to where the meter reads 0.8 milliamps. Make a mark on the outside of the meter plastic at this point. This is 100 percent modulation. Now you have a reference level for the meter. To save on the meter battery, disconnect the battery when not in use. Of course 1 milliamp of current draw will allow the two cells to last a long time.

In some cases a diode such as a 1N60 will work better than CM. Position it so that the arrow points toward the top line and away from the base of QM. In other words, the cathode of the diode must go to the line to C3 and the anode must go to the base of QM. With the meter calibrated you can watch the meter fluctuate as you talk, as others talk, as the tape plays and the recordings play. If anything is too loud, adjust the proper pot (R1, R2) to compensate. The peaks of talking should hit the 100 percent line you drew. The lower parts of the speech will fall much below. Sometimes you will need to have a recording on and talk over it. In this case usually the recording is music, and its level will be steadier than your voice. You will learn to let the peaks of your voice hit the 100 percent level while the recorded music is less.

Sometimes the poor man in the control room is asked to record on tape, for later broadcast, a program that is going on in the studio. He is also expected to be putting on a live show. Our setup is not fixed for such recording, but with another position on S3 to the mike input of the tape recorder it could be done. Often there is a jack panel at a station so that unusual connections can be made. In this case the control room man must watch the blinker light or meter on the tape machine and keep the pot (R3 in this setup) well controlled. He must also watch the on-air VU meter and keep the records spinning and talk as necessary.

If this work appeals to you, it will pay to practice hour after hour. Learn clever ad-libs to keep programs lively, but watch your expressions. Regulations still exist concerning vulgar words on radio and TV, believe it or not. If you consider TV station work, the audio section will be similar to the console of a radio station. Of course, there also are video boards and much more. In addition, if you aspire to be a DJ, announcer, or combo man, you would do well to study speech in high school, college, or at night school. Learning to speak plainly with correct but not elite grammar is the trick. Good language does not call attention to itself.

Quiz for Chapter 6

▶ 1. The process of changing a carrier in size to put voice or music on it is known as ______ .

▶ 2. Describe two early ways of generating a carrier.

▶ 3. What two items besides a tube or transistor are necessary to produce oscillations?

▶ 4. Name four types of microphones and explain how they work.

▶ 5. What stages in a transmitter amplify the oscillations and sometimes change their frequency?

▶ 6. What stage in a transmitter causes the speech to affect the size of the carrier?

▶ 7. Besides this stage the speech part might have several _______ amplifiers.

▶ 8. What type of coupling between final power amplifier and modulator would allow 100 percent modulation and allow both stages to get full supply voltage?

▶ 9. If a transmitter overmodulates the result that can be heard will be _______.

▶ 10. When one cues a record the turntable control should be in _______ position.

7. Signal Generators as Test Instruments

PROJECT: Generating Signals for Testing Radio and TV Circuits

IT IS hard to convince anyone who has repaired autos, power tools, and the like that a lot of test equipment is required to find the faults in electronic equipment before repairing it. With an automobile one can often, though not always, see the damaged part and put another in its place.

By this time, you should know that the bad part in a radio or TV set usually does not *look* bad. In rare instances in power supplies you can see a charred resistor or smell insulation burning in a power transformer. Most of the time, however, it takes considerable equipment to reveal faulty components. Using a voltmeter or ohmmeter, you can usually find the bad part or parts in a stage that is bad. Finding the defective stage requires other equipment. It would take much too long to test out every transistor, tube, resistor, capacitor, and coil in a TV, so the voltmeter usually is brought into play after you have a pretty good idea where the trouble is.

The faults in the picture on a TV screen usually give clues to the general section of the set that is malfunctioning. A flipping picture usually indicates trouble in the vertical sync

section. A weak picture could mean trouble in the antenna, tuner, IF, detector, or video amplifier, or even at the picture tube. The signal tracer you experimented with in Chapter 1 is of no use now, as it traced only audible, hearable, signals. It would not reveal anything of the picture content. On the other hand, you could inject your own signal into the set. Someplace between antenna and picture tube is a bad stage. It could be the tuner stage, the IF stage, and so on. Once you find the bad stage you can test it with voltage readings.

A signal generator is a type of oscillator used to put radio frequency into a radio or TV. The term "radio frequency" is not just frequencies on which radios operate, but anything above audio frequency range. With the assumed defective TV, you could set the signal generator at the IF stage frequency, 45.75 MHz for modern sets and about 25 MHz for older sets. (The IF was in the 20-MHz range in early TVs.) You would be putting in a carrier now, but you would need to modulate it to see anything on the screen. A 400-Hz tone generator could be added into the modulation input of the signal generator. This would make several bars across the screen of the picture tube.

You could find the last IF tube or transistor as a starting point for testing. Consult Figure 12-7. Some of the following may be hard to understand at this point, but it shows how a signal generator can be used to locate troubles. Note that the IF is low, as L18 is tuned to 25.6 MHz, L19 to 23.3 MHz, and L20 to 24.6 MHz, so set the signal generator for about 25 MHz. That's close enough. The shaded area on the picture is mainly the IF components. Note on the schematic that V7 is the last IF tube. The term *video IF* is used, because *video* means "picture" or "that which is visible." All this part of the TV is producing the picture. Some sets have an extra IF stage for the sound. With the set shown, the sound and picture travel through the same IF sections. Pin 1 of the tube, V7 (a 6CB6 pentode) is the signal or input grid. At this point you could inject your signal with modulation from the signal generator. You would have to get under the chassis board and find pin 1. You would then ground the signal generator to the chassis, and touch the test probe to pin 1. If bars came on the screen you would know that V7 is amplifying and in working order, and that everything from V7 back to the picture tube is good. Now the problem has to be somewhere between V7 and the antenna.

Since 25 MHz will work as frequency in all the IF stages, it is convenient to check them out before trying the tuner. In this case all three tubes are the same type, so all have pin 1 as signal grid. You can try the probe at pin 1 of V6 and V5 (not shown on the chassis picture). If, for instance, V5 produced no bars on the screen of the picture tube, you could try the plate of the same tube, pin 5. Assume you get a picture of bars now. The trouble must be in the tube itself. You try another tube and get no bars at pin 1. Now it is time for the voltmeter. The meter reveals no voltage at pin 5 instead of the 135 volts the schematic says you should have. You test at pin 6, the screen grid, and get the voltage.

What components must the plate voltage have in series in order to get the supply that the screen grid does not? R27 and the primary of L18 are the current paths. Since they are in parallel, if just one were bad, some voltage should still be at the plate. Since there is none, both must be defective. By close checking, you find a crack in the circuit board that breaks the circuit where R27 and the primary coil of L18 go to the socket of V5. This is a typical problem, and to solve it you would jump a wire across the break and solder at both ends or melt solder enough to remake the connection.

On the other hand, suppose you could get a signal through all IF stages but the picture is still very weak? Now you have to see whether the trouble is on all channels. If so, pick a channel that normally is good and set the signal generator at this channel frequency. See Figure 12-1 for the frequency to set. You

would have to go into the tuner and test there. Leave the modulation at 400 Hz. If you can even get the signal through at the antenna terminals, chances are the antenna leads are loose at the top or shorting, so get out the ladder. If you can't get any signal through the tuner, try setting the generator to the oscillator frequency. You can make the signal generator act as the oscillator. You would have to connect the antenna for regular signal from the station input. In such a case you would suspect that the oscillator is not working, and if you can get a picture now, that might be the case. If the oscillator is dead, you probably would have no picture at all.

Although a signal generator usually is not used with radios anymore, let's go through the procedure in order to learn more of signal generator use. See Figure 5-1 *b*. If you had such a pocket radio that produced no sound, you could set the signal generator in the audio range—at 1,000 Hz, for example. Then you would ground the ground clip and touch the probe of the generator to the center lug of the volume control. If you get a high-pitch, 1,000-Hz tone in the speaker, you know that everything in the audio section is OK. If you do not, you would try at the base of the driver transistor and then at each base of the push-pull transistors.

If, however, you get a sound at the volume control, reset the signal generator for 455 kHz and test the bases of the IF transistors. Finally, if everything else is good, use a radio station frequency such as 700 kHz and try at the base of the first transistor, or substitute the oscillator as before. Here is a tip in case you are fixing radios and suspect the oscillator of not working. Place the defective radio next to a good radio. Turn both on. Rotate the dial of the defective radio, and if its oscillator is working somewhere on the dial you will produce a whistle in the good radio. This is easier to hear if you have a station tuned in on the good radio near the high end of the dial—1600 kHz. The dial setting on the defective radio will vary, so rotate it back and forth. Chances are a low setting should whistle the good set if the oscillator is OK on the defective one. Such a system will work to check oscillators on TV receivers too, but the bulk makes it less practical. No connection between sets is needed.

Your laboratory radio can be used as a signal generator in a pinch. Although it doesn't operate on as high frequencies as TV channels or TV IF, it will create enough harmonics to be used to find faulty stages. Connect your radio equipment as shown in Figure 6-1 *a*, the code oscillator or transmitter. You will not need a telegraph key for this, but you will connect the negative lead to the battery as needed. Instead of going to ground, the extra positive lead will be a ground to the set under test. An alligator clip on the ground lead to set under test is useful. The antenna can be disconnected at the antenna capacitor, and a lead with a probe connected instead. Connect one TV VHF terminal to the ground from the oscillator and the other to the high side probe. The regular twin lead antenna is disconnected for this experiment.

With the TV operating and set on Channel 2, 3, or 4 (the lower the better), connect the negative wire to the battery. This should cause a blinking flash on the TV momentarily. Now turn the regeneration control potentiometer up until a whine begins in the earphone. At the same time first one, then two, then more horizontal bars will form on the TV screen. They may not be dark, but they will be there. Some TV sets reject such noise and the indication will be dim, yet dark enough to troubleshoot the set. As the pitch goes up, the bars will increase. Finally, if you can adjust it very carefully, the bars will cross over and become vertical before they disappear altogether. At that point the modulation is lost. You hear nothing in the phone.

Learn where to set the control for the clearest bars for generator work. You may find that changing the frequency of the oscillator will affect the bars also. If you had a little more experience you could test the IF stages, detector, and video amplifiers. It would be best, however, to learn how these stages work and then come back and perform the exercises.

The generator made from your laboratory radio is not in the same class as a high-frequency signal generator with modulation input and so on, but you can still use it to troubleshoot a TV circuit. In a pinch even a door buzzer will inject a signal into a TV receiver. For radio servicing the home-rigged signal generator is quite good. See Figure 5-1 *b* again. With the generator set so that it produces an audible tone in the earphone, you can test at volume control, first audio base, either push-pull transistor base, then back into the IF stages.

Because the oscillator produces a multitude of frequencies and self-modulates, you need not adjust it from audio to RF or IF servicing. Because it is protected by the capacitor above the feedback coil, the battery voltages don't mix. In some circuits you may find that the signal injected is so strong that a ground to the set being tested is unnecessary. Of course you can put signals in at collectors as well as at bases. Usually no signal can be injected at an emitter, but sometimes it can if the emitter doesn't go directly to ground.

This signal generator can also be used to test out amplifiers for phonographs, guitars, violins, stereos, and the like. If the signal injected is so strong that the set plays a tone even when the lead is just brought close to the transistor or tube, place a smaller capacitance between the feedback coil and the probe. Go to a 0.001 or even a 100 picafarad capacitor.

Several types of signal generators are available on the market today. One is called a *marker generator,* and the type you have used could be used as a *marker generator.* A marker generator puts a signal into another type of signal generator to mark a certain frequency on the second generator. The other generator is known as a *sweep generator.* It *sweeps* through a range of frequencies that are adjacent to each other.

For instance, if you wanted to send a signal

through a TV IF and cover the entire range of frequencies in that IF circuit, you would connect a sweep generator and set it to go between 41.75 and 47.25 MHz for a modern TV or 21.75 up to 27.25 MHz for an older one. This generator would constantly be going up and down between these frequencies, covering every one in between as it sweeps up and down. The output of this generator would cause an oscilloscope to show a pattern of the sweep of frequencies. The sweep generator would go in at the input of the IF stages and the oscilloscope would connect at the output, near the detector. When a marker generator is set to one of these frequencies, a little blip shows on the oscilloscope of the waveform at that point. Such a system is illustrated in Chapter 19.

Another type of generator makes a pattern of the screen of a TV set to converge the colors of the set. This is a *crosshatch, bar, dot generator,* a special type that we will cover in Chapter 18.

In this chapter you have learned:

1. A signal generator is an oscillator that can be set to a desired frequency.

2. A signal generator is a device that injects a signal into a TV or other electronic device.

3. By working backward from output until the signal's path is broken, you can find a stage at fault with a signal generator.

4. A sweep generator covers a set range of frequencies by tracing up and down through them.

5. A marker generator shows a particular place on a sweep generator's pattern.

6. A crosshatch, bar, dot generator sets convergence on TV.

QUIZ FOR CHAPTER 7

▶ 1. Why was it necessary to modulate the manufactured signal generator with 400 Hz?

▶ 2. Though a signal generator finds the faulty stage, what two items are usually needed to find the defective part or parts?

▶ 3. What visual indications might occasionally show a bad part?

▶ 4. What part of a TV often tells which stage has the trouble?

▶ 5. What is probably faulty in the following situation? You can put a signal in on an IF plate, but not at the grid, and all parts have their proper voltages (or almost, but at least all connections are OK).

8. Voltmeters and Other Test Equipment

PROJECT: Constructing a Voltmeter-Ohmmeter and Testing TV Circuits with It

A GOOD VOLTMETER and ohmmeter are just about the best friends a TV serviceman can have, because they tell you exactly the part or parts that need replacing in a faulty receiver. These two instruments are opposite in design and function. The voltmeter measures voltages dropped across resistances whether actual resistors, tubes, transistors, capacitors, or transformers. It must be used when the TV is plugged in, turned on, and adjusted as nearly as possible to normal operation.

The ohmmeter measures the resistance in ohms that a circuit component such as a resistor, transistor, tube, capacitor, or coil offers to that circuit. The set should be completely unplugged from the wall outlet for testing with an ohmmeter. The ohmmeter furnishes a small current of its own to test the resistance of the component this small current goes through. Note in Figure 8-1 *b*, that only a flashlight cell supplies the current for the ohmmeter in this case. Some ohmmeters use up to 4½ volts. Remember that voltage measurements are made with the supply voltage to the set, while ohmmeter readings are taken with the power supply disconnected from the wall outlet. Voltage from the set plus the battery could ruin the meter.

Since one meter movement can be adapted to measure ohms, volts, or amps, all three are often combined into what is known as a *multimeter*. A multimeter has switches and sometimes different probe connections to change its function. In my years of troubleshooting TVs and other equipment, I have very seldom needed to take a current measurement. Perhaps one reason a serviceman avoids current readings is that he must put the meter in series with the item through which he is measuring current, and this process calls for a lot of unsoldering and resoldering. Here is the point. If the voltage is correct across a resistor or transformer, the current is correct also. Voltage readings merely require placing the probes of the meter across the item whose voltage you wish to measure. Since current measurements are time-consuming and are avoided whenever possible a good voltmeter and ohmmeter take care of 99 percent of the work.

USING METERS

The meter movement is usually a combination of a permanent magnet and an electromagnet. The electromagnet is on a pivot. Flexible wires go from the electromagnet to the terminals in the back. As currents supplied by the circuit under test travel through the electromagnet, they produce magnetism that opposes the permanent magnet. Since the electromagnet is on a pivot, it turns to get away from the opposing magnetism of the permanent magnet. The north poles repel each other, as do the south poles. A hairspring keeps the movement within bounds. The more current that goes through the electromagnet from the circuit of the TV or whatever is being tested, the more the electromagnet is moved around, and an attached pointer indicates this on a scale on the face of the unit.

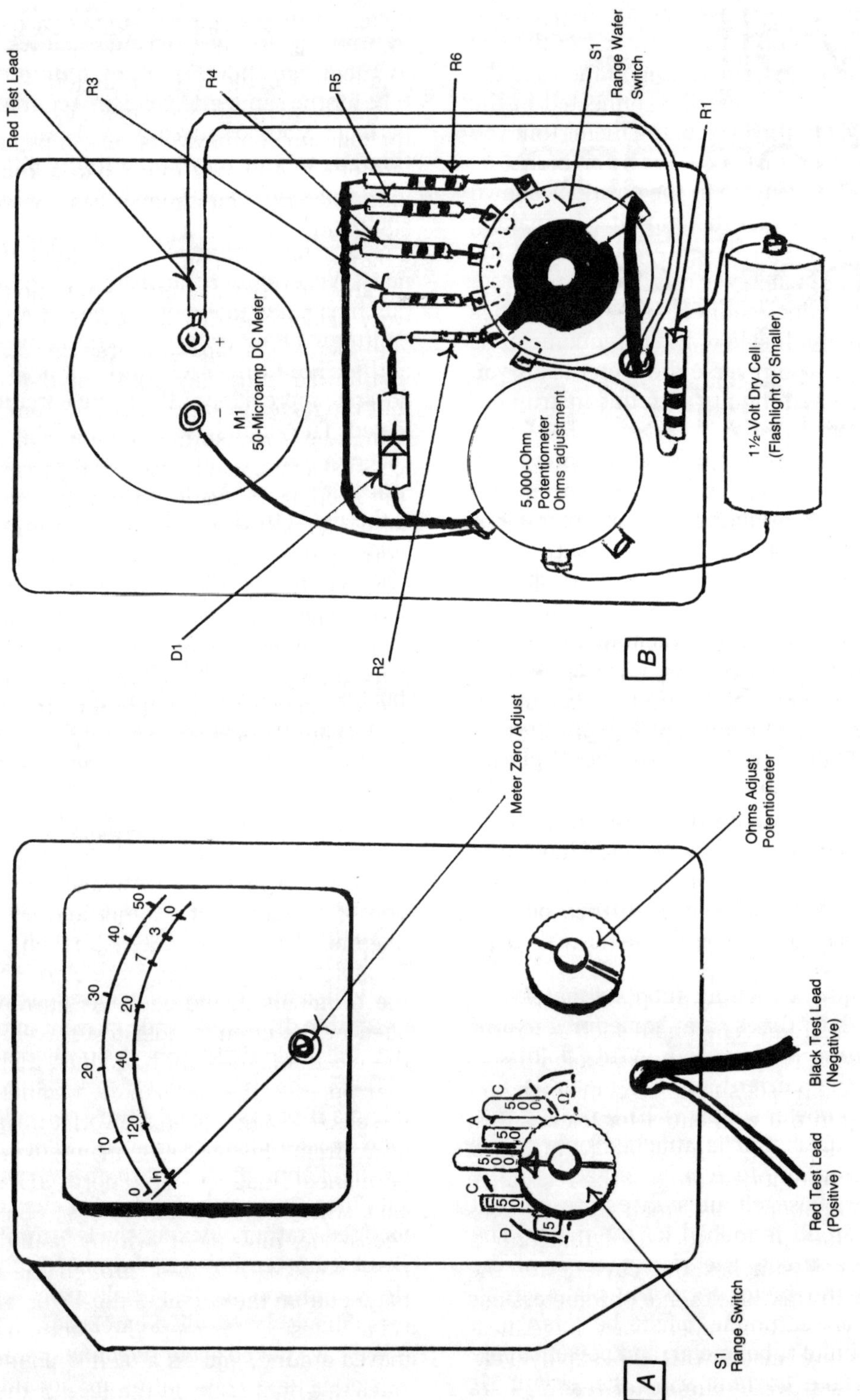

FIGURE 8-1
HIGH-IMPEDANCE HOMEMADE
VOLTMETER-OHMMETER

Since north pole must repel north pole, and a negative current makes that end of the coil negative, it is extremely important that the proper end of the meter be connected in the proper way. On the back of the meter unit you will find either just a (+) sign or both (+) and (−). They may be in the positions shown in Figure 8-1 or just the opposite. In making your own voltmeter, be *sure* you connect the (+) terminal, wherever it is, to the red lead. The (−) terminal connects to several things through the high side of the potentiometer as a tie point. If you reverse the meter leads you will have great difficulty with this instrument and may ruin it.

Many meters read full scale for one milliamp or 0.001 amp. This is the type meter recommended earlier for the VU meter of the transmitter console. Such a meter has a rather obvious drawback when used to test voltages, especially with transistors. Since many transistors draw only about a milliamp themselves when the meter is used, about half the current, or even more, is taken up by the meter, so the circuit can't act normally with this side path for current to flow in. You would get very incorrect voltage readings. Such a meter is referred to as a "1,000 ohms per volt meter." For every volt it measures, only 1,000 ohms of resistance are in its circuit, allowing the full milliamp of current to flow in the meter for full scale reading.

This problem can be solved in several ways. One is to use a vacuum tube voltmeter. Remember that tubes are sometimes called *vacuum tubes*. With a vacuum tube voltmeter (VTVM), the current for the meter is furnished by a power supply or battery. The voltage being measured is attached only to the grid and ground. Since only voltage is on the grid under these circumstances, no current from the circuit is robbed for the meter. The grid voltage affects the plate current of the tube, and the meter is in the plate circuit much as an earphone might be. Hence, a series circuit of tube, meter, and power supply does not take current from the circuit. It would have infinite ohms per volt. The trou-

bles with such a meter are: (1) it takes a while to warm up and give decent readings and (2) as tubes age they may draw a different current for the same grid voltage. Accuracy is not perfect. A lot of these VTVMs are around, though, so you may come into contact with one, especially if you work in someone's TV shop.

Solid state voltmeters are quite the rage now. With regular transistors, a tiny current (such as a few microamps) goes from base to emitter if a PNP transistor. This is the current of the circuit under test. Since it is so small it doesn't appreciably affect the circuit being tested. The small base current affects a larger collector current, and the meter can be in the collector circuit. Another solid state voltmeter is the field effect transistor (FET) type. Since field effect transistors act much like tubes, a small voltage on the gate allows a current to go from source to drain, and another transistor or the meter can be in series at that point.

Either way, these instruments are good in that they draw very little if any current from the circuit being tested. Of course, if the transistor or transistors are not quite linear (that is, if they do not draw current in the collector proportionately all the way up and down the scale, directly in relation to base current) they can give inaccurate readings. Field effect transistors are very easily damaged, and to be useful they should have built-in protection for overloads.

Another way to assure accuracy is to use a meter that draws very little current in order to give full-scale deflection. A 50-microamp (or 0.00005-amp) movement meter is quite desirable, as it offers 20,000 ohms per volt. With such a meter chances are you will not rob any circuit of enough current to give very inaccurate measurements. Such a meter can be obtained rather inexpensively from Radio Shack (the present part number is 22-051). Meter movements, especially the sensitive ones, should be handled carefully.

We have talked about the ohms-per-volt ratings of meters. These ohms are not offered by the coils of the electromagnet. After all, it is just a few turns of wire. The resistances have to be put in as resistors. The resistor cuts the current down to the point at which 50 microamps will move the needle all the way to the right side, where the scale will read 50. If 10 microamps goes through the circuit, with the same resistor connected, the needle will deflect one-fifth the way and read 10 on the scale.

Naturally it is desirable to make the numbers on the scale correspond to voltages. By the proper choice of resistor you can make the 50-microamp current flow when the input voltage is 50 volts. A series resistor of 1 megohm or 1,000,000 ohms will do this. By now you should definitely have learned the color code for resistors; the 1 megohm resistor would be brown/black/green. Since the current is so small in this circuit, you need not worry about wattage rating of the resistor. A one-fourth or quarter watt resistor is more than large enough, but you can use a large one if you have it, and it will fit into the case properly.

Look at Figure 8-2 *a*, the schematic. Current would enter at the black lead from the set under test. S1, the wafer switch or range selector, would be set at R5. Current would go through R5, the 1-megohm resistor, and drop the full 50 volts here, except for a tiny part of a volt that would drop across the electromagnet. Then the current goes through the meter from $(-)$ to $(+)$ side to work the electromagnet. Finally it goes out the red lead back to the circuit being tested on its positive end. This makes the meter give full-scale deflection, because a full 50 microamps went through it.

Now we leave the probes on the same part under test but change the range switch to the R4 setting to measure 500 volts of DC. The only difference is that the series resistor is 10 megohms instead of one. Now, instead of being dropped across the 1-megohm resistor, the 50 volts must fight its way across ten times the resistance. This will cut the current to one-tenth its former amount. Such a current of 5 microamps will lift the needle only a tenth of the way or halfway between 0 and 10

on the scale as shown in Figure 8-1 *a*. You are still measuring the 50 volts, but you must interpret it differently. When the switch is set for 500 volts DC, a full-scale reading of 50 means 500. You have to add a zero to each measurement you read. Since the meter is pointing to 5, you add a zero and get 50 volts.

You cannot turn the meter to R6, as it measures 5 volts and the 100,000-ohm resistor would allow too much current in and peg the needle past its normal place. If such voltage were left on very long the meter would be ruined. If you don't know what value of voltage to expect from a circuit, start with the high range. Then if the needle scarcely moves you can lower the range. Protect your meter!

In the 5-volt position, using R6, you must remove one zero from the reading in your mind to get the right voltage. A reading of 30 would really be 3 volts. Sometimes when you measure a voltage the needle will go down scale—that is, even below zero. If it does, remove the red lead immediately. Either you have your leads wrong, or you are measuring a negative voltage. For instance, with PNP transistors you would need the red lead on ground and the black to the collector. With NPN transistors you would need the black to ground and the red to the collector. If you get a down scale, reverse your leads. Don't ever leave a down scale reading on very long. Learn to beware of this situation and remove one lead quickly.

So far we have dealt with DC measurements, which are what you usually will measure. Sometimes, however, if the trouble is in the power supply, you will need to measure AC. Or you might want to troubleshoot a washing machine or another AC device in your house. AC meters measure AC directly. With a DC meter though, a diode can be used to rectify the AC, changing it to DC. The diode works the way it did with the radio signals; it cuts off half of every cycle, which then can go through the meter and be measured. The problem is that by cutting off the negative cycles (or positive ones, depending on which way the diode is set), some of the voltage is

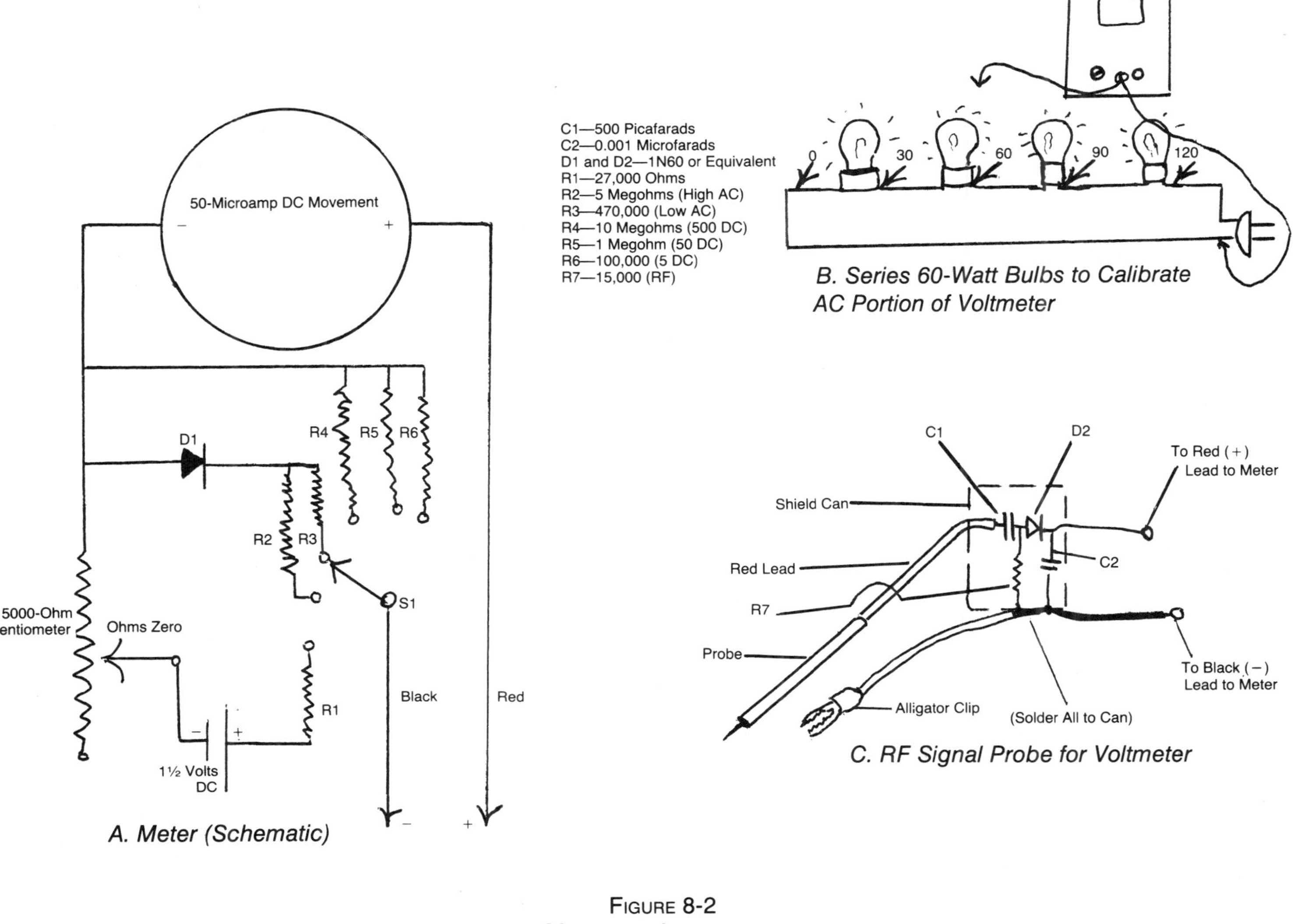

Figure 8-2
Meter and Accessories

lost. A different calibration of the scale is therefore required to be accurate for AC. Note that S1 selects the AC just by changing positions, and there are two choices, R3 for lower voltages and R2 for higher voltages. D1 must have its cathode toward the resistors. Current can go into the negative side of the meter but not back the other way. When you are measuring AC it matters little which way the leads are. After all, first one lead will be positive and then the other one, 60 times a second. With DC the way the leads are attached to the circuit is critical.

One way to calibrate the meter for AC is to rig up four lightbulbs of the same wattage in series. See Figure 8-2 *b*, which shows where you would measure the different voltages if the bulbs were plugged into a 120-volt AC outlet. You could note where the needle points for each measurement, and if the positions did not quite correspond to the DC readings, you could add a scale either by carefully removing the plastic from the meter and penciling it in or by drawing a scale on the panel just above the meter plastic.

The ohmmeter section of your multimeter is in the circuit when S1 is set at ohms or when R1 is in the circuit. In ohms position the current leaves the battery at negative side, goes through as much of the 5,000-ohm variable resistor or potentiometer as you set it for, through the meter, down the red lead to whatever you are measuring in ohms, up the black lead, through R1 and to plus side of battery. In ohms position the red lead supplies negative current and the black positive. This is a disadvantage, but it is common to many meters. The only way to overcome it is to have extra leads from the meter for ohms. Turning the battery around would ruin the meter movement. Just remember that in ohms position, the current goes out at the red lead. When you check transistors this fact is important; otherwise (with capacitors, resistors, coils, and so on), it is not.

This ohm circuit is set so that shorting the leads together will give a full-scale reading, if you adjust the ohms adjust potentiometer correctly. As the battery wears down, different settings of the variable resistor compensate for the difference in battery strength. With the leads shorted, set the ohms adjust potentiometer for full-scale deflection as shown in Figure 8-1 *a*. Now the meter is ready for ohms readings. Note that the range switch is set to the far right as seen from the front. The sign you see (Ω) stands for ohms. You will find it in many schematics.

Try a 5,000-ohm resistor and note where it fits on the voltage scale. Then try a 10,000-ohm resistor and measure it with the probes, then a 20,000, and so on until you have the meter fairly well calibrated. You can record the resistances either on the scale, carefully avoiding damaging the needle, or above. Remember these numbers as thousands of ohms and omit the zeros. Most numbers used in TV work will represent thousands of ohms. The spot where the needle normally comes to rest, zero volts, you should label *Inf.* for *infinity* or so much resistance that the current doesn't even move the needle. Note that even gripping the lead ends with your hands will give a small reading, because your body will pass a small current. It is good practice to leave the range switch, S1, in volts position. If it is set on ohms and the ends of the leads touch each other overnight, it would wear out the battery in time. When they are used only for measuring ohms, batteries can last over a year.

Some ohmmeters have several scales. The ohms $\times$ 0 is a low-ohms reading and is often used for troubleshooting motors. This position usually puts a healthy current out into the circuit under test and should be avoided when testing diodes and transistors. Then there will be an ohms $\times$ 10 position. In that case you add a zero to every reading. The ohms $\times$ 100 is usually the highest and you add two zeros to every reading. With the meter in Figures 8-1 and 8-2 you would add three zeros to any reading. This is an ideal scale for TV work. To have added other scales would have necessitated more positions on S1. With a factory-made meter, use the highest scale when measuring transistors or diodes.

Figure 8-2 *c* shows how the DC voltmeter section of the multimeter may be used to measure RF signals, such as the IF signal through a TV set. The red line to the probe from the shield can should be no more than 3 feet. The shield can may be any metal container, but it must be something that the ground lead can be soldered to as well as R7 and C2. (These resistor and capacitor numbers are strictly for this project and do not refer back to radio projects. Every schematic has its own R1, C1, and so on.) A 1-volt signal will not necessarily give 1 volt of DC reading on the meter. It will show relative signal strength, however, and can be useful in determining whether or not there is a signal at any point. In RF stages such as a tuner it may not indicate a weak signal. This is normal.

MAKING A VOLTMETER-OHMMETER

A multimeter should have a wood or plastic case, because metal may short to parts in time. True, it will prevent RF from coming in, but this is usually no problem. I personally like wood sides and a masonite front. The soldering of R2 through R6 will usually hold those parts as well as D1 in place. If not, a piece of insulated wire can be wrapped around for support and fastened to the back of the front panel with screws. The flashlight cell is soldered into the circuit. When you must change it (when full rotation of the ohms adjust potentiometer won't give zero ohms with the leads shorted), new soldering is necessary. Of course, a battery clip may be used instead.

The meter movement may not come to rest exactly at zero, but you can adjust it by turning a small slot in front with a screwdriver. Often a small plastic plug must be removed, or in some cases the plastic cover must be removed. The leads can terminate in probes or alligator clips with insulating sleeves on them. Solder all connections except at the meter.

Usually the meter has solder lugs, but if not, connect the wires and tighten them with nuts. This is not the world's greatest multimeter, but for the money it is very hard to beat.

We will not give step-by-step directions for construction of this project, because by now you should be able to plan your own steps for building the set. The pictorial drawings of Figure 8-1 should help considerably in layout for a neat arrangement. Figure 8-2 *a* will aid in the connections. Note the parts values listed at the lower right.

Once you have made the multimeter, spend some time considering how to use it effectively to find bad parts. We have discussed some of this before, but it just can't be repeated too often. Consult Figure 12-4, upper left, the schematic of a tube type TV tuner. Normally you need a schematic of the set in order to service it. Howard W. Sams publishes a photofact packet for just about any TV you will encounter. It includes schematics, parts layout information, and alignment and adjustment procedures. These packets are available at many TV parts centers for about four dollars. This material shows what value of voltages you should find at many places in the set. The TV manufacturer often has similar material available at low cost.

Suppose at the mixer plate of V2, pin 2, you don't get the 95 volts shown. Instead, you get well over 100 volts by clipping the black lead to the chassis and touching the red lead to pin 2. The meter is set on 500 volts DC. When a tube is dropping a very heavy voltage, like a resistor, less than normal current must be passing through the tube. Several things could cause this problem. If the tube is old the coating on the cathode may no longer give off enough electrons to travel to the plate. The grid bias could be so large that it is cutting down on electron flow to the plate because of changes in the value of grid or cathode resistors. The tube might not be making contact with some of its pins. You can pull the tube out by holding it with a tube puller or handkerchief—it will be hot. Measure plate voltage without the tube in. It should be very high now, as the tube is gone and the greatest of all resistances is offered—no path for current at all. In this case we find the same high voltage as before, meaning that the tube was not conducting. And replacing the tube doesn't help matters. (Of course, the substitute must be the same number tube.)

Now it is time to cut the electricity off and change from voltmeter to ohmmeter operation. With the ohmmeter measure from cathode pin 7 to ground. Note that the cathode should be directly grounded with this tube. You measure infinity or no path at all and find that pin 7 has a poor solder joint in its connection to pin 4 and ground. With no cathode current there was no current in the tube, so the plate voltage was very high. A bad tube would have given the same high voltage. Had this been a transistorized circuit with an ungrounded emitter, you would have measured a high collector voltage. With NPN transistors you would measure the same way but with a lower range, probably on the 50-volt scale. With PNP transistors you would secure the red probe to ground and use the black one at the collector. Transistors are subject to shorting, and in case of a short no voltage would be at the collector. Tubes seldom short from cathode to plate, but if one does there would be no voltage across the tube. In the case of a tube with a cathode resistor and the tube was shorted internally, a very small voltage would be at the plate.

Now with Figure 12-4, assume that pin 2 of the plate of V2, the mixer section, measures somewhat low, but not low enough to indicate a shorted tube. Still, when plate voltage is low, plate current is too high. What keeps too much current from going out the plate? The bias, remember? The grid, pin 5, calls for -2 volts. This is bias. To measure, set the meter on 5 volts DC and connect the red lead to ground and the black to pin 5. Since the pin specifies a negative voltage you know to do this. (You should know anyway, as tubes have a negative grid bias.)

If you are unable to measure the bias voltage, pull the plug and convert the meter to an ohmmeter. You check R15 and R16 and they are of proper resistances. You note that from pin 5 you can get a direct short to ground. Since the meter furnishes DC in ohms position, the capacitors C10 and C11 should block it. You can rotate the tuner and watch the meter with one lead on ground and one at pin 5. L14 is part of the tuner. If C10 is shorted, flipping the channel selector should break the circuit to ground momentarily between connections to coils. If this does not happen, then C11 probably is shorted. You would desolder the high side of it and test across it for a short. With capacitors, if an ohmmeter shows continuity (completely connected, zero ohms) the capacitor is shorted. Only an electrolytic capacitor in the $20+$ microfarad range will show resistance, and it should go very high after a second or so of touching the ohmmeter to it. In the case at point C11 was shorted. This would be a very small adjustable capacitor. If C16 shorted there would be no voltage at pin 2.

We have tried to show how a voltmeter and ohmmeter can be used to test tubes and transistors under operating conditions. Pulling a tube should raise plate voltage; replacing the tube should lower voltage. Another way to see whether a tube is amplifying or at least capable of it is to clamp the voltmeter to ground and red lead to plate. With a 100,000-ohm resistor—holding it by the body and not the leads to avoid shocks—touch between grid and ground. This would be the signal grid or the biased grid. The voltage should fluctuate at the plate, because you have changed the bias on the tube. If the tube were not capable of amplifying this would have little effect.

You should not hold the resistor to the grid and ground for any longer than absolutely necessary, as you are making the tube act differently from the way it was designed to operate in the circuit. If the tube has a cathode resistor, take one of equal value and bridge across the first one, thus lowering the cathode resistance and putting the grid and cathode at closer to the same potential, so plate current should go up and plate voltage should dip down a little. This is one way to test a tube if you don't have a tube tester. Of course, always look for obvious things like a tube not lit or not warm, cracked glass, and so on. You can look for shorts by taking a tube out of the socket and measuring across all pins with your ohmmeter. Naturally, the heaters should show very little resistance; actually this meter will show a short at heaters, and this is desirable. Tubes burn out and have heaters that don't heat or cathodes that put out weak electrons far more often than they short.

Transistors short more often, and you have learned how to test them for shorts. To find out whether a transistor will amplify in a circuit, connect appropriate meter leads to ground and collector with the power on. Now touch the 100,000-ohm resistor between base and ground. This will cut down on forward bias and the transistor will have less current. The collector voltage will increase. Briefly connect the resistor between base and collector. Collector voltage should go down, as you are increasing the forward bias and therefore the current.

These two tests usually tell whether a transistor is capable of amplifying or not. However, it does not tell whether or not the transistor can work in a high-frequency circuit. Sometimes you just have to remove one transistor and replace with another to be sure. When a transistor is out of the circuit, whether it is a new one or one you have removed, you can test it with your ohmmeter. Put your red lead tip to the base. Measure by touching the black lead first to emitter and then to collector. The readings should be either very high or very low. Now put the black lead to the base. Touch the red one to emitter, then to collector. If the readings were high before, they should both be very low now or vice versa. Finally, measure with red lead to emitter and black to collector, then with red to collector and black to emitter. Neither reading should be below 500 ohms for audio type or less than 5,000 ohms for RF types.

Resistors should show their proper ohm readings in a circuit. A coil usually shows a short if it is not of many turns, and this is normal. Capacitors should show infinite resistance. But can you find out whether a capacitor is open? The best plan is to try injecting a signal through it with a signal generator. The signal should go through either side. You can test a coil with an ohmmeter to see whether it is shorted to the chassis in the middle. Power transformers are best tested with a voltmeter set in AC position (we will cover this material in Chapter 11).

In your checking, be sure that no other component connected to the one you are checking is giving you a false reading. For instance, in Figure 12-4 suppose you were measuring the resistance across R9 and instead of 22,000 ohms you got nothing or a short. To the left is the coil in parallel with it, and the coil shorts out the DC current of R9. This is normal, as this is a tuner. Still, you might remove R9 and go to a lot of trouble for nothing. When other parts are connected to the spot where you must check, remove one end of the part under test with your ohmmeter. This will eliminate errors. Lifting either side of R9 would allow you to see whether it has the 22,000 ohms it should. It must be resoldered into the circuit after the test, of course, and such work makes TV servicing take time, although customers sometimes can't appreciate the hours it takes to fix a set.

OTHER TEST EQUIPMENT

Just about everyone in the TV repair business agrees that a repairman must have a tube tester, a transistor tester, and a cathode ray tube tester (a picture tube tester). It is much easier to pull and test tubes in a separate unit than to get the voltmeter in and change bias, measure voltage changes, and so on. Tubes might be on their way out as home electronic equipment amplifiers, but if so, their retreat is most gradual. Tubes will be around and will need testing for many a year yet, but since there will be fewer as time goes by it might be best not to stock every tube that could possibly be needed.

A serviceman used to carry two tube caddies to each repair job, as there was about a 70 percent chance that he would need one or more. He probably had three more caddies of less commonly used tubes in the truck. Today, the serviceman often knows whether the set is solid state or tube before he makes the call. It makes sense for an independent shop to carry a few of the most needed types of tubes and to order everything else for the job.

Tube substitution is by far the best check of any tube in a TV. Sometimes you can test a tube in a tester or checker and it will show "good" but it still will not work in the circuit. Then you can put in another of the same type and all is well. Even so, a good tube tester will aid in servicing. It should check 12-pin, 9-pin, 7-pin, and octal (8-pin) tubes. The old 4-pin and 6-pin types are seldom going to be needed, so if you buy a new or used tester, be sure it will test modern tubes. Appendix B includes a schematic of a tube tester that you can make for a few dollars and that will take care of many tubes.

Transistor testers vary in their ability to check transistors in the circuit. The tests described earlier with a 100,000-ohm resistor are more definite. Even so, these little beeping goodies that you connect to emitter, base, collector while you push buttons until you hear two beeps are much in demand. They do have limitations, and their testing is not absolute. Since more and more TVs are composed of integrated circuits (ICs) with several transistors in each IC, a transistor tester could be a white elephant to you.

The picture tube tester has certain advantages. If it checks each color beam system of a color picture tube separately, it can come in handy. Most of these testers not only check emission of electrons from the cathodes, but also can be used to rejuvenate a bad unit. (Color picture tubes are covered in Chapter 10.) If the red were defective, for example, and the blue and green OK, you could save

the very high cost of a new picture tube. Of course such rejuvenating is not always successful. It could last years, months, or days. The customer should understand that nothing can be guaranteed by rejuvenation. You are doing him a favor, trying to save him money. If you buy such a device, make sure you get it for color tubes. Directions for making a picture tube tester are given in Appendix B.

The oscilloscope is a rather necessary piece of equipment since the advent of color TV. It might sit on the bench for months and gather dust, but then it could need instant use. This device has a small cathode ray tube on which waveforms appear so you can see the carrier, modulation, and so on. The amplitude is the input to the oscilloscope. The frequency of the wave is made by an oscillator in the oscilloscope. Since the up-and-down motion comes from a wave from the TV, and the oscillator in the oscilloscope is set to the same frequency as the incoming wave, and since this oscillator moves the signal from the TV right and left at the same frequency of the wave, one big wave is traced over and over on the screen. Some oscilloscope screen pictures are shown in Figure 13-2, W20 to W34B. Each picture is two complete waveforms. To get this pattern the oscilloscope oscillator frequency must be set to half the frequency of the signal being taken from the TV receiver. More oscilloscope pictures are shown in Figure 15-3.

Oscilloscopes come with instructions for use, as do tube testers and signal generators. It always pays to study the operating instructions carefully. In time, use of the equipment will be easy, but at first, it is wise to read the instruction manual again and again. Note the information given with the scope pictures in Figure 13-2—W20 for instance. The 5V PP means that you would set the height of the picture for 5 volts peak to peak. If you don't allow for this the picture can go all the way off the screen at the top and bottom.

Some scopes have a calibrated screen that shows how large signals of 5 volts, 10 volts, 20 volts and so on would be. The final V means to set the scope at the vertical rate. Since it shows two pictures I think they mean half the vertical rate or 30 Hz. If you start your own shop you might be able to take sets requiring oscilloscope tracing to another shop for a fee. A scope is a pretty good investment, but its purchase might be put off a while. You will find many of the faults in sets with signal generator, voltmeter, and ohmmeter.

What the oscilloscope does is show whether a signal is of proper size, shape, and frequency. If it is small it will not form high enough. If it has a part chopped off or is too flat or sharp, you would know to adjust parts or replace some of them. If it is at the wrong frequency the picture will not be still, as it can't trace unless the input and oscillator are at either the same frequency or one is half the other. When you study color sections of a TV set, you will learn about *burst*. If burst doesn't appear in an oscilloscope picture at the proper place, you know the trouble to a certain extent, and a voltmeter would not show it.

In this chapter you have learned the following:

1. A meter is a combination of a permanent magnet and an electromagnet that oppose each other and move an indicating needle when current flows through the electromagnet.

2. External resistors in a meter keep current flow down to the point at which maximum current will move the indicator to the high side of the scale.

3. A meter of 20,000 ohms per volt or higher is necessary so the meter will not absorb too much current from the circuit under test and give false readings.

4. Vacuum tube voltmeters (VTVMs) and field effect transistor voltmeters (FETVMs) draw very little current but require a battery or power supply.

5. Power must be supplied the set under test to measure volts.

6. When using a meter, you should be sure the red probe tests at the most positive point.

7. You should measure on a high voltage range if you aren't sure what the voltage will be.

8. Ohms are measured with the set unplugged. A battery supplies the current for the meter movement.

9. You should zero the ohms before testing each time, as the battery strength will change.

10. You must be sure you are measuring only across the tested item with an ohmmeter; if necessary, lift one end to measure.

11. For troubleshooting, you should use a schematic that gives the proper voltages to expect.

12. Excessively high plate or collector voltage indicates a tube or transistor that is not conducting current.

13. Excessively low plate or collector voltage indicates a tube or transistor that is carrying too great a current. Bias (or possibly a short with transistors) is often the problem.

14. Capacitors, if good (not shorted) will not show any ohms but infinity instead.

15. Open capacitors can best be checked with a signal generator or by substituting.

16. A modern tube tester should measure 7-, 9-, and 12-pin and octal (8-pin) tubes. Do not buy one that is not so provided.

17. Oscilloscopes show waveforms when a light-forming beam is moved vertically by an incoming signal and horizontally by an oscillator in the scope.

Quiz for Chapter 8

▶ 1. A 1-milliamp meter would offer how many ohms per volt as a voltmeter? Would it be a desirable voltmeter?

▶ 2. Normal battery voltages for ohmmeters are _____ to _____ .

▶ 3. What is down scale in a meter, and what problem could it cause?

▶ 4. If a meter has a full-scale reading of 50 microamps and the current fed into it gives 20 percent deflection, what number of microamps would it show?

▶ 5. What item is added to an AC voltmeter that a DC voltmeter does not need? How must this item be put in so it will send current to the meter movement properly?

▶ 6. Explain how to use a voltmeter to measure bias on a tube.

▶ 7. If a tube were bad in that it sent no electrons from cathode to plate, what might be measured at the plate?

▶ 8. A multimeter is a combination of ______ , ______ , and sometimes (not in the set illustrated) ______ .

▶ 9. The best test for a tube in question is ______ .

▶ 10. Why should a meter not be left in ohms position after normal use?

▶ 11. Explain an easy way to calibrate an AC voltmeter.

▶ 12. How can you adjust a meter so it will rest on voltage zero if it doesn't?

▶ 13. VTVM means ______________ .

▶ 14. Explain how to measure voltages on a PNP transistor and on an NPN transistor.

9. The Four-Part Television Signal

PROJECT: Observing the Video Output Waveform of the Last IF Stage of a TV Receiver on an Oscilloscope

IN RADIO broadcasting only one signal is necessary to carry music, talking, and other sounds to the radio set. Television, however, is a combination of elements and requires four separate signals to produce an acceptable color program. Even a black-and-white picture requires more than one signal—one signal makes the screen light or dark in any area, and another traces out the picture on the screen. Color demands its own signal. And of course, TV, like radio, has sound. Actually the TV channel contains four signals that must be separated by circuits in the receiver to produce an unscrambled color picture with sound.

This chapter will describe the carrier that carries the four signals from the transmitter to the TV receiver, the signals themselves, and the way they fit on the carrier. First we will cover the carrier, the frequencies of the signals, and the color and sound signals. Then we will take up more complex video (black-and-white) signals with sync or synchronization.

Alternating current, AC, is the type of electricity that is supplied to your house. Normally it is referred to as 60 hertz or 60 Hz, meaning that 60 times a second the polarity reverses itself. The wire that was positive becomes negative and then positive again 60 times each second. This is also known as the frequency of the current, and frequency is expressed in hertz. If the speed of the AC is much faster, with the polarity reversing itself more often, the frequency may be in thousands of hertz, or kilohertz (kHz). As you have learned, radio stations produce their frequencies in this region. WSM, the Grand Old Opry station at Nashville, Tennessee, operates on a frequency of 650,000 Hz, or 650 kHz. The transmitter uses tubes and solid state equipment to make the very fast AC. High-frequency AC will produce radio waves that travel many miles.

Television has been assigned much higher frequencies in the millions of hertz, called *megahertz* or *MHz*. The TV station has some very expensive and exacting equipment to generate electrical frequencies in the MHz range. Remember, however, that TV stations have to transmit four separate signals, and if they were all on one high frequency, they would be jumbled together so that no TV receiver could separate them. The solution is to have a wide band of many frequencies so that the picture and sync can be put on one, the color on another, and the sound on a third part of the band of frequencies.

The wide band of frequencies is a channel just like the one you choose with the channel selector on the front of the set. The band of frequencies is known as a *carrier*, and it carries the four signals from the transmitter to the receiver many miles away.

Refer to Figure 9-1 as you continue with this lesson. The TV carrier is 6 MHz wide. This is as the complete channel carrier appears in a receiver. Both ends tend to have

Video, Horizontal and Vertical Sync Signals
FM Sound System
Red
Horizontal Sync Pulse
Video
Equalizing Pulses
Burst
Blue
Sync Level
Black Level
White Level
Vertical Sync Pulse
Color Subcarrier
Carrier Unmodulated
Carrier Modulated Below
Amplitude
−1.25 MHz
0
3.58 MHz
4.5 MHz
6 MHz

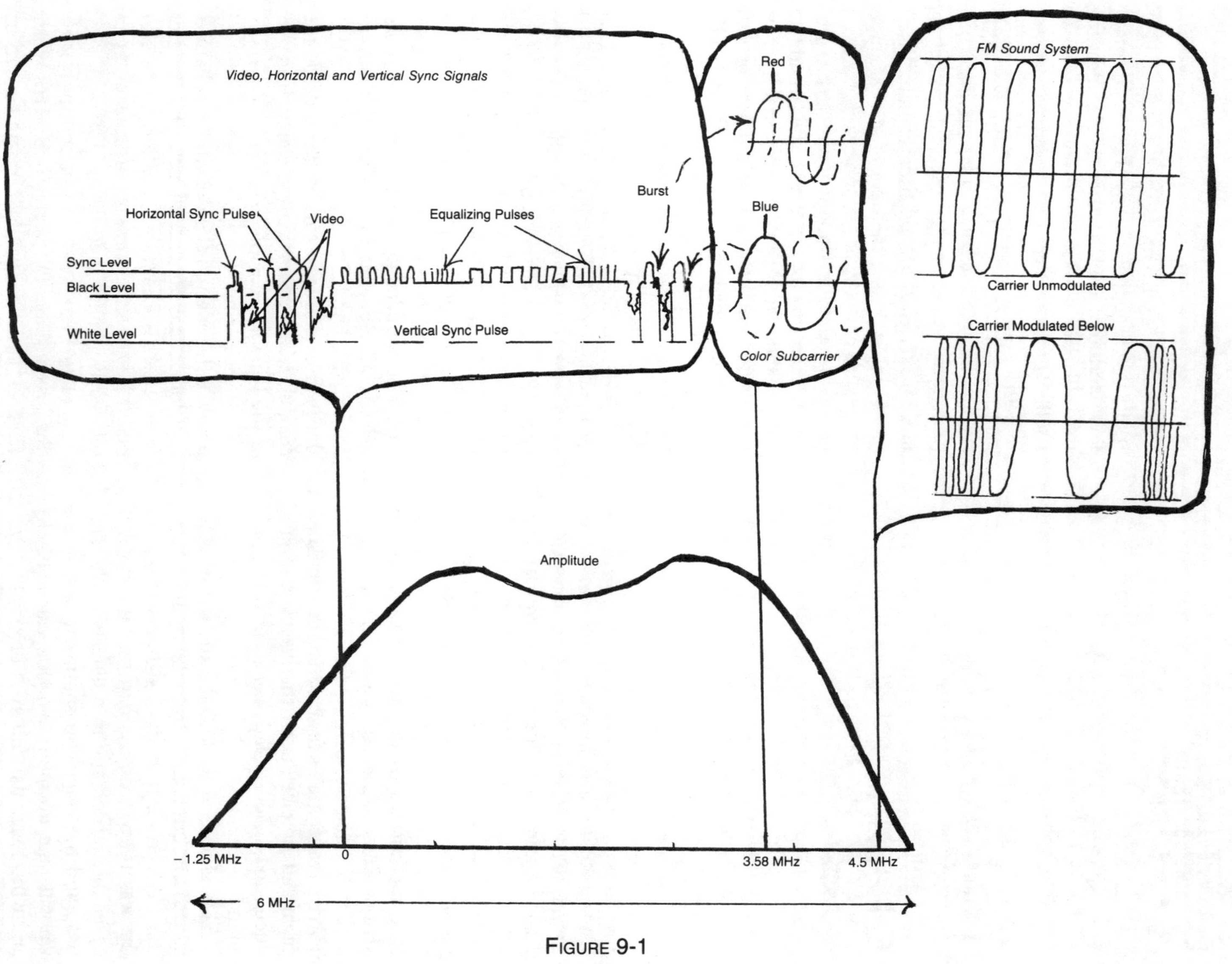

FIGURE 9-1

very small *amplitude,* which is the amount of power the carrier has. Note that the carrier is more powerful in the middle than at the ends. The best is a double-humped carrier as shown, with greatest power at the frequencies necessary to transmit video, sync, and color. The sound can have less power at its frequency and yet produce plenty of volume.

The frequencies shown are always the same with relation to each other. That is, the sound is always 4.5 MHz above the video and sync. Actually, the TV channels start with Channel 2 at 54 to 60 MHz, then Channel 3 at 60 to 66 MHz, and so on. If this were a sketch of Channel 2, then the video would be 1.25 MHz above the 54 MHz bottom frequency, or 55.25 MHz. The sound would be riding on 59.75 MHz.

As Figure 9-1 shows, the video and sync ride on the carrier 1.25 MHz above the bottom of the 6-MHz bandwidth. You have not learned about sync yet, but you will read much about it later. For now, you only need to know that sync is the signal that moves the beam that traces out the picture on the screen across (horizontal sync) and down (vertical sync). The color signal is always 3.58 MHz above the video and sync.

At this point you need to understand how these signals ride on the carrier. The carrier is a group of continuous alternations, AC, all just alike in shape and in size, or amplitude. The process of putting a usable signal on a carrier is known as *modulation.* Three types of modulation are used in this case: amplitude modulation, frequency modulation, and phase modulation.

The video-sync signal is *amplitude modulated* on the carrier. We use the initials AM to stand for amplitude modulation. AM is a process of the video and sync, in this case, allowing the carrier to be large in power or amplitude, or small in amplitude, depending on what is desired at that time. Thus the sync or video controls the amount of carrier. Note in the upper left-hand part of Figure 9-1 when the sync is low, at white level, the modulation

would prevent much power from getting out of the carrier at zero frequency. Of course, the carrier would be larger in the 3.58 and 4.5 MHz regions. They would not be affected. Now, going up that first sync pulse to the sync level, the modulation would allow the carrier to increase greatly in power. This would be enough signal at zero level (remember that zero is really 1.25 MHz above the bottom of the channel) to blacken the screen of the picture completely at that instant. Amplitude modulation then determines the amount of signal.

The sound for TV is frequency modulated, FM, and does not change the size of the carrier. Rather, it bunches or spreads the actual alternations of the high-frequency AC. Note the upper right-hand part of Figure 9-1. When no sound is needed, the carrier is continuous. When sound is present, the carrier is compressed and expanded in proportion to what is being said or played.

On one end of the carrier the amount of carrier is being increased by AM, and on the other end the speed of the carrier is being changed slightly by FM. For TV FM sound, the sound can move the carrier as much as 100 kHz in either direction. On a carrier of at least 59.75 MHz (for Channel 2—and it is higher for all other channels) a full swing is very little proportionally. The sketch exaggerates a good deal to be sure you can grasp what happens.

In the middle is *phase modulation,* but some background is required for understanding it. The color signal really adds to the video signal in the picture tube, so black-and-white video is necessary in color sets too. The color travels in two parts. Note that on the back part of each horizontal sync pulse in Figure 9-1 is a little mark labeled "burst." This burst signal, although illustrated with the video and sync signals, really is at 3.58 MHz. It is drawn with the video-sync, however, as it occurs during the end of each horizontal sync pulse, the part of the sync pulse called its *back porch.* The burst must always come at exactly the same time; it must always be of the proper frequency, 3.58 MHz; and it must have a uniform size. All this must be attended to at the TV station, but sometimes a receiver can lose this burst signal.

The burst is a reference signal. It causes circuits in the TV set to produce an exact 3.58-MHz waveform. A second signal traveling at 3.58 MHz is going through other parts of the TV. Look at the top of Figure 9-1 in the color subcarrier balloon. The solid lines represent the signal created by the burst. The dashed lines represent the subcarrier signal often called *chroma signal*. They are both of the same frequency. Note that they are not together. When they are 90 degrees apart a voltage is created that produces red in the picture tube. When they are 180 degrees apart a blue picture would result. This lead and lag is known as *phase*. Thus the color is sent in two parts from the TV station as phase modulation.

To review, AC at very high frequencies produces radio or TV waves. TV must have a band of adjacent frequencies 6 MHz wide. The video and sync signals are always 1.25 MHz above the lower limit of the channel. The color subcarrier is always 3.58 MHz above this. The sound signal is always 4.5 MHz above the video and sync. The video and sync are amplitude modulated in that their changes affect the size of the carrier. The sound is frequency modulated, meaning that the spacing of the waves that make the carrier is changed. The color is phase modulated by a combination of a burst and subcarrier signals.

FORMATION OF A PICTURE ON THE SCREEN

The only successful way to date to transmit and receive pictures is by scanning them. There is no known way to put a whole picture into electrical pulses all at the same time. The scanning principle used in commercial television consists of tracing a line at a time from left to right as you see the picture tube from the front. Starting at the top the trace is from left to right, down slightly, then left to right again, and so on to the bottom of the picture tube screen. Inside the picture tube are three guns that send out streams of electrons—tiny negative particles of electricity. These beams or streams travel across the picture tube to the inside of the screen. Where they strike it, light appears. You will learn more about this later.

The sync pulses move this beam in their pattern across the screen. A color set has three beams, one for each primary color, that must be held very close together in their trace. There are two types of sync pulses—horizontal ones that move the trace across the screen from left to right and return it quickly to the left again, and vertical ones that move the beams slowly down the picture tube so the lines won't go on top of each other. The vertical sync pulse also returns the beams to the starting position after a complete trace. See Figure 9-2 *a* and *b*. The horizontal sync pulse is much shorter in duration than the vertical. After all, it doesn't take nearly as long to trace one line as top to bottom across the entire picture. An entire frame consists of 262½ lines. That means that 262½ line traces must occur before a vertical pulse returns the beam to the top to make a second frame.

See Figure 9-1, upper left. Note that the vertical sync pulse is broken into little serrations or equalizing pulses. During the time the vertical sync pulse is operating, the horizontal sync circuits in the receiver might get out of time. These equalizing pulses keep the horizontal circuits working during the movement of the beam back to the top.

Even though the video and sync are on the same frequency, they do not interfere with each other because they don't occur at the same time. The sync pulses come between lines of the picture, and they are shorter in duration than the lines of video, so the picture seems to the eye to be on all the time, even though there is a short period for the pulses between traces. Remember that this all happens very fast. There are 60 complete frames

Sync Level
Blanking Level
Black Level
White Level
A. Horizontal Sync Pulse
B. Vertical Sync Pulse
C. Video Signal Intensity and Effect on Picture
D E F G H I J K L M N O P Q R S T U

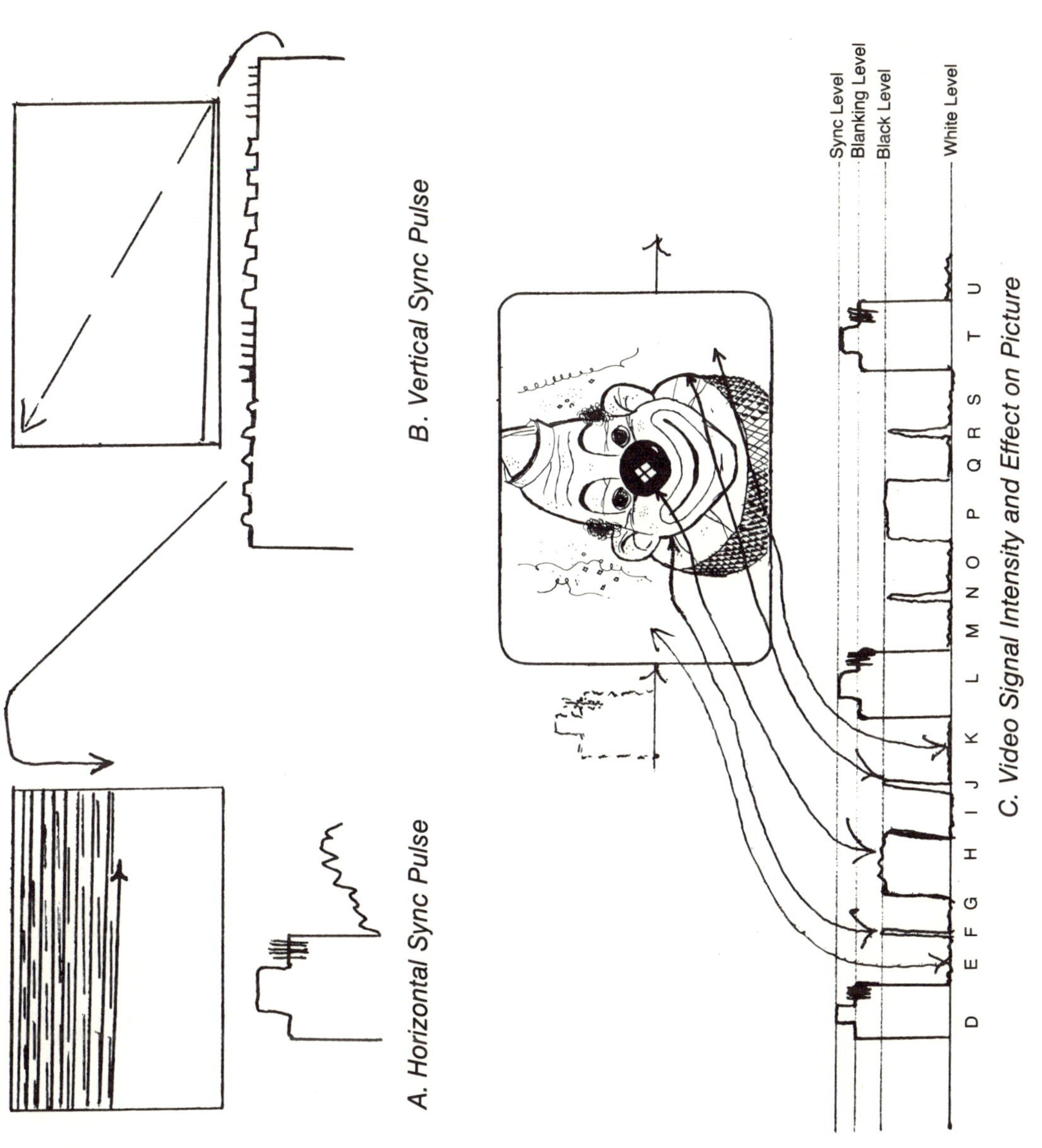

FIGURE 9-2

of picture a second, so the eye doesn't notice the very short time that the sync pulses consume. Actually, the trace is such that there are 30 complete pictures a second. Every other trace of the 262½ lines goes between the lines that were just shown. This is called *interlaced scanning,* and it produces less flicker. All in all, 525 lines make up a complete picture, but half of them go from top to bottom, then the other half are traced halfway between the spots where the previous ones had been. Although this will come later, the sync signals work circuits that develop great magnetism, and electromagnets are used in a device called a deflection yoke. This yoke sends the magnetism inside the picture tube, and this magnetism bends the beams of electrons to form the scanning you have just studied.

Now let's turn our attention to picture forming. Although the horizontal sync pulse comes before the picture trace, it triggers a circuit in the receiver that moves the beam across after the pulse has been received. During this time the video signal can make the beams of electrons light or dark. The more signal, the darker. Study Figure 9-2 *c* carefully. The white level at the bottom of the signal is the way the screen would be with no channel tuned in. When no channel is coming in but the screen is completely white and lighted, the white light on the screen is called a raster. Black level is a signal strong enough to make the screen completely black at the point where the beam or beams are at that time. The blanking level is beyond any light being shown. It is the "back porch" level of the horizontal sync pulses. Note that the burst for color is shown there, but we are not concerned with it now. Sync level is the top of the sync pulses. This area is even stronger signal and no light could be on the screen; no light would be there anyway because the sync pulses occur between traces.

Here is the way one line of video helps make up the face of the clown in the TV picture in Figure 9-2. The horizontal sync pulse

at *d* is the one represented by the dashed lines at the left of the trace. It comes before the picture, but it will carry the picture line across and even return the beam to the left for the next trace. Then the trace begins. At time interval *e* the signal is very low, so the screen gets bright white in this area. At point *f* in the video a spike of signal is received. This darkens the screen just long enough to make a part of the side of the face of the clown. Then *g* is low signal, allowing the screen to become bright again to make part of the cheek. At *h* a longer duration high part of the video signal makes the screen black for the mouth. (Now, of course, the color subcarrier and burst might trigger the red gun to go strongly and make red lips, but for now we are concerned only with light and dark.) At *i* the signal again is low in amplitude so the screen brightens up and gives a white countenance to the clown. At *j* the other side of face outline is produced, and *k* gives white screen because there is no background.

Then another horizontal sync pulse comes up for the next line of trace. Since it would be so close to the one before, it would be little different. So *e* and *m* are similar; *f* and *n* are similar, and so on. You can see that the size of the signal determines how dark the screen will be. A process of this kind is called *negative picture transmission* and is used in the United States. Some countries, at least some years ago, had positive picture transmission, in which stronger transmission produced a brighter screen. Negative picture transmission has the advantage that lightning and other disturbances blacken the screen, producing less annoyance to the viewer than large bright flashes.

Though the sync and video occur at different times and aid rather than interfere with each other, they still have to be separated in a receiver and sent to their proper sections to work. Since the sync signals are greater in amplitude than the video, transistors or diodes can be adjusted to take only the top of the sync, or to reject it and take only the lower video. It is very important that the sync pulses be higher than the video for that reason.

In retrospect, then, the sync signals trace the picture out on the screen of the picture tube. They occur between traces, but because of the great speed of transmission, no flicker occurs in the picture. The video signal makes the screen bright or dark at any point, depending on the strength of the video signal. The reason we see a complete picture instead of the separate traces is partly the speed of the trace and partly the fact that human eyes have what is known as *persistence of vision*. The eyes continue to "see" light for a fraction of a second after it is extinguished.

If you have access to an oscilloscope, set the sweep at 30 Hz, connect to the ground of a set and to a test point at the output of the video detector. Perhaps someone more familiar with TV sets will help you find this point on the TV receiver. You should be able to observe a trace similar to Figure 9-3 *a*. Naturally, the TV set must be operating and have a station coming in. Watch how the lines of video rise and fall, darken and lighten as the scenes change on the picture. The places where big gaps occur are the vertical sync pulses. This is a much compressed trace, and each horizontal line is one or more lines of picture run up and down instead of across. You do not see the actual picture, because the scope shows average light condition for that particular line, not its components.

Now change the scope sweep speed to 15,750 Hz. You should see about half of Figure 9-3 *b,* one horizontal sync pulse plus everything that makes up one complete line of a picture. Now set the sweep to about half that frequency or 7,875 Hz. Now you should see something similar to 9-3 *b* with two horizontal sync pulses and the video between that makes the screen light and dark as the picture is traced. This is the same as what is drawn in Figure 9-2 *c*. Note that the video is constantly rising and falling. Watch until the TV picture goes black, perhaps just before a commercial. What is the level of the video on the oscillo-

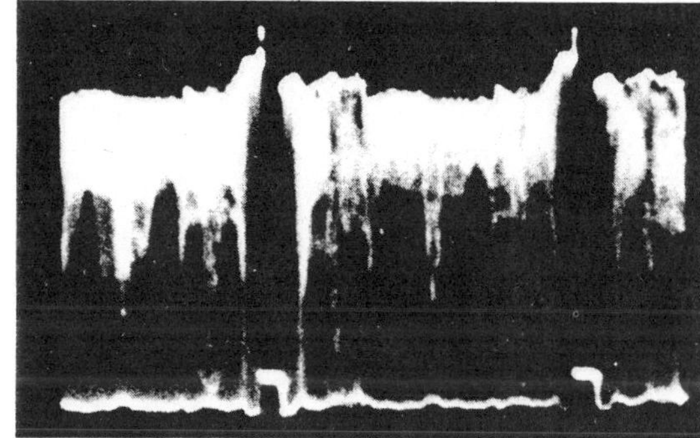

A

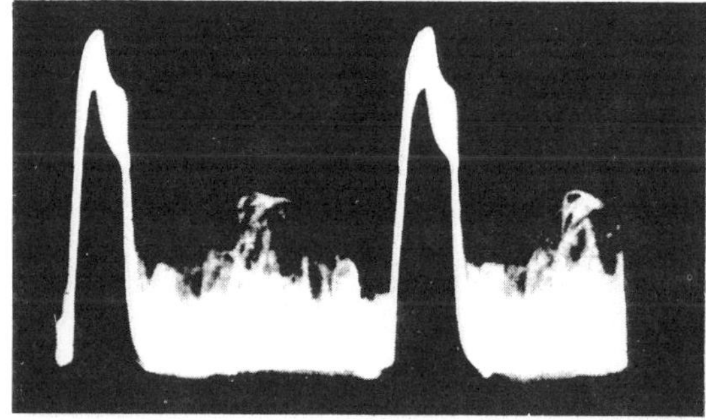

B

Figure 9-3

scope trace? Note also the video on the scope when there is a very bright picture.

Important points of Chapter 9 are:

1. A TV carrier is 6 MHz wide. Such a wide band of frequencies compose a channel.

2. The width of the channel permits video, color, and sound signals to be separated from each other so they can be transmitted at the same time. Sync shares the same frequency as video, but the video intervenes between sync pulses.

3. Video or light forming signals vary the amplitude of the carrier. The darker the picture is to appear, the greater will be the amplitude modulation at that instant.

4. Sound TV signals travel at frequency variations of the carrier in the general region of 4.5 MHz above video signal frequency.

5. Color information has a burst signal that is a reference. It is exacting in frequency and phase. A second color signal goes through other circuits and is slightly ahead or behind the burst cycle. This is a phase difference, and the amount of phase difference between the second color signal and the burst reference determines color of the picture.

6. The horizontal sync signals trace the lines of picture across the screen of the picture tube. There are 262½ of them for each vertical sync signal. In that way a whole picture will be traced by the combination of vertical and horizontal sync pulses.

QUIZ FOR CHAPTER 9

▶ 1. What four signals make up a complete TV transmission?

▶ 2. Three of these signals are separated from each other by ______ .

▶ 3. The two parts of a color signal are the ______ and ______ .

▶ 4. Horizontal sync produces one ______ of picture, and vertical sync makes ______ .

▶ 5. The part of the signal that makes the screen light or dark is known as ______ .

10. The Stages of a Color TV Receiver

PROJECT: Identifying Parts in Several Color TV Sets

Now THAT you have studied the chapter on the composite signal of video, sync, color, and sound for each channel, you will see how the TV set makes use of these separate signals to produce a good picture and sound. You will recall that a superheterodyne radio tunes in the desired station, converts the signal to an intermediate frequency, IF, and detects it before amplifying the sound for the speaker. Although a TV does somewhat the same thing, it is far more complicated because it receives four signals.

First the TV must select the one channel the viewer chooses and reject all others. Next it must amplify everything strongly at an IF so all signals will work in the rest of the set. It must separate the different types of signal and use each to perform its desired function. Because a picture tube is much larger than most other tubes, large voltages are needed to operate it. And large signals are needed to operate it properly. The signal that will operate a radio loudspeaker is not nearly strong enough for the picture part of TV. For this reason a TV set may have twenty or more tubes or that many transistors, whereas a radio needs only four

tubes and six transistors. You will find black-and-white TVs that seem to have as few as eight or nine tubes, but what you see are tube envelopes. Inside each glass envelope will often be two or even three working tubes, each with its set of cathodes, grids, and a plate. A tube that has several working units inside is called a compactron. And each working unit is really a tube by itself. It takes many amplifiers, tube or transistor, to make a TV work.

Figure 10-1 and the text that follows will give you a fair idea of what takes place in a color TV receiver. Each of the blocks represents a stage, including the tubes or transistors, resistors, capacitors, coils, and everything it takes to make that stage work. In old-fashioned TV sets, even early color ones, most of the stages were all together on a metal chassis just under the back of the picture tube. It was hard to tell where one stage ended and the next began. Although often a drawing was attached near the rear showing each tube location and its function, the parts associated with that tube could be scattered all over the TV. You very likely will be required to service such sets. With a manufacturer's pictorial you can find the parts for each stage, or you can get a *Sams Photofact* of the set.

Today each stage may be on its own module or circuit board, which may hang sideways, over the tube, under it, or almost any way. Many circuit boards are labeled, aiding considerably in servicing. Some circuit boards contain more than one stage, and sometimes a part of a stage will be on one circuit board and part on another. And some parts can't be on the circuit board. The sound volume control, for example, must be on the front or side of the set so the viewer can adjust it. Long leads are therefore needed from the sound module to the potentiometer on the set. Similarly, to allow the color to be adjusted from the front panel, a color or hue control must be placed there and leads must go to the color module.

Let's go through the stages as shown in

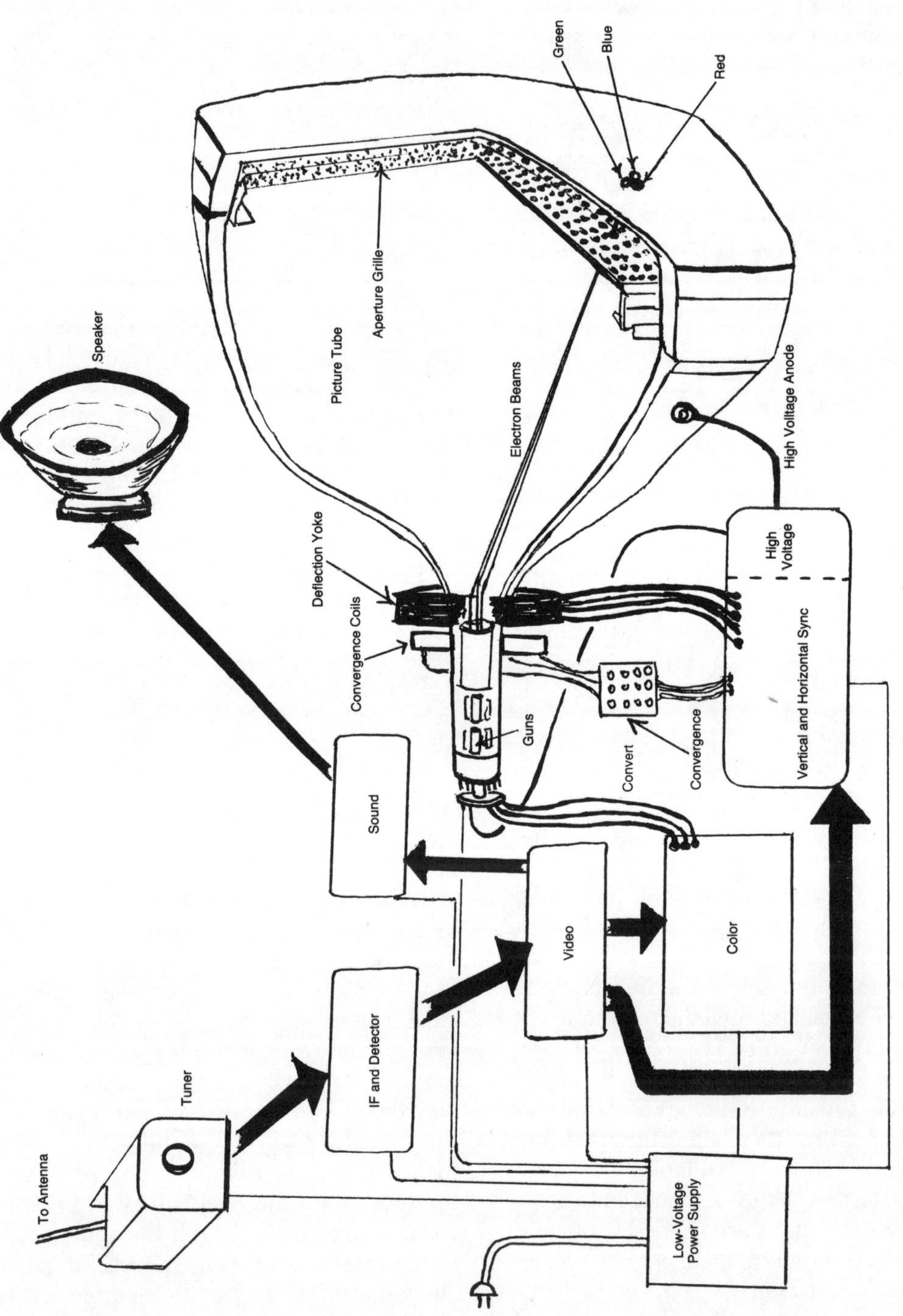

FIGURE 10-1
BLOCK DIAGRAM, COLOR TV

Figure 10-1 and learn how a set operates. In several later chapters you will study each stage in detail with schematics and layouts of the parts on the module boards. For now, you should learn the function of each stage but not yet concern yourself with how it works. The heavy arrows show the path of the signal from one stage to the next. Note that the video has three outputs to other places. (This is a simplified block diagram. Later you will study one that divides the color circuits—there are several—so that you can see how all of them affect each other.)

POWER SUPPLY

The low-voltage power supply gets AC from the power outlet in the wall and creates low-voltage AC for heaters in the set if it is a tube set, and high-voltage DC for the tube plates, screen grids, and negative for cathodes. If the set is completely solid state, the power supply provides low-voltage DC for the transistors, higher voltage for the output transistors, and a heater supply for the picture tube. At any rate, the low-voltage power supply provides from 6 volts to possibly 300 for parts of the set.

Though 300 volts doesn't seem low as far as shocks are concerned, it is low compared to what the high-voltage system under the picture tube handles (over 20,000 volts for the anode of the picture tube and from 700 to 5,000 volts, depending on the model, for the focus of the picture tube). You can see the lead to the high-voltage anode in the side of the picture tube. The focus voltage lead goes in at the neck of the picture tube on the wafer socket. Remember the difference between the high-voltage system and the low-voltage power supply. The low-voltage power supply supplies all stages, so thin lines are drawn to all stages on the diagram. Note that a line goes to the picture tube socket for heater voltage.

Naturally, if the low-voltage power supply is defective the entire TV is affected. Remember this if you do servicing in the future. If a set is completely dead, chances are the low-voltage power supply is not functioning. One type of set can be a fooler, however—a portable with series heaters. The heaters add up to the 115-volt AC input, so there is no power transformer. Any tube that has a broken heater will cut off the entire TV. It might be the power supply rectifier or it might be a video output tube. These sets are a real joy to service. Troubleshooting and repairing take more time, but because the sets are usually cheap, the customer doesn't want to spend much on them. Most color sets, even the better portables, are not of the series string heater type, so with color sets this should not come up too often.

ANTENNA AND TUNER

The antenna brings in all the channels available in the area, favoring those that have wavelengths nearest that of the particular antenna used. An antenna can be cut exactly for Channel 4, 8, 13, or any other, but most of them represent compromises that provide fair pickup of all channels. The tuner selects the one channel chosen by the viewer. It has an RF amplifier that strengthens this channel and then feeds it to a mixer. This is a heterodyning function. The tuner also contains an oscillator. As the channel selector is turned, the oscillator remains 44 MHz above the channel selected. (This varys a little from set to set. Some oscillate at 43.5 MHz above the channel, but it remains in this vicinity so that the heterodyning produces an output in the IF range of 41.25 to 47.25 MHz.)

You will recall that the video and sync were 1.25 MHz above the bottom frequency of the carrier band width and that the sound was 4.5 MHz above this or near the top of the bandwidth (see Figure 9-1). After the oscillator combines with the mixer (antenna) signal, the output is reinverted. The sound is at the bottom of the IF, or at 41.25 MHz. The color is at 42.17 MHz, and the video and sync ride at 45.75 MHz. This is the IF composite signal, and the frequencies listed are those of the various signals in the composite. Although the tuner converts these to the IF and inverts their frequencies, it does not bother their amplitude, phase, or frequency modulation, so the information that has been modulated on them is still there.

The IF tunes in the new frequency bandwidth of from 41.25 to 47.25 MHz and amplifies it. Since the IF must pass through several tubes or transistors and several tuned circuits, they can't each be tuned to the whole bandwidth. Instead, one will be tuned high (such as to 46 MHz) and another to the middle and another low (to around 42 MHz). All in all the IF amplifies the whole bandwidth well. The detector is usually on the same module and it detects much the way a radio detector does.

SOUND

In some sets the sound leaves at the IF; in others it goes through the detector; and in still others it goes to the video before going to sound circuits. The sound does not have to be detected by the detector, as it is FM and requires a different type detector to change the frequency shifts to voltages that represent sound. These detectors are of two main types: the *ratio detector* and the *Foster Seely discriminator*. This sound detector along with amplifiers would be in the sound circuit, and the amplified audio would go to the speaker, or speakers, in more expensive sets.

VIDEO

The video stage often has three or more amplifying devices. The actual video signal is the part of the composite waveform that makes any part of the picture light or dark. See Figures 9-1 and 9-2. Since the video and color are on different frequencies the part of the video stage that amplifies both of them must have a very broadband tuning mechanism. Af-ter the color goes on to its stage the video can be more narrowly tuned. Once the video is isolated it is strengthened considerably and added to the information for the picture tube in one of several ways.

In some sets the video is put in at the cathodes of the picture tube to make them emit more or fewer electrons and control brightness. In others the video is recombined with the color in the color circuits, and the combination of red and video goes to one grid of the picture tube, green and video to another control grid, and blue and video to a third. Remember that the color signals are separated in the first part of the video section or stage from the video signal and go to the color stage (the color is really several stages but for now we will think of it as one), and after the video is amplified more, it also goes to the output part of the color stage or to the picture tube. This will be clearer when we cover video amplifiers in Chapter 16.

COLOR

The color stage gets its signals from the video and control voltages that turn parts of it on and off from the horizontal sync circuit. This will be shown in the more complete block diagram, Figure 15-1. Two signals of color come to the color stage. One is the burst signal that occurs at the end of each horizontal sync pulse. The other is the chroma signal that occurs during line tracing.

The blending of these two signals, after the burst has been changed to a steady reference signal, makes the different color signals. If the chroma and reference signal are 90 degrees apart one color will result, and if they are 180 degrees apart, another will result. See Figure 9-1, upper middle. The circuit that mixes these color signals is called a *demodulator*. After the demodulator the color signals—red, blue, and green—are amplified and then may or may not be mixed with the video. At any rate each color signal goes to the socket of the picture tube to either grids or cathodes, and each controls a gun emission so that at any moment the picture could be red, blue, and so on at any one point. The color and video stages separate the signals by frequency by the use of coils that tune in the desired parts.

SYNC

The video and sync are on the same frequency, and they are really part of the same signal. The high parts are the sync and the low parts the video. See Figures 9-1 and 9-2. When the sync leaves the video stage it takes a healthy amount of video with it. At the beginning of the sync system is a sync separator that is so biased that only the sync pulses get through. The lower video is not amplified. Sometimes this separator is called a *sync clipper*. The clipped sync goes to both the vertical and horizontal sync stages, which are often on one circuit board. If there are tubes, the horizontal output tube usually is mounted near the high-voltage cage and not on a thin circuit board.

The purpose of the sync (short for *synchronization*) is to synchronize the tracing of the picture at the receiver with the tracing of the picture by the camera at the TV station. When the camera is picking up the top line of the picture from left to right, the receiver must be tracing it out on the screen. In the early days of TV, the 1930s, synchronization was difficult and often out of whack. Small wonder TV never really amounted to anything until 1947, with the picture tube and magnetic sync that is broadcast by the station along with the video.

You have seen in Figure 9-2 *A* and *B* that the horizontal sync traces out the picture line by line, and that the vertical sync moves the tracing down so that each line traces just a little below the previous one. The problem for the sync circuits is to separate the vertical from the horizontal sync for each circuit. The horizontal sync accepts everything, as there

are little equalizing pulses in the vertical sync so they act like more horizontal sync pulses and will not upset timing. The vertical sync system has a network of resistors and capacitors in its input that let only the longer vertical pulses through. The short horizontal ones are rejected.

Each sync circuit, horizontal or vertical, reshapes its signal, amplifies it, and sends it to the deflection yoke, which is composed of many coils of wire. The signals of horizontal and vertical sync produce strong magnetic pulses that go through the glass of the picture tube and send the electron beams of red, green, and blue information to the proper place on the screen. In Figure 10-1 they are bending it quite low and in the middle, so these beams would be in a trace across the lower part of the picture. The vertical sync holds the beam low as the horizontal is moving it left to right.

High Voltage

The horizontal sync also serves another function in generating the extremely high voltage for the picture tube. This is another output for the horizontal sync and has nothing to do with beam tracing.

The color beams tend to spread and not hit dots of color phosphor on the picture tube properly, so a system of converging them is necessary with a color set. Samples of vertical and horizontal sync are fed to a convergence board, where the adjustment of resistors and coils sets the beams for converging properly. The output of this board is the convergence coil assembly on the neck of the picture tube. This also sends out magnetic charges that bend the beams slightly to make them hit the proper dots.

If this were a black-and-white set the video would go directly to the picture tube neck socket with no color or convergence circuits or convergence coils. In addition, the picture tube would have one electron beam rather than the three. Otherwise it would be the same.

SUMMARY

Let's summarize the color circuits:

1. The low-voltage power supply furnishes AC for tube heaters and DC for tubes and/or transistors.

2. The high-voltage power supply furnishes very high voltage for the picture tube B+ anode and focus grid. The high voltage is a by-product of the horizontal sync.

3. The tuner amplifies the channel selected, heterodynes it with a local oscillator to produce an intermediate frequency of bandwidth 41.25 to 47.25 MHz (in modern sets).

4. The IF amplifies this bandwidth and the video detector detects the IF.

5. The sound is separated from either the IF, the detector, or the video by frequency difference, amplified, detected by a frequency-sensitive dector (ratio detector or Foster Seely discriminator), amplified at audio frequency, and sent to one or more loudspeakers.

6. The color is separated from the video by frequency separation. One signal becomes a reference with unvarying phase; the other varies according to the color desired. A demodulator converts both into color signals that are amplified and go to the picture tube.

7. The video beyond the color separation point is amplified and mixed with the color again either in the picture tube or near the color output. The color signal determines red, green, and so on, and the video determines the amount of light in that color.

8. The sync is separated from the video by amplitude variation (the sync being taller). The vertical sync input has a circuit that rejects the short pulses of horizontal sync. The horizontal sync runs during vertical sync time too, by use of equalizing pulses.

9. The sync signals are reshaped, amplified, and sent to coils in the deflection yoke on the neck of the picture tube. Magnetism developed there travels through the glass neck of the picture tube and bends the three beams of electrons to trace out a picture.

10. The color beams would not hit their appropriate color dots at any given instant except for a convergence board having extra coils that send corrective magnetism to keep the beams together better. Samples of sync pulses are fed to the convergence board.

FINDING STAGES IN A TV

Although a block diagram of the stages of a TV shows it logically and in proper sequence, the parts in an actual TV set may be arranged entirely differently. The set will be arranged for best operation, ease of manufacture, and in some cases, ease of servicing. Small board module sets often are labeled on the boards. The older sets with one big chassis for everything are the hardest to find stages in.

You can make use of some guidelines in finding stages, however, and without any other information (but with time and luck), you can find any stage. Some of the parts named may be confusing now, but concentrate on the general idea of finding stages. First, the low-voltage power supply will usually include a big black power transformer (see Figure 11-2, T800). It will also have one or more large electrolytic capacitors to filter the DC and make it smoother (see the same illustration, C800 and C803). Now you know where the power supply is in general, although some parts may be a distance away (L800 and L805 also are part of the low-voltage power supply). At any rate, look for a big power transformer and big can-type capacitors to find the low-voltage power supply stage. The power cord may not be close at all, because it must go to the on-off switch before anything else. The circuit breaker should be near the power cord.

The tuner will have twin leads from the antenna terminals, and the UHF tuner will usually be next to, below, or above the VHF or Channel 2 to 13 tuner. The knobs on the front of the set also clue you in as to the location of the tuners. In Figure 11-2, the VHF tuner is just under the top bracket under the hue and intensity controls. The UHF tuner is under it.

Normally the IF stage can be found by finding a coaxial cable that connects the signal from the tuner to the IF stage. There will be several slug-tuned coils in the IF circuit. The IF stage can be confused with the sound stage if both are on one big chassis, as both have several slug-tuned coils, and it can even be confused with the color circuit. The sound circuits will have two important connections, however. One set of wires will go to the speaker and another to the volume control. The three side lugs are the volume control. The two back terminals on the control are an on-off switch.

The color circuits will have a 3.58-MHz crystal in the general area. This item is usually easy to spot. It is elliptical on top, and it plugs in. Figure 10-2 is a top view, and the crystal is on the subcarrier regenerator module, upper left of the module. In Figure 15-4, which is not a picture of the color module but schematically represents the layout of parts, see if you can find the crystal. It is the fourth component from the top in the middle, Y1. If this were a picture it would show up larger than several of the other parts. Out of the color circuits will come the three color feed wires to the neck of the picture tube. Often their colors correspond to the colors they represent. You have to look carefully, though, as the screen grid leads to the picture tube socket also may be colored red, blue, and green. The hue control will have leads to the color circuits.

The video is often, but certainly not always, under the neck of the picture tube. The contrast control and brightness controls will have leads to the video. In several sets the video will have leads to the picture tube.

The high-voltage supply stage is easy to find. The anode lead to the picture tube will go to the cathode of the high-voltage rectifier tube or solid state rectifier. The flyback transformer will usually be in a cage nearby. A cap

will come out to the horizontal output tube and connect to its top. All this helps you find high-voltage and horizontal output, and the vertical stage can't be too far away. In some older black-and-white TVs the vertical might be some distance away, but you can find the vertical by tracing from height control or vertical linearity or vertical hold controls to the vertical circuits. The horizontal hold control also will lead to the horizontal oscillator.

Try to get the idea of tracing from controls to parts or stages, and later, after you learn how the stages work, reread this and it will make part tracing more meaningful.

The convergence board with approximately twelve service adjustments that make the beams trace well is usually mounted near the top of the set. The convergence coils will be on the picture tube just before the deflection yoke. In Figure 10-4 it is labeled "HA." Now with Figure 10-5, the pictorial, you should be able to spot it. It is above the brightness, hue, and intensity service adjusts. In this position the convergence board is set up so the serviceman can set the controls on it conveniently while watching the picture on the screen in a mirror or by watching from the front with a hand behind to adjust. When convergence is complete, the circuit board, "HA," is lifted off and fastened down in the chassis so the back cabinet may be put on. Note that part of the wires go down into the set. These are to pick up control horizontal and vertical sync pulses. The other set of wires goes to the convergence assembly on the neck of the picture tube. This same illustration as well as Figure 10-4 shows well the high-voltage cage, lower right, and the horizontal output tube behind it.

Chassis identification by obvious clues is fine if you have no other guidance. Fortunately, TV manufacturers produce a manual for each TV they make. If you decide to work for a shop, chances are the shop will represent one particular brand—or perhaps two in smaller communities. The shop should have service books for all the late sets of the type it sells and services. Figures 10-2 through 10-5 and

DISCRIMINATOR COIL
QUADRATURE COIL ADJUSTMENT (TOP)
INPUT COIL (BOTTOM)
SUBCARRIER REGENERATOR MODULE 9-27
INPUT COIL
AUDIO OUTPUT
AFC SWITCH
AFC TEST POINT
AFC OUTPUT
CHROMA DEMODULATOR MODULE 9-37
A.C., VOLUME and COLOR COMMANDER SOCKET
CONVERGENCE SOCKET
BRIGHTNESS CONTROL
CHROMA LEVEL CONTROL
AFC 150-6
AFC CHASSIS
150-6
150-206A
SOUND MODULE
To I.F. OUTPUT on VHF TUNER
VERTICAL HOLD CONTROL
150-166 I.F. CHASSIS
121-712 OR 121-713 SOUND OUTPUT (ON HEAT SINK) Q202
3.58 MHz CRYSTAL
SUB-CARRIER (BLACK)
22-6073
22-6071
40 OHM 15WATT RESISTOR (MUST BE MOUNTED IN CLIP)
VERTICAL OUTPUT TRANS
24 V. FILTER CHOKE
V202 6U10
121-744
Q205
Q204
121-695
AFC B+
MUTE
24 V.
AFC GRD
ALIGN
NORMAL
CHROMA (ORANGE)
COLOR COMMANDER
GRD
APC CONTROL
ACC CONTROL
BEL-FUSE 2.7 AMP
BEL-FUSE 500 MA.
SPK'R
A.G.C. (YELLOW)
Q401 121-671
Q403 121-447
Q404 121-447
Q402
121-446
Q406 121-699
Q405 121-671
Q210
Q209
Q211
121-743
121-776 OR 121-777
121-587
Q207
Q206
121-822
Q203
Q208
V201 10JA5
V204 19CG3
V203
20LF6 or 26LX6
TEL.
FILTER CHOKE
POWER TRANSFORMER
SILICON RECTIFIERS
C.T.A. (CROSS TALK ADJUST)
A.G.C. MODULE 9-23 (DOTS REFER TO COLLECTORS)
GROUND STRAP
FUSE .5 AMP
IN EARLY PRODUCTION MODELS only
FOCUS CONTROL MODULE
TRIPLER SWEEP TRANSFORMER
YOKE SOCKET
123-4150 E
CRT TUBE SOCKET
BLACK and WHITE SET-UP SWITCH
KILLER THRESHOLD
AGC LEVEL
AGC DELAY
PEAK PIX
BRIGHT
NOISE LIMITER
NOISE GATE
VERT LIN
NORMAL
SET-UP
BLUE
GREEN
G2
RED
RED GAIN
BLUE GAIN
GREEN GAIN
VERT SIZE
HORIZ CENT
HORIZ HOLD
VERT CENT
PIN AMP
PIN
PHASE
NOISE GATE CONTROL
A.C. INTERLOCK
VIDEO GAIN ADJUSTMENT
HORIZONTAL HOLD CONTROL
Courtesy Zenith Radio Corporation

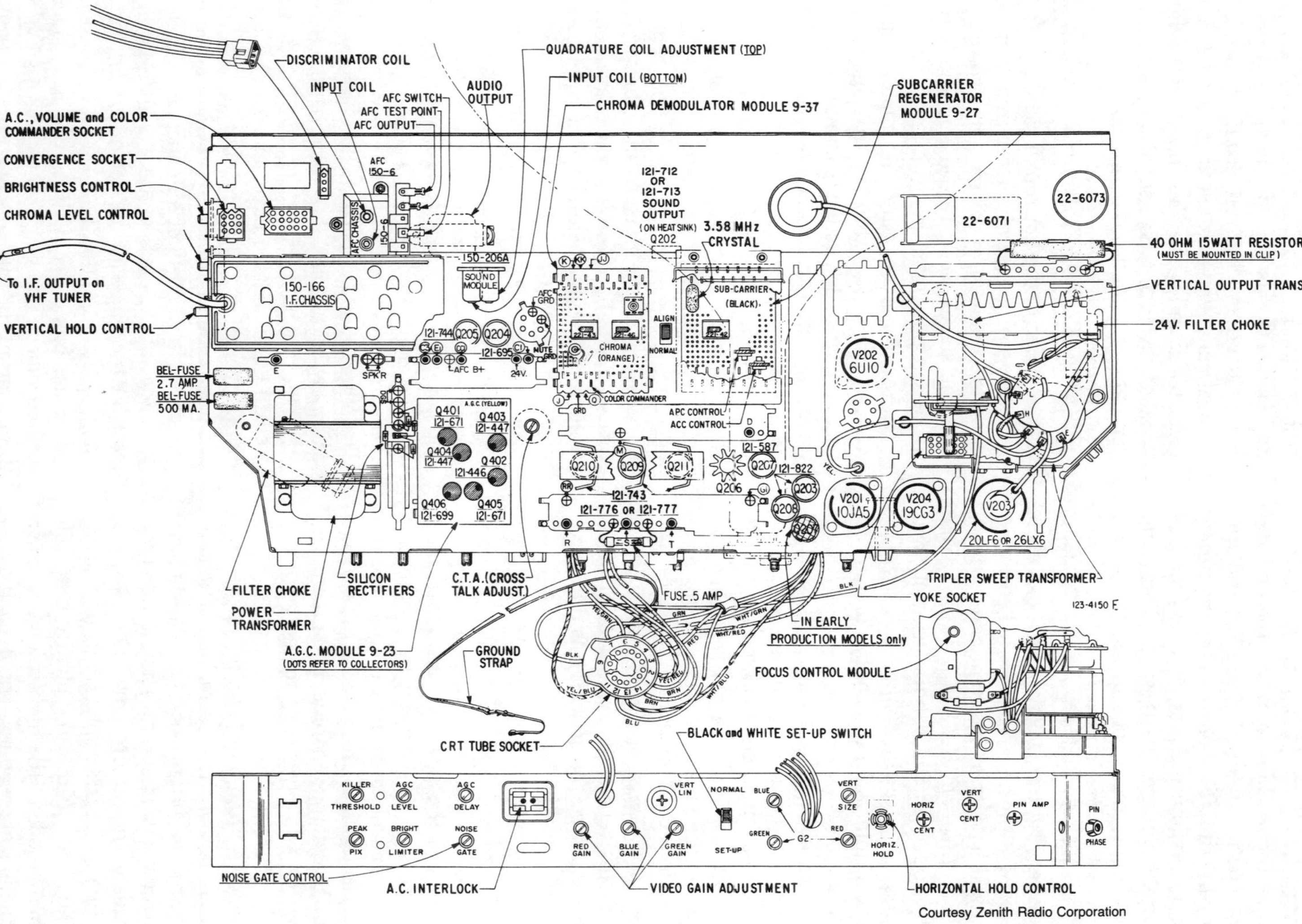

FIGURE 10-2

many schematics in this book were taken from such manuals or service books. By using these manuals, you can quickly find the stages of a set and find the individual parts on a circuit board or chassis. The books also have schematics of the set and voltage readings at plates, grids, bases, collectors, and so on to aid in servicing. Always use these manufacturer's service manuals if you can. They are available at small charge or in some cases are free. You will have a certain advantage if you service just one brand of TV. You will learn the general pattern of the physical arrangement of the parts, of typical operation, of troubles with the set. These will not vary much from year to year until a radical model change occurs.

Actual pictures of the set are more help than you might imagine at first. You could read the control names off the set itself, of course, but if you are in a hurry and are looking in the back of the set wondering what the control is on the front that you spot from the rear, a quick look at Figure 10-5, upper left, will tell you without your having to look around. A drawing such as 10-4 (the same set as 10-5) is even more help. Note that each panel has a letter or two-letter identification. The IF is "BA"; the convergence is "HA"; the color is "SA"; the deflection is "FA." Another part of the manual tells how to replace an entire panel according to the troubles in the set. For instance, it might say, "Weak or no video? Place screwdriver on delay line input or output, with finger inject noise. Does picture change significantly? Yes? Replace panel 'BA.' No? Replace panel 'SA.'" This is fast servicing, and we will take it up in Chapter 17.

Let's find the stages you have studied with the aid of Figure 10-4. The tuner is not labeled but is easy to spot at upper left. The IF lies on its end a little toward us from the tuner and shares the circuit board with the audio. Figure 11-2 shows this IF module well.

You located part of the low-voltage power supply earlier, but the DC regulator panel, "ZA" (Figure 10-4), is an important part of it,

as you will learn in Chapter 11. It fits behind the side panel, but the arrow indicates that it is shown swung out to reveal the 20-volt adjust. Continuing with Figure 10-4, note where the power line comes in and connects to the AC in plug. Often this is connected to the back cabinet in such a way that removal of the cabinet removes the AC. You can use a cheater plug for servicing or remove the power cord from the cabinet, sometimes by drilling out rivets, and use it. It would be impossible to service a set with the back panel on. The circuit breaker can be tripped, and resetting it may be the only trouble with a TV receiver if it is "dead." Unless there has been a big surge on the power lines, such as lightning hitting them, why did the circuit breaker trip? Chances are there is other trouble if there have been no storms.

The "SA" panel is a combination of color and video. It is conveniently placed so leads can go up to the neck of the picture tube at the socket. The socket has been removed in Figure 10-5 for inspection of other parts. It is in place in Figure 11-2. The deflection panel, "FA," is what we called the "vertical and horizontal sync" circuits in Figure 10-1. All of the vertical circuit is there, including the tube. The horizontal circuit starts on the "FA" panel and ends up to the right. High voltages would not work well on a phenolic panel.

A page from the RCA service manual for CTC 44 is shown as Figure 10-3. The parts lie on the set horizontally, but they are mostly in modular units nevertheless. Looking down at the set, you see that the power supply (low-voltage) is at the left with the power transformer at the rear and filter capacitors and choke farther toward the front of the set. The tuner is not shown. The IF, video, and sync separator module, PW300, is near the front of the set and under the picture tube to some extent. The chroma panel, PW700, contains all color circuits including the color signals. PW500 is the vertical sync, and most of the horizontal sync is on PW400,but some of it is

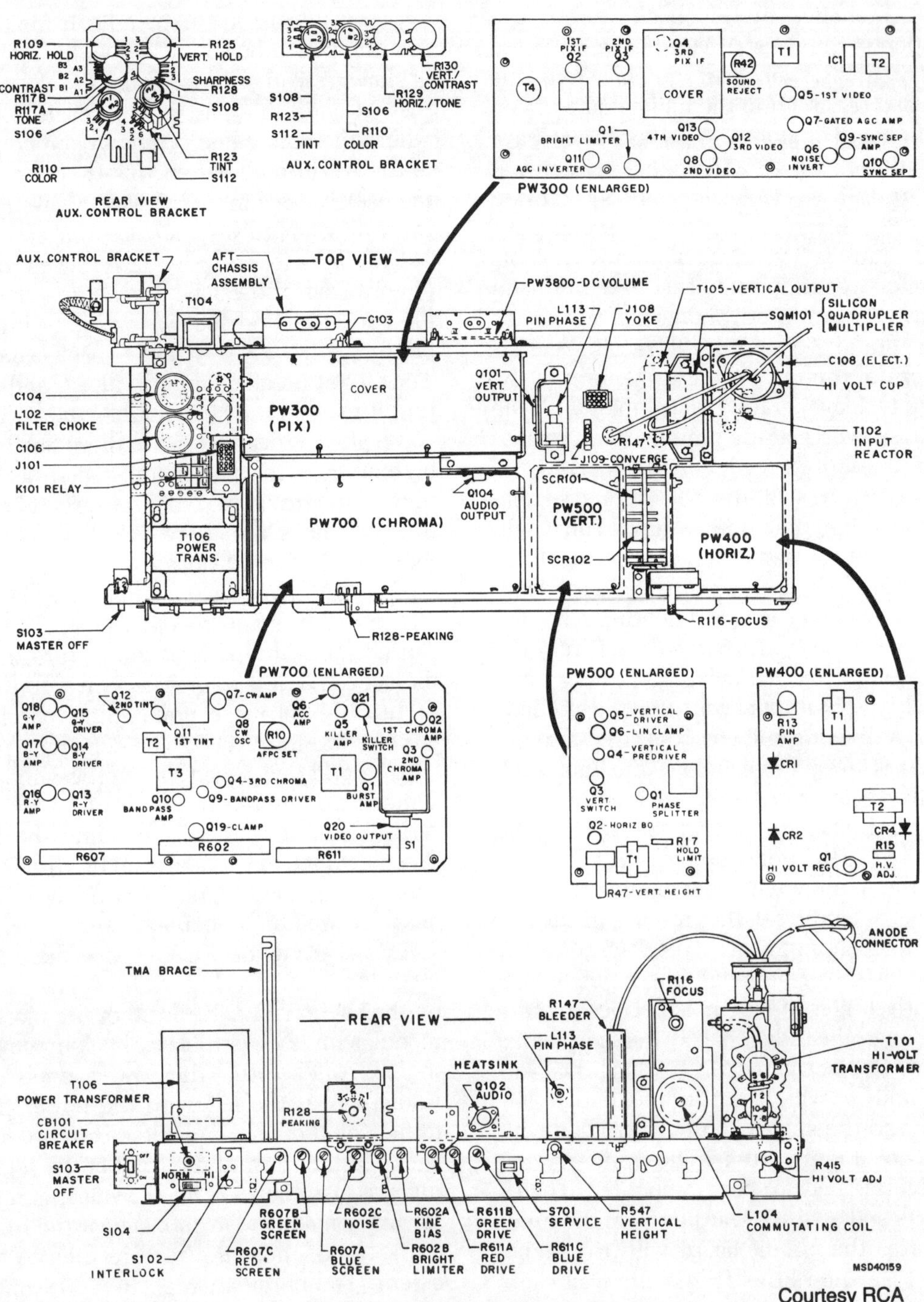

FIGURE 10-3
CHASSIS LAYOUT

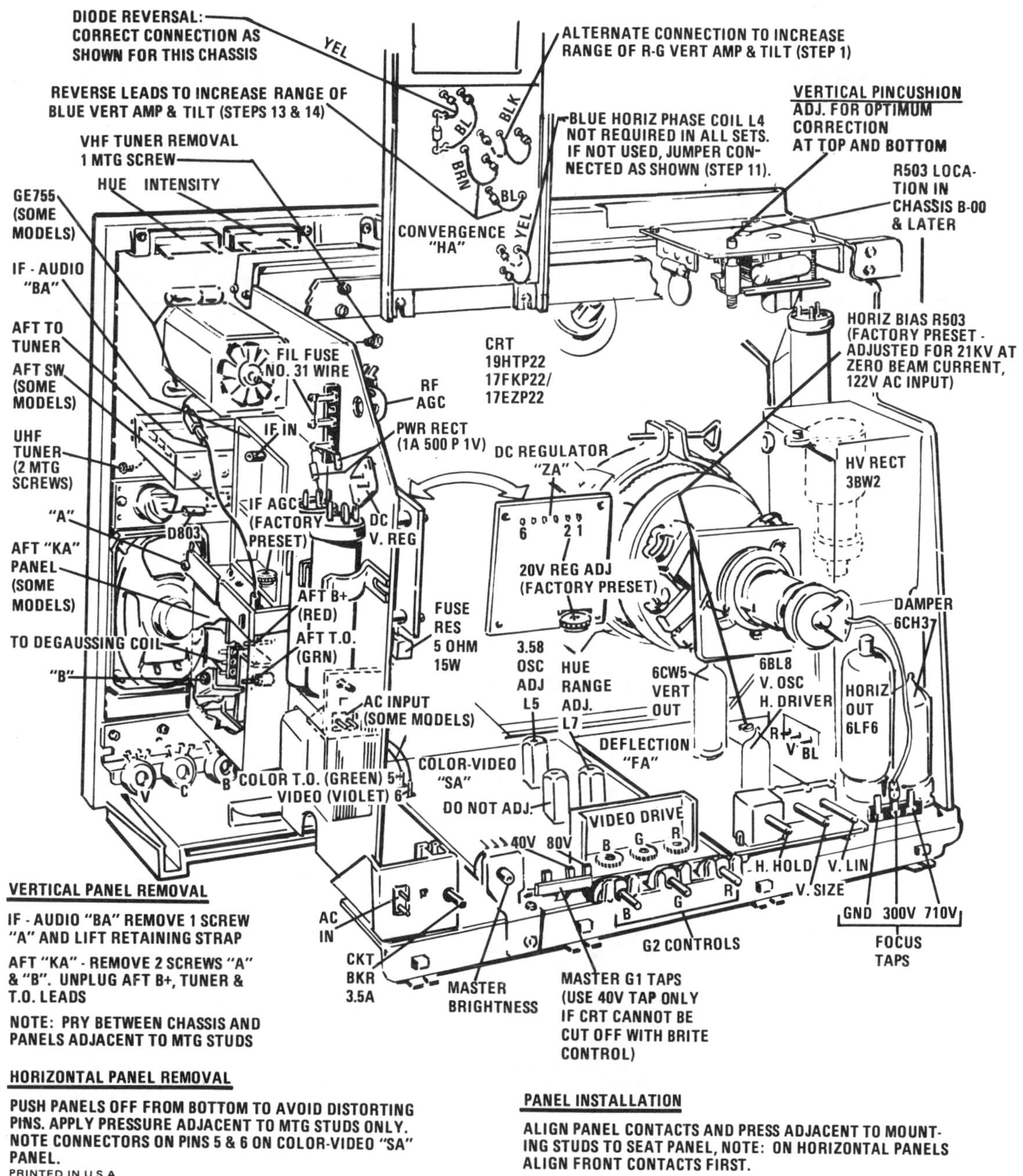

Reprinted with permission of Quasar Electronics Company

FIGURE 10-4

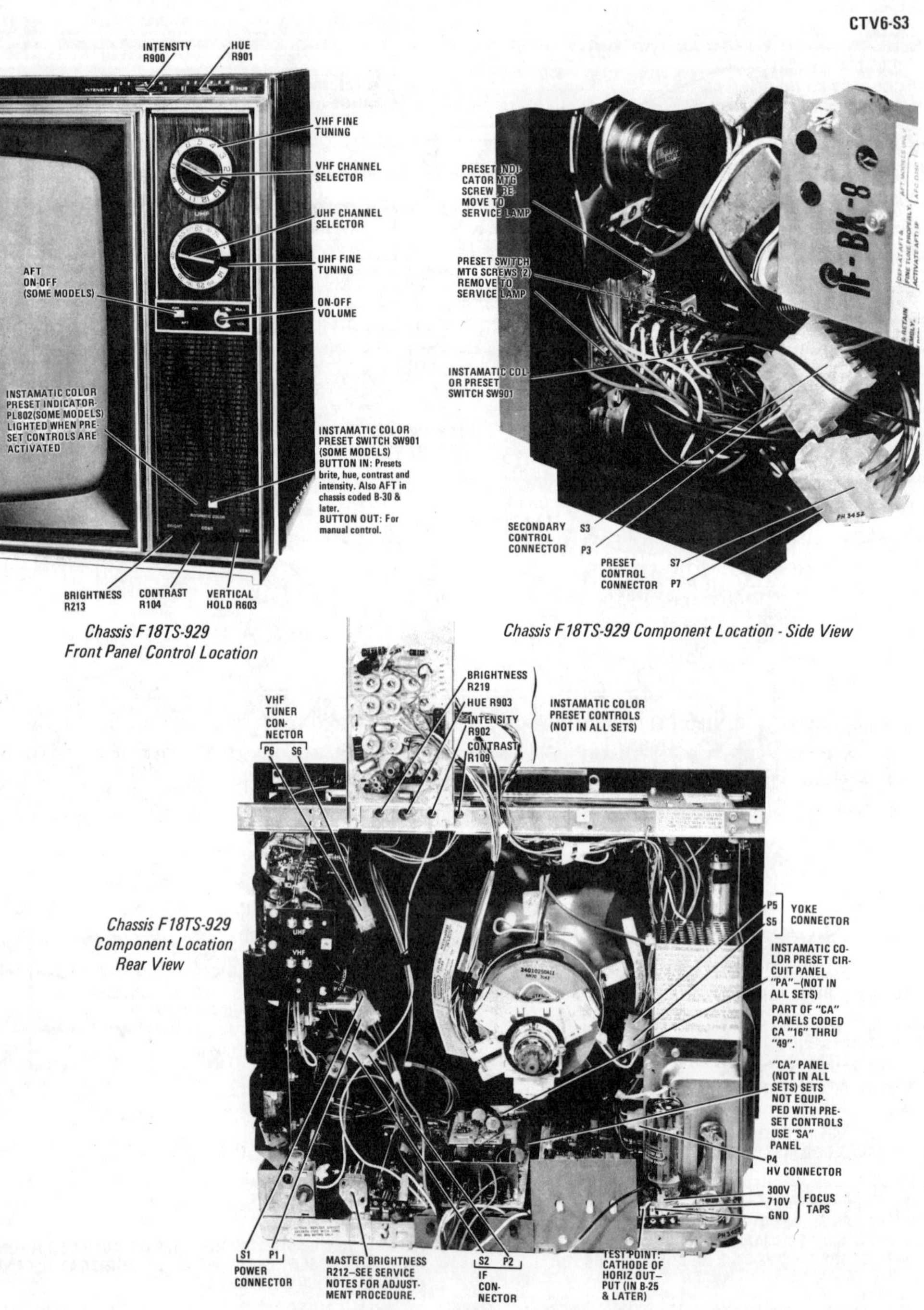

Reprinted with permission of Quasar Electronics Company

FIGURE 10-5

between PW400 and PW500. SCR101 and SCR102 replace tubes as the horizontal output.

The high-voltage circuit is directly front of PW400 with T105, SQM101, C108, and a quadrupler multiplier. Note how the enlarged drawings with arrows help identify individual transistors, transformers, large resistors, and so on. You could look at a schematic and still not know where the parts were unless you traced part by part, line by line. Manufacturer's service manuals speed up servicing greatly.

Now let's consider the Zenith service parts placement drawing, Figure 10-2. Study these drawings from time to time to increase your general familiarity with color TVs. As you go through the chapters that follow, it is good practice to note the credit on each figure so you can identify the set with each schematic. If it is a Quasar set, refer back to Figure 10-4 to see where on the actual set the schematic parts would be. This kind of orientation has been all but lacking completely in the study of TV servicing in the past. Often a person finishes TV school with a head full of theory and schematics but without the foggiest idea of where the parts are in the set. Complete separation of lab and lecture study help confuse the issue. If you will just take the time to refer back, however, you will be way ahead of the game when you start working with actual sets. If you can find sets similar to the ones covered in this course, studying them, adjusting them, and measuring voltages on them will also strengthen your knowledge in a most practical way. For the RCA schematics later, refer to Figure 10-3, and for Zenith refer to 10-2.

In Figure 10-2 you can see that the IF chassis is mid-left. It shows the coaxial lead from the tuner, though it does not show the tuner. This IF chassis enlarged is Figure 12-8, and Figure 12-6 is a schematic of it. The sound module is to the right of the IF. See Figures 14-3 and 14-4 for a schematic of this module and a photo of it. The color is separated by the

modules, chroma demodulator, and subcarrier regenerator. The chroma is the bandpass phase shifted signal that is out of phase with the burst, reference signal to carry color information. The subcarrier is the reference one. Three color output transistors are not mounted on the chroma demodulator circuit, but rather at the back of the set a bit. They are Q210, the red output; Q209, the blue; and Q211, the green. In this set the video is already mixed with each color at these output transistors. The video circuit is scattered. Q204 and Q205 are between the IF board and the chroma board. Q206, another video, is back with the color outputs.

The horizontal and vertical sync circuits are at the right. In this TV, only these circuits use tubes. The power transformer is at the back left. Then comes an automatic gain control (AGC) circuit. AGC is somewhat like automatic volume control in a radio, but it controls gain in the tuner and IF so they will not overload later stages with too strong a signal, and so they can build up signal on weak stations. Note the socket at the back of the set. It has 14 connections, though some are not used. This socket goes on the back of the picture tube. The video, color, heater, screen grid, and focus voltages are put in the picture tube here. Note the lower drawing, which shows the back of the chassis. Many of the controls are adjusted here by the serviceman.

Again, the manufacturer's service manuals for the set you are going to repair are your best source of information. A general repair shop that doesn't specialize in one make is likely to have few service manuals, but you do have an alternative. The Bobbs-Merrill Co., Inc. (4300 West 62nd Street, Indianapolis, Indiana, 46268) puts out folders for any TV or radio you are likely to want to service. The trade name is *Sams Photofacts*, the publications mentioned several times before. Each set of pictures, layout drawings, schematics, and probable troubles is for one or usually two TVs or radios. The cost is a few dollars a set. Most radio-TV supply houses sell *Sams Photofacts*, so it is not necessary to order from the publisher. If you start to work for a general service shop, chances are they will have many sets of *Sams* material. As new sets are brought into the shop, the owner or manager will purchase more *Photofacts* and file them after use.

If you start your own shop, you can buy these as needed, and even though they will cut into the profit to be made per set fixed, you will probably need each one again later. Some distributors of TV equipment and parts allow the TV serviceman to buy recent TV *Photofact* series folders on monthly installment payments. It is good to meet and become friends with your parts distributor. In some cases he will let a repairman come in and look at a *Photofact* free if he buys all his parts from that distributor.

Now a few words of caution on the circuits might be in order. The circuit boards are often made of phenolic material, which can crack with rough handling, and the printed circuit on it can be lost. In addition, the connections to the transistor or integrated circuit can become loose with abuse, heat, or with time alone, so sensible handling is necessary. Also with these circuit boards, suspect the printed circuits and contacts to parts as well as wiring harness as being bad as often as you might suspect a resistor, capacitor, transistor, or tube. The copper circuit material may have cracks or other lack of completing the circuit.

TV CONTROLS

The controls of a TV fall into two categories—the ones the customer should adjust from time to time and the ones the serviceman should adjust as needed. With early TV sets the general rule of front controls for the customer and rear controls for the repairman worked fairly well. Then came the portables, and it was convenient to hang controls on the side, rear, or anywhere. Controls that the customer needs to adjust began to appear at the rear of the set. Like controls, customers fall into two groups—those who are afraid of their

TV sets and will call a serviceman for almost anything except channel selection and those who believe in turning everything that can be turned as often as it can be turned and think they are fixing the set.

Normally the following are customer controls, and it occasionally takes patient explanation by the serviceman to show the customer how to manipulate them correctly. Look at Figure 10-5. The on-off volume control is either a conventional control that is snapped on and then turned farther clockwise as desired for volume, or it is a pull-out for on and either a rotation or a separate control for volume. Usually this operation needs to be explained only when a person purchases a new set.

The channel selector for VHF is a rotary channel switch that snaps to each channel position. It is not continuous, but either has each channel or doesn't. The fine tuning control is often a slip ring around the channel selector that improves the tuning. The contacts of these controls often get dirty, so when a picture comes in poorly it pays to rock the channel selector back and forth a bit, and the picture may clear. In some models an automatic fine tuning (AFT) control can be turned on, and the AFT will do the fine tuning as channels are changed.

In many sets the VHF tuner must be put in *U* or *UHF* position between channels 2 and 13 on the indicator before the UHF can be tuned in. Often the UHF tuner is a continuous control that does not snap into and out of channels. Some sets have no fine tuner, and careful adjustment of the UHF tuner is necessary. Others have a gear drive that produces more control by allowing a greater spin for small movement within the tuner.

The brightness control adjusts the amount of light the picture tube gives out into the room in general, while the contrast control adjusts the blackness of the blacks and the whiteness of the whites. An engineer usually sets these as follows. First, with a picture on the screen, the contrast is turned all the way down, counterclockwise, for disappearance of picture. Then the brightness is set by flipping to an unused channel and turning the control until, with average room light conditions, the picture tube looks the way it would with the set off. That is, the picture would be darkened by the brightness control just to the point at which it doesn't appear lit. Then the first station is found again, and the contrast is turned up until a desirable picture emerges.

This is the recommended way to adjust brightness, and everyone hates it. The problem is that many dark scenes are not visible at all with such a setup. Dark alley gunfights are completely lost. Instead, you need a "setup" condition in which blackness in the picture will still leave slight light on the screen. The viewer understands that this means darkness, but he doesn't want his screen to go completely blank. The best bet might be to follow the procedure outlined above and then bring the brightness just up to the point of pleasant viewing.

The vertical hold is a customer control. So many things make a picture flip vertically that the customer must operate this control now and then. All the controls mentioned so far would be on the front of either a black-and-white or a color TV set. Black-and-white sets sometimes have the horizontal hold control on the front too. This is not usually true of color sets because the setting of this control is critical to the picture and is best left to the serviceman.

Color controls on the front usually consist of a hue control and a tint or intensity control. The hue control adjusts the color of the red, green, orange, brown, and other hues in the picture. The tint or intensity control adjusts the shade of the color, determining, for example, whether the red will be a dark rosy red or a light pastel pink or whether the green will be a dark, swampy color or a light springtime leaf color. The intensity control is set for pleasing allover color conditions, and the hue control is usually set for flesh tones. Since atmospheric conditions, strength of signal, in-

terference, and skill of the station engineer during broadcast all affect color, it is frequently necessary for the hue control to get a slight touch-up. If the service controls are set properly by a qualified serviceman, the customer should be able to make the minor adjustments necessary in the hue control.

The instamatic color preset switch defeats the customer brightness, hue, contrast, and intensity controls and in their place connects such at the rear of set, just under the convergence panel. The serviceman has preset these controls for best operation, and an automatic color control voltage system in the set shifts colors depending upon their strength. This feature helps bring color movie reprints up to live broadcast level.

Look at Figure 10-2. Many back-of-set controls will be taken up in Chapter 18, but some of those that appear on black-and-white sets and that customers are used to adjusting will be mentioned now. The vertical size or vertical height control used to be anyone's plaything, as did the vertical linearity control. With color sets the adjustment of these controls affects convergence and purity as well. They should be set by a competent repairman before he converges and otherwise sets the color beams. The height or vertical size control determines how well the picture fills the screen from top to bottom. The linearity adjusts so that all parts of the picture will look in proportion vertically—no big heads and short bodies, for example. Vertical size and linearity are best set with a crosshatch generator connected to the antenna terminals. Adjust for an even set of marks from top to bottom, using both controls. While you have the crosshatch generator on, it would be well to set the horizontal hold, horizontal centering, and horizontal drive, if any, for even marks horizontally. Once these have been set and the set has been converged they should be left alone unless you want to converge the set again. Converging is also taken up in Chapter 18; it is quite complicated until you get the hang of it. Some sets will have other names for all these controls, but the general function of each one has been covered.

THE PICTURE TUBE

Now you have seen the general way signals go through a TV set and what they do. You have identified the stages that work with these signals on actual TV receivers. You have learned about basic customer controls. Now let's see how the picture tube works.

The general idea of a cathode ray tube, or picture tube, is the same as that of any tube. Cathodes are heated by heaters and give off electrons. A positive charge on a plate attracts them. In between, grids control and affect the flow of electrons to the plate. Refer to Figure 10-1 throughout this discussion. A regular tube with a solid plate would be of little use, as the picture could not be seen through the metal plate. Instead, the electrons are focused into thin beams and directed with such intensity that they don't just go to a plate, but they hit a screen of phosphor dots that give off light in colors when the electron beams hit them. Then the electrons bounce off and go to a high-voltage anode, which is the metal inside the picture tube near the viewing screen, and out the high-voltage connector as current to the high-voltage circuit. The high voltage (+) charge on this anode brings the electron beams across the long distance, but they hit the viewing area first, as they travel in a straight line. Since the electrons must travel many inches in a picture tube rather than the half inch or so they go in a normal tube, there has to be a very high voltage on the anode. For a black-and-white TV, 17,000 volts is common, and with color, over 24,000 volts is sometimes used.

Notice the aperture grille. Without it, the beams would hit not only their dot but some of each adjacent dot as well; the red would slop over and light the blue and green a little. The aperture grille has many tiny dots (much magnified in the drawing) that sharpen the beam as it goes through. This helps, but for the picture to be as bright as a black-and-white TV picture, the high voltage has to be higher to get the beams through the aperture holes. A black-and-white TV has no aperture grille, as all phosphor dots just glow white when hit with electron beams.

The color tube is made of sets of different tiny dots of phosphor, often arranged in triangles. When one kind of phosphor is hit by an electron beam it will glow green; another will glow red, and another blue. These triangular combinations are all over the front of the picture tube on the inside of the glass. They are much smaller than shown. If any beam strikes a phosphor dot, the color of that dot will show on the screen. That is why it is so critical for the red beam to hit only red dots, the green beam to hit only green dots, and the blue beam to hit blue dots. The beams are certain quantities of electrons and determine how much of a color will be shown; no actual color flows in the beams. It takes a gun, or a combination of cathode, control grid, and screen grid, to make up each color beam. The cathode emits the electrons as usual; the control grid has bias and perhaps a signal to determine how much of a given color is going to show at any instant; and the screen grid pulls the beam into action toward the anode.

Sometimes, to be sure, the video is put in on the cathode and the color signal is put in on the control grid or vice versa, or both are put in on the cathode or control grid. There are many manufacturers, and each has his pet arrangement. The screen grid always increases the speed of the electrons. A focus grid then shapes the beams more thinly and also helps pull them along. Of course, the thousands of volts at the anode will pull much harder. Without the shaping and accelerating by the screen grid and focus grid, the electrons might go lazily right to the anode and out, and not make a picture at all. Of course, the deflection yoke will aim the beams too. Figure 18-2, lower left, shows the assembly of the guns with glass removed. The cathodes and heaters are inside the first grid and cannot be seen.

Accelerated beams are now going out, shaped by the focus so they are thin and reasonably close together so they will strike the adjacent dots of phosphor. The convergence coils help assure accuracy. The deflection coils by action of sync move the beams across from left to right (as viewed from the front of the set), then down slightly, left to right again, down slightly, and so on until 262½ lines of each color have been traced. Of course, all colors may not have been active all the time. If a red ball appeared in the center of the picture, the blue and green beams would have been nearly off during this part of every trace during the tracing of the ball. Suppose someone in the picture is wearing a violet robe. At that point the red and blue gun beams would be strong and would strike the red and blue phosphor dots, and the viewer's eyes would blend the colors so he would see violet.

If yellow is called for, the red and green beams would send lots of electrons to make yellow. At school you probably learned that yellow is a primary color. That is true with opaque colors on which light is shining from a bulb or the sun. When light is coming through the colors and they are a part of the light, however, green rather than yellow is the primary color. Also remember that the video is lightening all the colors with white, putting a certain yellowish tone in them.

Now let's trace current superficially. Currents from the low-voltage power supply would enter transistors or tubes and go out to the cathodes of the picture tube. Emitted electron beams now represent the current. In three beams they travel through grids to the picture tube viewing screen, bounce off, and go out the high-voltage anode (as current now) to the high-voltage circuits. Through this circuit the currents find their way back to the low-voltage power supply.

The aperture grille is not the anode. Remember that the inside of the picture tube around the edges near the viewing screen is metal, and this is the high-voltage anode. The aperture grille, again, shapes the beams sharply enough to hit only one dot.

One type of picture tube has the phosphor material lined up in strips instead of in triad dot groups. The aperture grille is also in strips. This system, by Sony, is called the *Trinitron picture tube*. It uses only one gun, but has three cathodes, one for each signal. Deflection plates bend the beams to strike the proper slot for each color. The tube still has three beams, but it has the advantage of a small neck on the picture tube (if that is an advantage), and dynamic convergence is simpler.

SUMMARY

More important things to remember are:

1. Whenever possible, you should use the manufacturer's manual to find stages and parts when servicing a TV.

2. If you don't have a manual, *Sams Photofacts* can do the job.

3. Tracing from a control to a circuit can sometimes help you identify the circuit.

4. Normally front-of-set controls are customer controls. These include on-off, volume, channel select, fine tuning, contrast, brightness, vertical hold, hue, tint or intensity, and automatic color and tuning switches.

5. Adjusting vertical and horizontal controls affects convergence on a color set.

6. Each color system of cathode and grids in a picture tube is called a gun. A color TV has red, green, and blue guns.

7. The guns do not send colors across the tube. They emit electron beams that form colors as they strike phosphor dots of the proper color.

8. The color information can be put to a gun at the control grid or the cathode.

9. The video information can be put to a

gun at the control grid or cathode or can be combined with the color before entering the picture tube.

10. The screen grid accelerates beam flow for each color.

11. The focus grid shapes the beams and gives them general stability so the electrons stay together.

12. The deflection yoke controls the main course of the beams. The convergence coils help keep the beams on their own color dots.

13. The high-voltage anode on the side of the picture tube collects the electrons after they bounce off the phosphor dots.

14. Combinations of beams at a certain strength make up the different colors (red and blue make violet, for instance).

QUIZ FOR CHAPTER 10

▶ 1. The _______ stage develops signals that make areas of the picture bright or dark and usually supplies signals for the color and sync stages.

▶ 2. As a by-product, the horizontal sync stage also develops the _______ and _______ .

▶ 3. The vertical sync stage rejects horizontal sync pulses because they are _______ .

▶ 4. The sync is separated from the video by virtue of its _______ .

▶ 5. The color is separated from the video by virtue of its _______ .

▶ 6. The three main parts of a tuner as far as function is concerned are the RF amplifier, _______ , and local oscillator.

▶ 7. For a modern TV the IF bandwidth is from _______ to _______ MHz.

▶ 8. The horizontal and vertical sync pulses are applied to the beams to bend them magnetically by the _______ .

▶ 9. The _______ furnishes important voltages to all other TV stages.

▶ 10. The tuner signal is often coupled to the IF by a _______ .

▶ 11. The best source of information on a TV parts location or schematic is the _______ .

▶ 12. The electron beams hit _______ to form the actual colors.

11. Television Power Supplies

PROJECT: Measuring AC Input Voltage Heater Supply, and DC High and Low Voltage

WITH THE exception of very small portables, TV receivers plug into 115-volt AC outlets. Because very few television components will operate on AC or 115 volts, it is necessary for the set to have a power supply to (1) change the alternating current to direct current, which is needed for tubes and transistors, and (2) change the 115 volts to lower voltages for transistors and even higher voltages for the plate circuit of tubes.

We sometimes refer to a high-voltage system in a TV set. This is not the power supply, however, and must be distinguished from it. The high voltage is developed only for the high-voltage anode of the picture tube and sometimes for a screen grid on it. The lower voltage power supply develops voltages for all tubes or transistors of the set. If this power supply is not functioning properly, no part of the TV can work well; in fact, usually the set will not work at all. Power supply problems make up a great proportion of troubles in TV receivers.

Many TV sets still in use (and often in need of repair) have tube type power supplies. These sets are all tube. Some other tube type sets have solid state diodes in the power supply. Still other sets are called *hybrids*. They are part tube and part transistor. The Quasar set we will study in this chapter, Figure 11-2 and 11-3, is such a hybrid set. Then there are all-transistor sets. Some may have integrated circuits, but these are combinations of transistors. Despite all these kinds of sets, there are really two basic types of power supplies: (1) the old-fashioned tube type with a tube rectifier and (2) the solid state diode type.

TUBE TV POWER SUPPLIES

We will take up the tube type first. Figure 11-1 shows the schematic. If you have an older TV set around it would be well to measure voltages as you study this material. Use a quality voltmeter and be sure to have the meter set for AC when you are measuring AC and DC when you are measuring positive voltages. Since these sets usually have an interlock, it may be necessary to use a cheater cord to measure voltages or work on the set so you can put the power on as desired. The set may have to be approached from underneath or turned on its side to get at the voltage spots. Be very careful where you put your hands while the power is on. Safety is most important. If it is necessary to move or remove the high-voltage lead to the picture tube, see Chapter 13 before proceeding.

The power transformer, T1, is usually a big black or silver color metal-covered box near the rear of the set. In Figure 11-2, part T800 is a typical power transformer. Some cheap portable TVs don't have power transformers but drop the 115 volts across the heaters of all tubes by arranging the heaters in series. If one tube is burned out everything is dead. Such a transformerless set is hard to repair because fixing it is so time-consuming. You have to pull one tube at a time, look up its heater connections, and check it with an ohmmeter. Such sets are usually black-and-white or monochrome; few color sets are without a power transformer.

When a transformer type TV is dead—no lights, no noise, no voltages, no tubes lit—chances are the trouble is with the line cord, the on-off switch, or the primary of the power

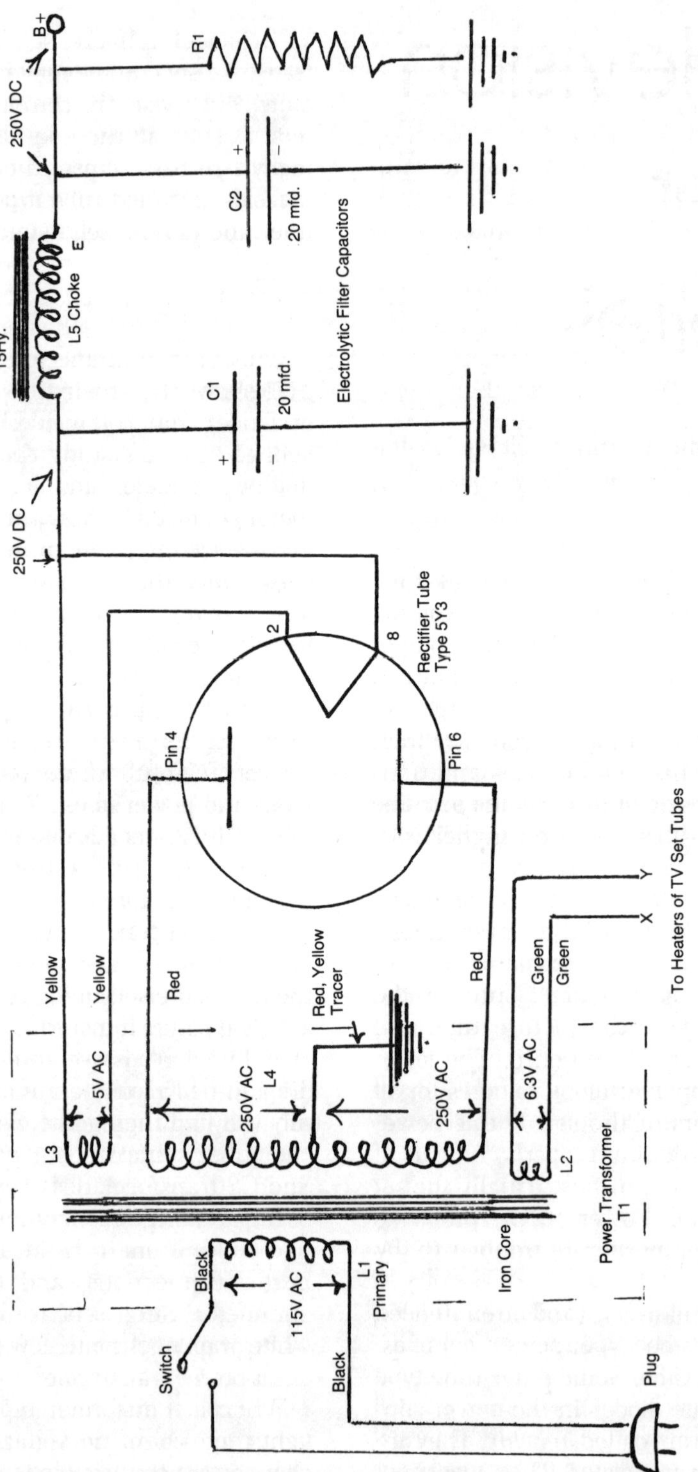

FIGURE 11-1
TV TUBE TYPE POWER SUPPLY

transformer. You can test this all at once. With the cord out of the wall, connect an ohmmeter, set on low ohms scale, across the plug terminals. Now turn the TV switch on and off a few times. If the meter goes to infinity on "off" and to about 10 ohms on "on," then the switch, power cord, and primary of the power transformer are good. There may be a fuse or circuit breaker in series somewhere in this primary circuit, so look for one before ordering another power transformer. If L1, the primary, is broken internally, then a new transformer is the only cure.

Sometimes a power transformer is so hot that it smokes. If it does, turns have shorted inside and the whole thing will short with use. It is normal for transformers to run warm, but not extremely hot. This transformer has three secondaries. L3 supplies 5 volts of alternating current to heat the filament of the rectifier tube. Some sets use a rectifier tube that can operate on 6 volts of heater, so they would not have this extra winding but would get their voltage from L2. This L2 winding supplies 6.3 volts to the heaters of all tubes in the set. The designations X and Y show where the leads go to all other tubes. Remember that a transformer passes AC, not DC, from one coil to the other. This is fine in these circuits, as heaters can work on AC, but all other parts of tubes must have DC. If they had AC they would pick up the 60 Hz hum, which would make ripples in the picture and hum in the speaker. These low-voltage windings, L2 and L3, have just a few turns in proportion to the L1 winding, so the voltage is lowered.

Winding L4 has many more turns than L1, so the voltage is raised to 500 volts. This winding has a center tap that is grounded to the chassis of the TV, so there are 250 volts in each half of the winding. So far this is still AC. The 5Y3 rectifier tube will change it to DC. Here's how. At a given instant the voltage at pin 4 of the tube will be (+) and at pin 6 it will be (−). The heating of the filament at pins 2 and 8 causes tiny electrons to be given off. Electrons, being negative, are attracted great-

ly by positive charges of electricity. At the instant we are concerned with, pin 4 plate has a positive charge, and the electrons stream to it, form a current, and go on around through the top of L4 to ground, then out through the chassis and up each tube's cathode, through the tube as electrons again, then out the plate and its load as current and come back on the schematic at the top of R1, left through coil L5, and back to the heater.

This is a continuous circuit. But, as you will remember, AC changes its polarity 60 times a second. (That statement applies to household AC, of course, not high-frequency RF.) So a fraction of a second later pin 6 is positive and pin 4 is negative. Now the heater sends electrons to pin 6, up to ground through the bottom half of L4, then out to the cathodes of the tubes again and back as before. In this way current flows in one direction for either half of the supplied AC. When current flows in only one direction it is DC. This DC would ripple very badly, though, because between the half cycles of AC change it would lower in voltage. This problem is solved by a filter system. C1, L5, and C2 smooth out the DC and end the hum it would otherwise create in the TV set.

The filters are large electrolytic capacitors. They have a polarity that must be observed in replacing them. The outside metal is normally negative and is soldered or bolted to the chassis. The inside lug or lugs (some have several sections) are positive and connect to the coil L5 and heater of the rectifier tube. You usually will find these electrolytic capacitors near the power transformer. In Figure 11-2, C800 and C803 are electrolytic capacitors.

Capacitors store electricity. When the voltage builds to its highest level these electrolytics charge up. L5 cuts down on the large surge of current at peak voltage. At the time when the AC is neither positive nor negative at either pin 4 or 6 but is changing polarity, neither plate attracts many electrons. To prevent the current through the whole set from dropping, the capacitors slowly release some of their charge; thus, they boost the electricity

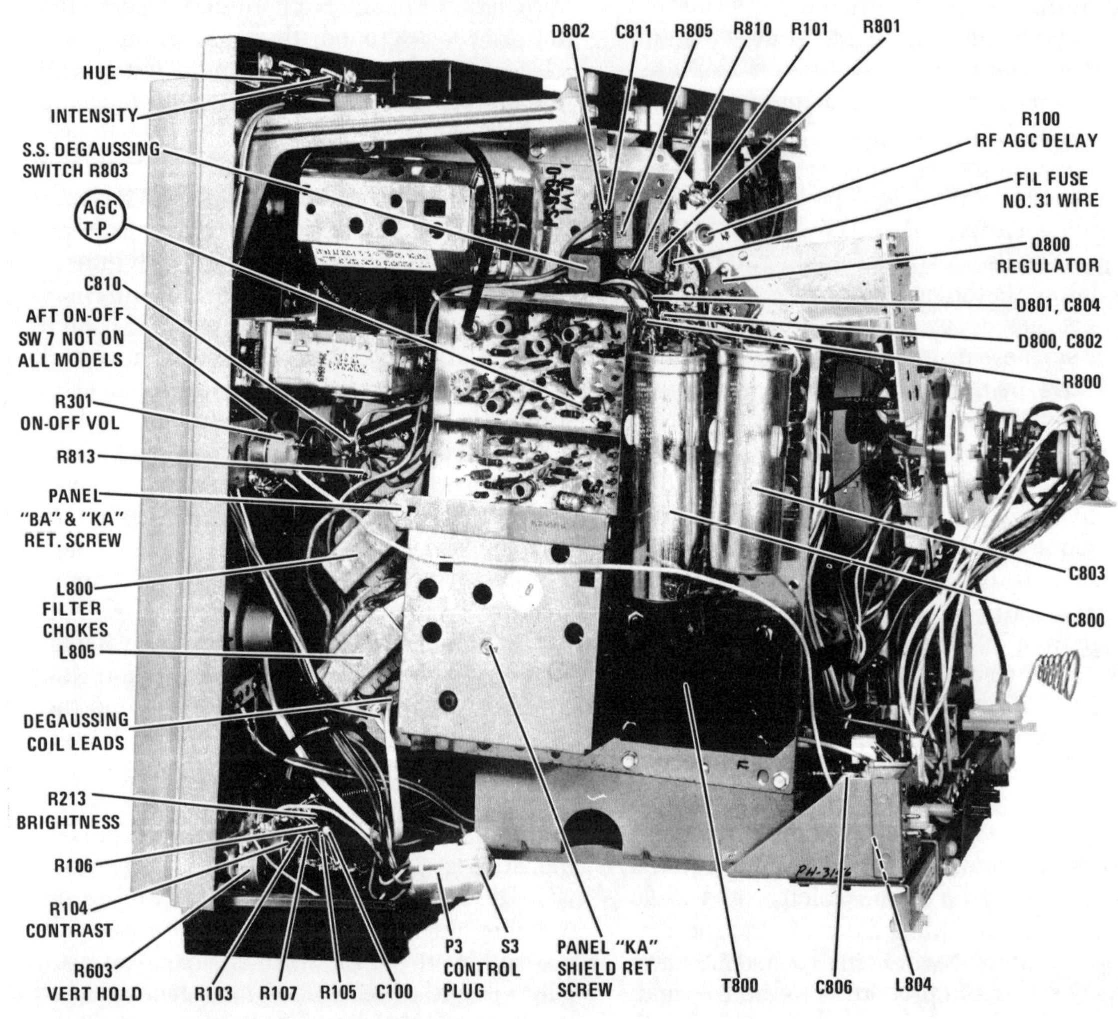

C18TS-929 Chassis - Parts Location - Receiver Side View

Reprinted with permission of Quasar Electronics Company

FIGURE 11-2

POWER TRANSFORMER, FILTER CAPACITOR, CHOKE, AND OTHER PARTS IN THE POWER SUPPLY

during half-cycle lulls and store it during the big surges. This smooths the electric flow, and pulses of direct current become almost, but not completely, pure DC.

Using a tube manual, check the plate numbers of the rectifier tube in your TV. Set your voltmeter on AC high scale and measure across the pins (4 and 6 with a 5Y3). If you can't measure DC in a power supply, see if you can get the AC. If not, perhaps L4 or the corresponding secondary, is broken. If you have AC but still no DC, try replacing the rectifier tube. Now, with the lower AC voltage setting on the meter, measure the heater AC for the rectifier. In this case measure pins 2 and 8. You can't develop DC if there is no heater voltage for the rectifier. Usually you would know if the rectifier had no heater voltage, as the tube would be cold and dark; if it is a glass type you can see whether it is lit. Check the voltage with the negative probe connected to the chassis, the meter set on high DC scale, and of course the meter set on DC function. Touch the red or (+) lead to either pin 2 or 8 or the center probe of C1. Any of these places should give you the DC voltage.

With a tube set you may have DC voltage, sometimes referred to as B+, but the set is completely dead. Don't forget the heater winding for the tubes of the set. Tubes won't work at all without heater voltage. Any time you hear a lot of hum or see wavy lines in the picture, suspect that the filter capacitors are open. If L5 were open, however, you would not have wavy lines or hum. Why? Because if L5 is open the complete series circuit of the DC path is broken, and the receiver would be dead. Some sets have no filter choke like L5. Instead, a large resistor is inserted between the capacitors.

If the filter capacitors short you will know it. The rectifier will light up like a Christmas tree momentarily and then you will get nothing. Say that C1 shorts. The path from ground to the filament of the rectifier tube is a complete short. Before the short the circuit had the

tubes of the entire television set and their loads in the circuit. Now there is nothing, and the current from heater to plates, through transformer secondary to ground and directly to heater again, is great enough to burn out the rectifier tube. It could damage the power transformer, but often it doesn't. Shorts could exist at other places in the set, but check out the electrolytic filter capacitors first. With an ohmmeter (and the set unplugged), you would get a direct short between (+) DC and ground. You could pull out the tube and see if the short persists. If the tube shorted internally (from pin 2 to 4, for example), you would have a similar situation. But if the short still shows without the tube, you must check further.

To be sure whether the problem is a filter capacitor, you would have to desolder all connections from the center terminal of the capacitor. Now measure only the capacitor between the lug and ground. The meter needle should first go nearly to zero ohms and then swing back up to a decent reading of 50,000 ohms or more. Try reversing the leads of the ohmmeter. The higher reading is the one to count. If the needle indicates zero ohms and stays there, the capacitor is shorted and must be replaced.

The bleeder resistor, R1, drains the capacitors of large currents when the set is off. If R1 is defective, you could get a nasty shock touching the B+ or (+) DC line. If you are going in with an ohmmeter or fingers it is good practice to short each electrolytic capacitor to ground. If you get a spark or a snapping sound, chances are the bleeder resistor is not doing its job. Of course, you only short to ground, measure ohms, or touch electrolytics with the set off and unplugged. Remember the safety first admonitions you read in this book.

Solid State TV Power Supplies

Now let's consider a solid state power supply. Refer to the schematic Figure 11-3, but try

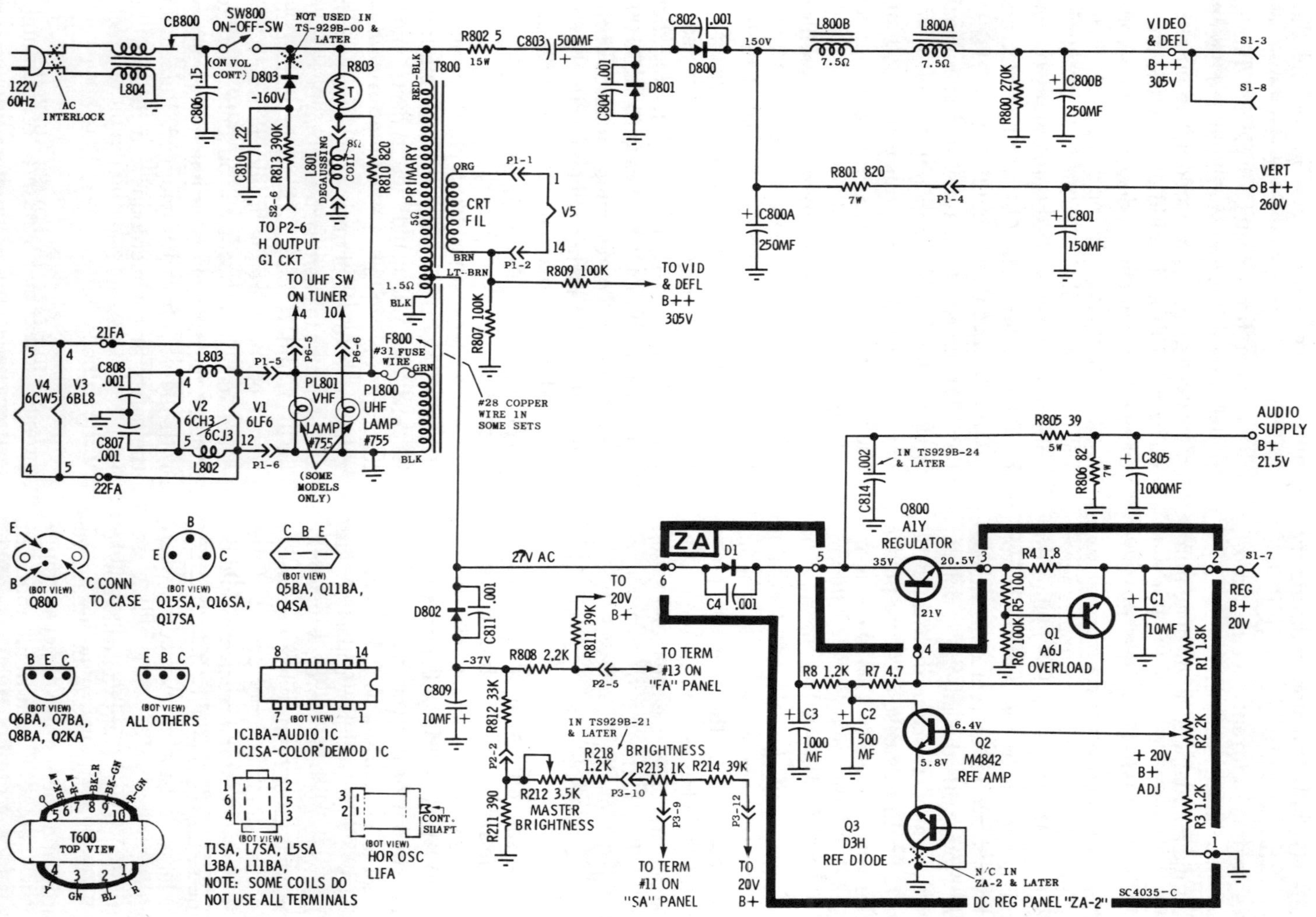

Reprinted with permission of Quasar Electronics Company

Figure 11-3

Regulated Power Supply

to find as many of these parts as you can in Figure 11-2 to learn how to recognize them in an actual set. Let's start with the AC supply. From the plug, current goes through circuit breaker, CB800, to the on-off switch, then through the red-black lead at the top of the primary of the power transformer, T800. Finally the path goes to ground, through the frame of the chassis, and back up to L804 and out the plug again. That is half of one cycle, and 60 full cycles occur every second. The primary places voltages on two secondary coils by induction and is tapped to place a low voltage on still another circuit. Also the top of the primary feeds to a high-voltage power supply.

We are considering only the AC supplies first. The voltage must be lowered, as these supplies are needed to heat cathodes of tubes in the vertical and horizontal sections and the picture tube cathode. The picture tube heater voltage supply is a separate one with orange-brown color codes for the wires. Note that the wires from the secondary winding go to pins 1 and 14 of the picture tube, V5.

The other AC supply is a secondary with green and black leads. This winding has the proper number of turns in proportion to the primary to produce 6.3 volts of AC for its circuits. Tubes V1, V2, V3, and V4 are fed in parallel from this secondary winding. Heaters do not give off electrons, so AC can be used, as it will not get into the signal, providing there is no heater-to-cathode short in a tube. Coils L802 and L803, with capacitors C807 and C808, aid in getting higher-frequency signal that might be on the line to ground. The VHF and UHF lamps are attached here merely for the convenience of getting the 6.3 volts they need to operate. They receive a ground on the tuner switch, so as the tuner is set on VHF, the VHF light would function. These lights have nothing to do with the heater supply circuit.

The horizontal output tube, V1, a 6LF6 type, draws a lot of current when operating. Should it short internally from heater to B+, a large load would be put on the secondary winding of T800. Therefore, fuse F800 would blow out and break the circuit before real damage could be done. The circuits we have just studied need the same 60-Hz AC that comes in on the power line. The voltage must merely be dropped to an acceptable level. The difference in the turns ratio between the primary and the two secondaries in the power transformer, T800, made the difference in voltage.

Now we will consider the two DC voltage supplies in this system, one being a fairly high-voltage supply for the tube plates and parts of the picture tube, and the other being a low-voltage system for the transistors.

Let's consider the higher-voltage system first, the one attached to the top of the primary of T800. There are two diodes, D800 and D801. The current can enter easily at the cathode, represented on the diagram by a straight line, and go out through the arrow, which represents the anode. Current could easily flow from L800B to C803, but it would be all but stopped from flowing the opposite way. A rectifier diode is a diode that can handle large currents, and D800 and D801 are rectifier diodes.

The current tries to go both ways as all AC does, and it has little difficulty passing C803, as capacitors pass AC but block DC. An upward flow of current would go through R802 and pass C803, but it could not continue through D800. Instead, it would go through D801 as it has its cathode in the direction to accept an incoming current. This half cycle goes to ground, then through ground to the bottom of the primary of T800, also to ground, and out at L804. This puts a large positive charge on the right side of C803.

The current reverses itself, however, and goes downward, out into the circuit at L804. At the cathodes of the tubes (not shown) are connections to ground. The current flows up through these connections directly or through resistors, as electrons, and flows through the tubes to plates. Wires or printed circuits carry the plate currents either to S1-3 or S1-8 or to the B+ connection. The currents now return through either R801 or L800 and are able to go through the diode rectifier, D800, because they are going in the proper direction, and back through C803, R802, and so on. This way every cycle or alternation of the 60 Hz puts a current in the right direction to supply the plates of the tubes. The set changes AC to DC this way, but it is very pulsating DC. As mentioned before, the filter capacitors, C800B, C801, and C800A, change these pulses to fairly smooth DC. A capacitor not only can smooth out between half cycles as in the tube power supply we just studied, but can even smooth out between whole cycles. Coils or chokes L800A and L800B help keep the discharge current steady.

If a diode rectifier shorted, then AC would be in the DC circuits. If a filter capacitor shorted, the diode could burn out. As stated before, an open filter capacitor would be indicated by hum and ripple in the picture.

The low-voltage power supply section for the transistors—and note that these are NPN type transistors with (+) collectors—has an audio supply similar to the ones for the tubes. Note, however, that the filter capacitor, C805, has a very high capacitance. This is often used in low-voltage systems as they need as ripple-free a DC as possible for the transistors. The rectifier, D1, converts the AC to DC for both the audio and the regulated 20-volt supply. A tap very low on the power transformer primary at the light brown lead brings 27 volts of AC to rectifier D1. D802 rectifies the cycles that can't go through D1.

A special board or panel, ZA, has a rectifier and some of the voltage regulator circuit on it. The actual regulator transistor, Q800, is a power type and needs mounting separately so it can dissipate more heat. Note the connections drawing of Q800 at the left. The collector is the case of the transistor. Always learn the lead identification. This entire panel uses NPN type transistors.

All the regulated 20-volt supply must return from the parts of the set where it is used through the emitter-to-collector circuit of Q800. The more current that can flow through Q800, the lower the voltage will be at its emitter. If the current flow is controlled, the voltage can be kept constant.

In color TV sets it is imperative that voltages that supply the collectors and bases of transistors be constant. Any change in current from base to emitter of Q800 will affect its current from emitter to collector. A small sample of the 20-volt system voltage is developed across R1, R2, and R3. This voltage is applied to the base of Q2, and any change of this sample would affect the current flow through Q2. This in turn will affect the voltage developed across R7 and R8 resistors. This voltage difference affects the amount of voltage, and thus current, at the base of Q800. The base current of Q800 will control the current from emitter to collector. Again, these are all NPN type transistors, so current flows from emitter to collector rather than the opposite, as it does in the PNP transistors.

All this determines how much current can go along the regulated line and keeps a constant 20 volts for the supply system. R2 is adjustable so the entire regulator system can be set for the right voltage to begin with. Q3 is a transistor used as a diode. Because of the way it is attached, this reference diode tends to make Q2 not conduct until a certain voltage is reached. Q1, the overload transistor, would bypass the regulator, Q800, on hard spikes of current. It would help lower the voltage quickly by conducting.

A regulator system helps keep a constant voltage, but it does not smooth out the pulses of voltage, that is, it does not filter. Capacitors C1, C2, and C3 do this job just as C805 does in the audio supply. One advantage of a panel such as ZA is that if you suspected poor regulation or filtering in the 20-volt regulator supply, you could insert another ZA panel. If that didn't help, you would know that the regulator transistor Q800 is at fault. Note that it is not on the panel. This would be a quick way to isolate the trouble if you had another ZA panel to substitute.

Laboratory work for Chapters 11 through 19 will vary with your situation. If you are studying this book as a television school text and are in residence at the institution, the laboratory instructor will have sets provided so that you may identify parts of color and monochrome TVs, measure voltages, repair similated faults, and align the sets. Naturally, at the end of each section of study, before the quiz, you should do work pertaining to that particular chapter. If, on the other hand, you are studying this text away from a technical or other institution, you will need to adapt the lab work to your own TV set and to those of obliging neighbors.

For the TV or TVs available, first study the placement of parts of the power supply. You will need to obtain a schematic and parts layout drawing of each set. Measure AC volts with either the multimeter you made in Chapter 8, or any high input impedence unit of 20,000 ohms per volt or more. Measure this at the wall outlet, at the transformer primary if you can find terminals; if not, skip it, as it is poor policy to remove insulation from wires of a working set.

Measure heater voltages at tubes, if applicable. Measure with the meter set to DC function, the plate, screen grid, or if solid state, collector, and base voltage between the named unit and ground (metal of set). Take precaution not to shorten base and collector. It would be fine to measure these voltages for several transistors and/or tubes in the set for the practice. Locate the voltage regulator, if the set has one, and measure voltages between ground and emitter of regulator. Then measure voltage between ground and collector of the voltage regulator transistor. Depending upon which way the circuit is connected, one voltage should be considerably more than the other. This indicates that the regulator is active. Figure 11-3, Q800 regulator is on the high or positive voltage side, so the greater voltage is at the collector. The correct voltage for transistors out in the TV is the same as the emitter of Q800 voltage or 20.5 volts in this case.

In order to get an idea of what happens when filter capacitors are open, if it can be done without disturbing too many other items, temporarily desolder one end of capacitors that compare to C805 and C1. Be sure every other connection is complete. It may be necessary to tape some to prevent shorts during this test. With the set on, you should hear a motorboat noise in the sound and see wavy lines of picture. These waves are 60 Hz modulation that the filter capacitors are not smoothing out. Return the set to normal operation.

Summary

Summary for Chapter 11 is as follows:

1. Power transformers convert 115 volt AC to appropriate AC voltages for a TV receiver.

2. Solid state or tube diode rectifiers convert AC to pulses of DC.

3. Filter capacitors and chokes smooth out this pulsating DC to nearly pure DC.

4. Voltage regulator circuits maintain voltage levels by sampling the voltage and correcting it by controlling the bias of a transistor that allows full power supply through its emitter to collector circuits.

5. A shorted electrolytic filter capacitor will cause way too much current through the rectifier and normally ruin the tube or solid state rectifier.

6. An open filter capacitor will cause motorboating and distortion in the picture.

7. An open choke will block all current flow.

Quiz for Chapter 11

▶ 1. What type of capacitors are used in filtering for power supplies? What precautions must you observe in replacing them?

▶ 2. What two units convert AC to DC?

▶ 3. What parts of a TV set need low-voltage AC?

▶ 4. What usually happens if a filter capacitor shorts? If it is open? How can you check a filter capacitor with an ohmmeter?

▶ 5. What happens if a filter choke coil is open? What is sometimes used in place of this coil?

▶ 6. Without using an ohmmeter or voltmeter, what condition might suggest a defective power transformer?

▶ 7. Describe a quick way of testing AC line, switch, and primary of power transformer all at once.

▶ 8. Tell the purpose of a voltage regulator and briefly describe how one works.

12. Circuits That Pick Up, Select, Convert to IF, Amplify, and Detect

PROJECT: Injecting Signals at Antenna, Tuner, and IF to Learn to Signal-Trace Troubles; Observing Oscilloscope Waveforms; Taking Voltage Readings at Various Points

THE ANTENNA, tuner, IF, and detector stages of a color TV carry all four signals—video, sync, color, and sound. If any of these stages is dead, none of the signals will reach its destination. You might see a raster, a complete trace of light, on the screen, but there would be no incoming sync to time it. Usually if a set has a lighted screen but no picture on any channel and no sound, the trouble is in either the tuner, the IF, or the detector, or the antenna system is faulty. Since all parts of the TV reception depend on the tuner, IF, detector, and antenna, the student should learn all he can about such stages.

TV ANTENNAS

Consult Figure 12-1. All transmitters and receivers—radio, TV, citizen's band, or whatever—require an antenna. Radios once had long, high, elaborate antennas, but with modern high-power broadcast stations, the receiving antenna can be nothing more than a ferrite rod with the tuning coil wire wrapped around it.

Not so with TV. Although TV stations often put out tremendous power in the thousands of watts, the 6-MHz channel is high enough in frequency that it travels best in a straight line from the station antenna. Just a few miles from the station the signal goes straight out into space because of the curvature of the earth. To get any reception at all, tall antennas are needed. Special parts strengthen the signal, and specific lengths of antenna make it pick up the signal better. In this case the signal means the whole channel, including the sync, color, and so on. Even in suburban areas an antenna above the building height is preferable for color reception, not just to reach up where the stronger signal is, but to get away from manmade interference such as auto ignitions, motors with brushes, and power lines.

You have learned that high-frequency alternating current becomes radio and TV carriers and that each carrier has a wavelength. Channel 2, for example, not only has from 54 to 60 MHz frequency of alternations, but each of these alternations has a length that it will occupy on a wire, metal frame, or other antenna. The current can travel best across this specific length, and the voltage that develops across this length will radiate into the air better than if the TV station had a wire shorter or longer. For best reception the receiving TV antenna must be the same length to pick up the maximum voltages radiated at the proper channel frequencies. Not only must it be of proper length, but it must be aimed properly to receive the maximum carrier along with its signals.

Figure 12-1 *a* gives the frequencies of the VHF channels and half their wavelengths. The frequency is expressed in megacycles, but they mean the same as MHz. Note the antenna below the chart. Most TV antennas are *dipoles;* they are made of two poles, each one about half a wavelength long. The figures of half wavelength are expressed in inches. Note that for Channel 2 the whole length would be about 100 inches. That would be a lot of antenna for the wind to blow against on the roof. The solution often is to compromise and have a shorter length that won't pick up Channel 2 as well but gets other channels fairly well. About the only place this won't work is in fringe areas, areas far from stations where the best pickup possible results in barely acceptable viewing. In fringe regions antennas need to be very tall, aimed precisely, and of exact wavelength of the desired channel. Color TV requires a better antenna than does black-and-white TV.

Note that the higher the channel number the shorter the wavelength. A plain dipole requires a 72-ohm coaxial cable lead-in to the set. Since the set is ordinarily designed to receive 300-ohm twin lead, you would need a small impedance-matching transformer, 72–300-ohm type, at the input to the TV set. An easy way to overcome this problem is to have a folded dipole as in part *b* of Figure 12-1. The folded dipole takes the 300-ohm lead more easily and can be attached directly to the antenna terminals on the receiver.

About the only time you will have to concern yourself with coaxial cable is with community cable TV antenna systems. Usually the supply to the house is with 72-ohm cable. A small transformer must go between the cable and the TV set antenna terminals. One other case might be in an area with a lot of electrical machinery. To keep the equipment from putting static on the lead-in wires, the antenna would be quite high, away from the disturbance. Either a plain dipole or a folded dipole and transformer at the top would be used. The lead-in wire would be coaxial cable.

Then at the receiver another transformer would be used to get the matchup back to 300 ohms.

Be sure you install the transformer with the proper side to its impedance. *Impedance* is the term for the ohms of resistance in antenna lines. These ohms would not be like those in an actual resistor, and you could not measure them with an ohmmeter. On the other hand, you certainly can use an ohmmeter to test out an antenna. With a folded dipole, after removing the antenna from the set you should get almost no ohms or a short indication on the ohmmeter by measuring at the twin lead that goes to the TV receiver. If you get infinity or very high ohms, there is a poor contact or no contact somewhere up there. One of the leads may be broken. You can also test for shorts to the pole. Connect one lead of the ohmmeter to the metal of the pole and the other lead to one of the twin lead wires. If you get any indication at all, the antenna unit is shorted to the pole that supports it. Check for worn lead-in shorting, or in rare cases the driven element insulators (the driven element is the dipole) allow the pole clamp to touch the driven element.

The TV station drives the driven element with signal, developing a very tiny voltage across it. A dipole alone would not likely pick up enough energy for good reception except in suburban areas. Figure 12-1 *c* shows the most standard type of antenna system. The elements before the driven element, the dipole, are directors. They help bring the very weak voltage to the driven element, and they must be shorter than the driven element. Behind the driven element is a reflector. It bounces signals back to the driven element, and it has another function.

You have a Channel 4 in your area, but there are more Channel 4 stations all over the United States. One is very likely to be in just the opposite direction from your Channel 4. The reflector is longer than the driven element and reflects a possible interfering channel away. There is little the TV receiver could

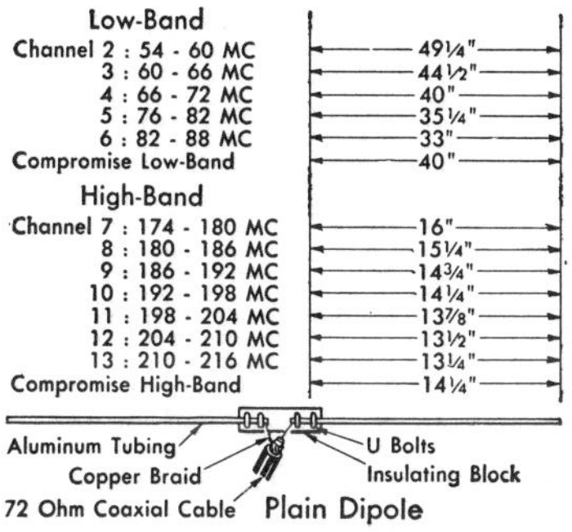

A. VHF Channel Wavelengths

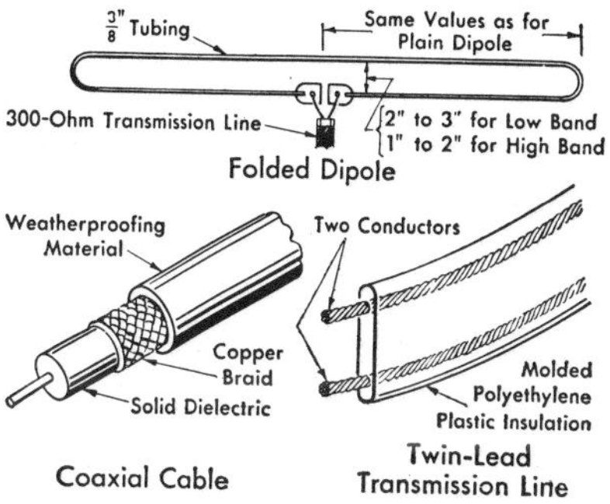

B. Folded Dipole and Lead-in Examples of Transmission Lines

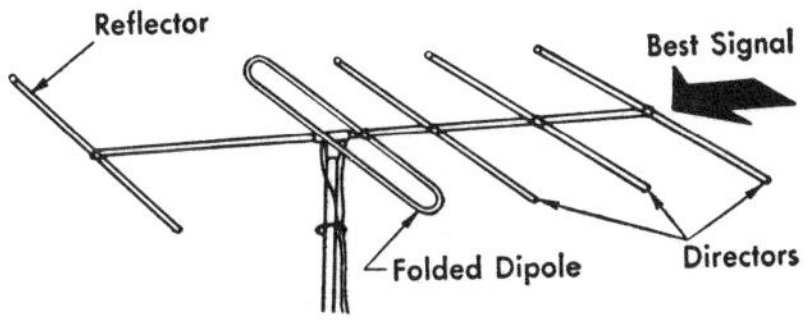

C. Standard Broadband Antenna Elements

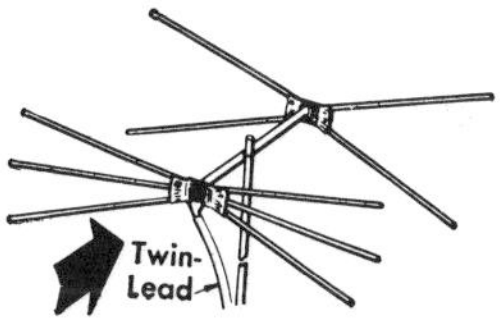

D. Conical Broadband Antenna Elements

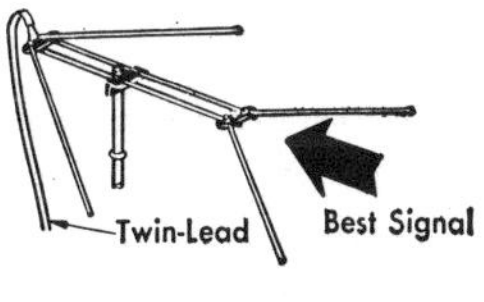

E. Double V Antenna

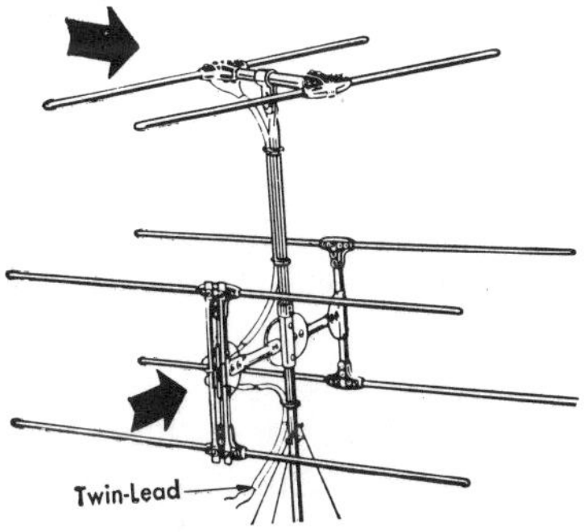

G. Piggyback Antenna

FIGURE 12-1
TELEVISION ANTENNAS

do to separate one Channel 4 from another; only a well-directed antenna can do this. The best reception is always broadside to the driven element, so when directors are placed before and a reflector behind the driven element and it is aimed as shown, the stations in the direction indicated will come in best.

Figure 12-1 *d* shows a conical type antenna. More active elements are added to pick up more signal, and all are cut to proper bandwidth wavelength. Note that this antenna has no directors. Sometimes extra driven elements are better. These elements are plain dipoles, yet a twin lead is shown. This is not technically correct, but sometimes oddball plans work better than engineering theory. The standard broadband and conical broadband antennas are called *broadband* because they will receive stations in the general direction they are aimed toward, and they pick up quite a few channels decently in areas not too far away. In some areas where interfering stations might be to the side of the desired channel, a double V antenna can be used (see *e*). The elements are cut to bandwidth, but are bent forward to receive from a very narrow area. These antennas have to be aimed carefully, and a wind can often move them enough to ruin their effectiveness. If the antenna must get one station that has other channels in nearly the same direction on the same wavelength, however, the V antenna is a good compromise.

In some areas the various stations come from different directions. One can invest in an expensive rotor system with a motor, or one can put up two antennas on the same pole (Figure 12-1 *g*). Preferably each should have its own lead-in, and a flip-over switch should be installed at the receiver input. You can, however, install a matching transformer on the pole for two separate channels with separate antennas and bring it all down with one twin lead to the TV set. The piggyback arrangement of two antennas is better than the rotor in several ways, one of which is that each can be cut for exact wavelength of its channel, whereas the rotary antenna will have

to be a compromise and never will bring in as clear reception.

You will seldom make antenna elements, so this information is presented as a guide to purchasing. You probably will install, replace, and repair antennas a great deal. The most important element of installation is clean, good contact of lead-in to antenna terminals. Securing the lead-in with standoff pieces on the way down is a good idea. In far fringe areas the use of antenna amplifiers near the driven element are useful, although they are subject to lightning damage.

Planning an antenna installation ahead of time is important. The customer seldom has a ladder, for instance. If you can, arrange the elements on the ground, fold them out, attach the lead-in, then go carefully up the ladder you have brought and attach the antenna to the pole. Attach guy wires on all sides and erect the pole if it is a telescoping type. The antenna elements are very easily damaged, as they often are made of light aluminum. UHF antennas are very short with lots of directors. It is best to get one made exactly for the channel in your area. Adapters can be installed to allow VHF and UHF to go down one lead-in, but if the customer can stand the cost, it is better to have one lead-in for the VHF Channels 2 to 13 and another lead-in for the UHF Channels 14 to 83 and attach each to its own connectors on the back of the TV. The so-called "all-channel" antenna may work on UHF if you are within a few miles of the station.

Experiment with a rotary antenna, either motor or hand-driven. Tune the TV set to a channel that is quite strong and rotate the antenna each way until the signal weakens or disappears, depending upon how close the received station is to the antenna. Note the approximate swing necessary for this to happen. Now tune to a weak channel and try the same procedure. Is the angle of decline of picture any less?

Note the type of antenna driven element. Use your ohmmeter to measure from the ground for shorts or opens. First disconnect the lead in from the TV. Try the ohmmeter across the lead in wires. If the antenna is a folded dipole driven element type, you should get full continuity, or zero ohms. If the antenna is the plain dipole type, you should get infinity or no current able to pass at all. Now test between one lead in wire and the support pole or tower. You should get no reading at all unless the lead in or dipole is shorted to the tower or pole.

Try the TV set with only one wire attached. Note the amount of reception. Then attach both properly and note the difference. Sometimes if you suspect the antenna you can tell if it is defective by this test. With a good antenna and lead in wires, attaching the second wire after trying the set with just one should increase picture clarity, sound, color, etc.

Here are principles to remember:

1. TV antennas are of the dipole type, with each pole being half a wavelength long.

2. The driven element picks up the signal and, with twin lead for 300-ohm systems or 72-ohm coaxial cable, relays this signal to the tuner.

3. Directors in front of the driven element bring the signals in from the proper direction. They are shorter than the driven element.

4. A reflector reflects the proper station back to the driven element and also reflects stations in the opposite direction away from the driven element. It is longer than the driven element.

5. An ohmmeter can be used to check for shorts to the antenna pole.

6. Separate lead-ins for more than one antenna system on a pole or for VHF and UHF are best.

7. Clean, tight connections to elements by the lead-in are very important.

Tuners

In very old television sets you will find continuous tuning mechanisms that do not select channels by snap type switches, but rather by continuous movement of a dial like a radio tuning dial. These dials have been replaced almost universally by switching mechanisms that flip in a channel by number and fine tuning devices that allow the viewer to select the point of best reception. Automatic fine tuning (AFT), has been added, and once the channel has been selected by a switching system, the AFT takes over and tunes in for the best reception of the bandwidth. The latest innovation is tuning without switches that select the channel but rather with *varactors*, diodes that vary in capacitance according to the voltage placed on them. Of course, someplace the set must contain a switch or switches to select the voltages to put on the varactor to change channels.

A tuner in a TV receiver must perform three functions: (1) it must preselect the channel the viewer wants to watch; (2) it must give some amplification to that channel; and (3) it must heterodyne the channel to present it at the proper frequency for the IF stage.

The radio frequency (RF) amplifier transistor and its properly selected coil on the wafer switch (or coil and varactor in new sets) amplifies only the channel that the coil or varactor is set to receive. Note in Figure 12-2 that the first wafer switch, SW5, is set to receive Channel 13. Only coil L12 is in the circuit to ground out pin 2 of the wafer switch. The other coils are shorted out.

Lower-frequency channels than 13 find an easy path to ground through coil L12. Higher-frequency channels find a path to ground through the distributed capacitance of coil L12. But the channel that is just right, 13 in this case, resonates or cannot get to ground easily, so it travels on through L13 and other items to the emitter of Q1, the RF amplifier. This is a little review of the theory of tuned circuits. Since only Channel 13 gets to Q1 with much power, it is the only channel that will be amplified. Other channels enter to a smaller degree, but the heterodyning will eliminate them. The RF amplifier selects and amplifies one channel strongly before the heterodyning. This process is known as *preselecting*.

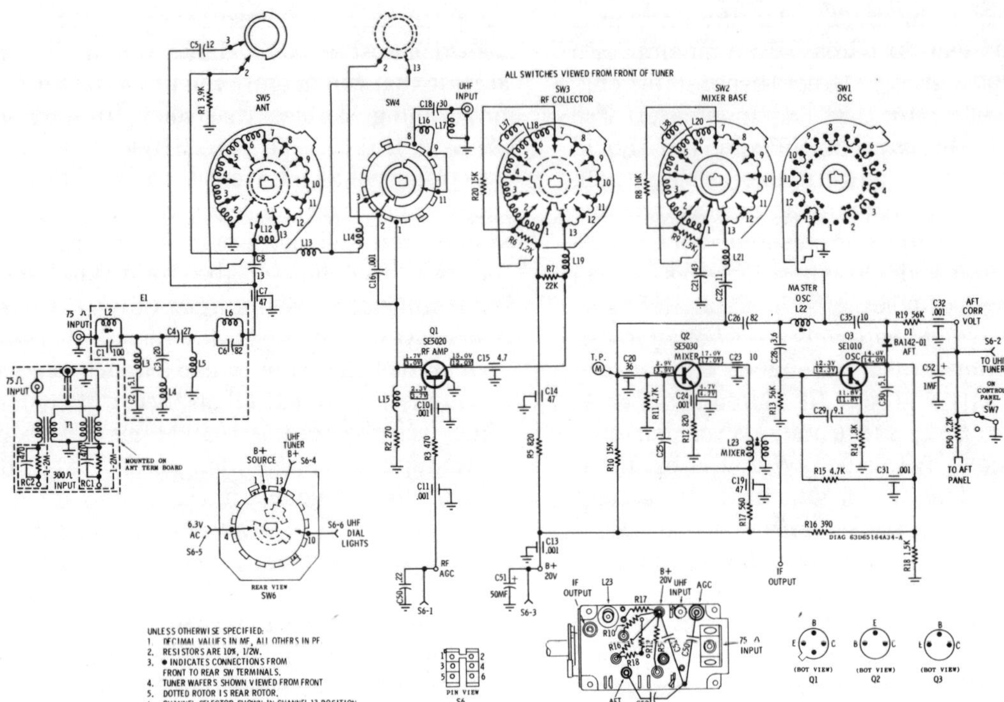

VHF Tuner Schematic - CPTT-429

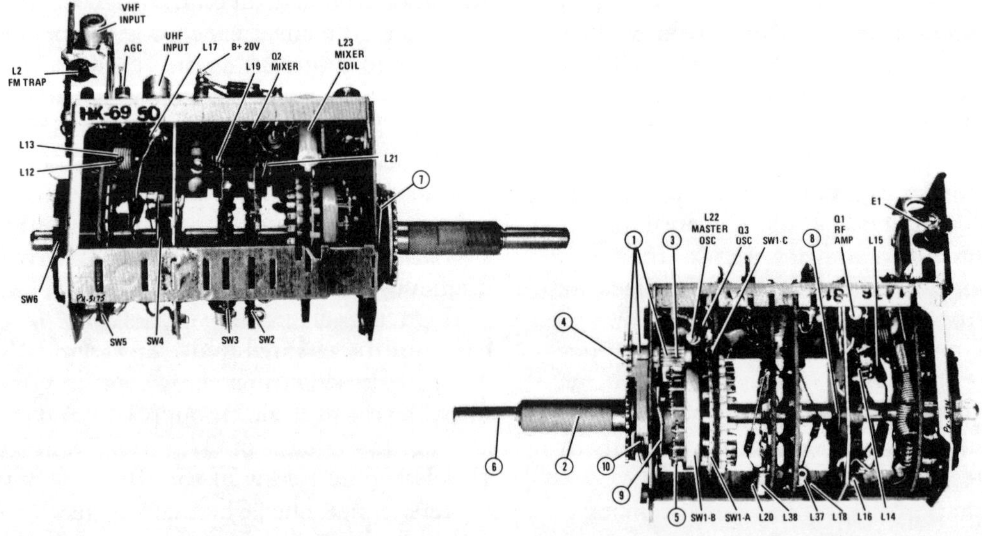

VHF Tuner CPTT-429 - Component Location

Reprinted with permission of Quasar Electronics Company

FIGURE 12-2
WAFER SWITCH TUNER, SCHEMATIC AND LAYOUT

The heterodyne process consists of creating a carrier much like the incoming channel, though without signals, and beating this created carrier against the incoming carrier. The created carrier is produced in the oscillator stage. The beating of the two carriers together is done in the mixer stage. The created carrier is at a frequency that will always be 44.25 MHz above the video center frequency. If you were receiving Channel 2, 54 to 60 MHz, the oscillator would have to create a 101.25 MHz. center frequency. The difference must always be 44.25 MHz. This is for a modern TV with an IF of 41.75 to 47.25 MHz. With older TVs with an IF of 21 to 27 MHz, the oscillator would be only about 25 MHz above the incoming channel center frequency.

Naturally, the oscillator must be changeable at the same time that the RF tuning is changed. In Figure 12-2 this change is accomplished by SW1. Switches SW1 through SW5 are all on a common shaft that goes to the channel selector knob on the front of the TV set. Thus all tuning takes place at once.

The mixer takes both the incoming channel with carrier and signals and the oscillator carrier without any signals and puts them together. When two bands of frequencies are put together in a stage, a choice of four outputs can be developed for the next stage: (1) the input channel, (2) the oscillator channel, (3) the sum of frequencies, and (4) the difference between the frequencies. It is the difference in frequencies that is used in TV. This difference channel—always from 41.25 to 47.25 MHz in modern sets—still has sync and video at 45.75 MHz, color at 42.17 and sound at 41.25. It is not the incoming band of frequencies from the antenna, nor is it the oscillator frequency. It is an intermediate band of frequencies called *intermediate frequency* or *IF*. The IF is amplified in its own stage. This is a repetition of material in Chapters 5 and 10 to some extent, but it bears repeating.

Now let's examine the wafer switch tuner circuit in Figure 12-2. This Motorola tuner has three transistors. Note their connections drawings to the middle right. At one time all transistors had a tab just before the emitter as in Q1, and going clockwise you came to emitter, base, and collector. Some had a red dot at the collector and you went backward for lead identification. Now the leads are in all kinds of combinations, so if drawings such as this appear on a schematic, pay attention to them. Some sets have *E, B, C* printed by the leads on the circuit board.

Note that this set is actually designed for a 75-ohm coaxial input. The circuit at lower left is necessary to adapt it to 300-ohm twin lead. T1 changes the input impedance to 75 ohms. Most of the parts in board E1 such as C6 and L6 are designed to trap out noise, amateur and CB signals, and the like. C7 is a feedthrough capacitor and goes through the metal of the tuner housing. Antenna switch SW5 has two sections. The main section places L12 in the circuit to tune in Channel 13, as stated before. Resistor R1 is always in parallel with the coils or coil selected to broaden the bandwidth received to a full 6 MHz. In some channels C5 is in parallel with C8 to couple more channel through, but in the case of Channel 13, C8 alone couples the incoming channel to the tuning coil. If the TV were set on Channel 2 then all the coils in L12 as well as all those around to pin 2 would be in the circuit in series, and much more inductance would result, thus tuning in a channel lower in frequency.

The channel carrier with signals travels through L13 and C16 to the emitter of Q1. In other channel settings it would have gone through L14. Q1, the RF amplifier transistor, is a four-lead transistor. The fourth lead must be soldered firmly to ground. This is a shielding network and helps keep stray signals from being amplified by Q1. L15 allows DC to get to the transistors but keeps the channel carrier from escaping to ground. R2 is an emitter biasing resistor to place the proper bias on the transistor.

There is another biasing method too. The

automatic gain control (AGC) voltage is placed on the base of Q1 to control its amplification. Thus, if too strong a signal is being received at the video amps or detector, a voltage is sent on the AGC line that lowers the forward bias of Q1, so Q1 will amplify less and reduce the overall signal. Q1 is a grounded base amplifier. When the signal or carrier is put in on the emitter, then the base is grounded (not DC grounded, but grounded as far as signal is concerned). C50, C10, and C11 place the base on signal ground. A signal can be put in at emitter instead of base if the emitter is not grounded directly. It is the emitter-to-base current that is important to trigger emitter-to-collector current. Note that this is an NPN transistor. Signals can be put in on the emitter of PNP types, too, with a signal-grounded base.

The collector signals are tuned again by SW3. This time L19 does the tuning for Channel 13. The swamping resistor, R7, keeps the bandwidth broad to 6 MHz. Pin 1 of switch 3, through R5, supplies positive 20 volts for the collector of Q1. This same source supplies collector voltage for Q3, the oscillator, through R16. R5 and R16 are load resistors. They keep the signal bandwidth from escaping into the power supply and they lower the voltage to the collectors.

The Channel 13 information gets to Q2, the mixer, by inductive coupling between SW3 and SW2. They are close together. After again tuning of Channel 13 by SW2, the channel goes through C22 to the base of Q2. Feed-through capacitor C20 is not a capacitor between parts for coupling purposes, but rather couples a very small part of the signal to ground and allows wires to go in and out of the metal frame of the tuner switching mechanism. The oscillator, Q3, supplies its carrier through C26 to the same base of Q2, the mixer. The bandwidth of the difference signal, 6 MHz, is fed out the collector through L23 to the IF circuits. The collector voltage is through R17. Note that R10 from the same positive source supplies a forward bias to the base of Q2. The AGC does not control the bias

of this stage. C51 keeps signals out of the power supply by routing them to ground. If signals got into the power supply they could modulate the set at other points and cause much distortion. R12 is an emitter bias, a protective bias that keeps the transistor from thermal runaway.

Now for the oscillator circuits. For an oscillator to make a carrier it must have a tuned circuit, and some of the output must be fed back to the input. The tuning is accomplished by the combination of L22 and whatever coil to ground is selected by SW1. L22 is a service adjustment and would be set by the TV repairman to get it to track properly over the different channel settings.

Small slugs of ferrite material in the coils of SW1 are adjustable for each channel to set the oscillator properly to be precisely 44.25 MHz above the incoming video frequency. The farther into the coil the slug or core is turned the lower the frequency of the coil. The feedback path is from collector through D1, C35, and L22, and back to the base. This feedback of the output develops the carrier. This carrier is fed to Q2 through C26, as stated before. C29 and C30 send a small amount of carrier to emitter, but this is not the oscillator feedback. It helps keep steady oscillations and removes unwanted signals. The automatic fine tuning (AFT) voltage acts with D1 in such a way that the feedback is varied.

This varying of the feedback actually changes the frequency of the carrier slightly. Thus, a voltage difference affects frequency and can keep the frequency of the oscillator correct in respect to the incoming channel. Without this AFT, L22 probably would be the fine tuning device and would be adjustable from the front of the set. Since a diode such as D1 allows current to travel only from cathode to anode, or into the arrow, the voltage on the anode can allow more or less signal to get through.

The UHF tuner is usually separate from the VHF with channel selection all its own. Figure 12-3 shows the actual UHF tuner with the case bottom removed and a schematic. The signal input takes a 300-ohm twin lead antenna lead-in. L1 inductively couples this signal to the "1st line," which is in series with C1A. This line and C1A form a tuned circuit that tunes to the desired channel. A single line instead of a coil is used for tuning because of the extremely high frequency. Through slots in the housing the incoming composite signal is inductively coupled to "2nd line" and C1B, which tune to the same channel.

UHF Tuner TT-640 - Component Location

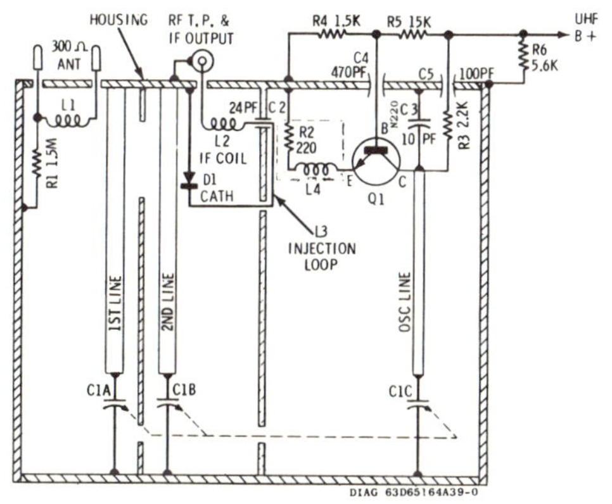

UHF Tuner Schematic - TT-640

Reprinted with permission of
Quasar Electronics Company

FIGURE 12-3
UHF TUNER

Q1, a grounded base transistor, is the oscillator. C3 provides feedback to produce oscillation. C1C tunes to the oscillator frequency. L3, the injection loop, picks up the oscillation and mixes it with the incoming signal that resonates on "2nd line" and C1B. D1 detects this mix, so actually the mixing is done without amplifying the incoming channel. The UHF output is an IF that is fed to the oscillator-mixer of the VHF tuner. See Figure 12-2. S6-2 input is the entry place. The VHF tuner reheterodynes this UHF-IF to the lower IF frequency for the IF stages.

R4 and R5 (Figure 12-3) are bias resistors for Q1. R3 is the collector load resistor. Needless to say, any moving of parts in the UHF tuner will affect tuning greatly. The gear drive for the UHF tuner has such a ratio that C1A, B, and C are moved less than the tuning knob. Otherwise just touching the knob would take the tuner off frequency. Such are the high frequencies of UHF.

A varactor-tuned system often is said to have no moving parts. In the actual tuner this can be true, but somewhere in the set there has to be some kind of channel selection. A varactor tuner would have the RF preselector, mixer, and oscillator stages, but instead of the coils being shorted in part or changed by a wafer or turret tuner, the coils would remain stationary. See Figure 12-5, which is a simplified diagram of the mixer stage of a varactor-tuned system. The RF and oscillator stages would have similar varactors. As mentioned, a varactor is a diode that changes capacitance between cathode and anode to a noticeable degree as a varying voltage is applied to it. With AFT in the preceding example a diode had varying capacity to change the oscillator frequency slightly. In this case, though, the varactor, X101, is the capacitor of the tuned circuit. Changes in its capacitance will determine which channel will be tuned in.

On the front of the TV are pushbuttons. Pushing a button connects a varactor in the RF in the oscillator and in the mixer (shown) to a preselected amount of voltage. This is set by connecting a positive and negative voltage across a variable resistor.

Figure 12-5 shows the pushbutton set for Channel 3. The potentiometer or variable resistor is set so that just enough voltage is placed across X101 to give it the capacitance to work with the secondary of T1 to tune in Channel 3. When the set is new, all these potentiometers have to be set for their respective channels, and they may have to be set again as the parts age. For tuning purposes, it is necessary only to push the channel selector button to connect the right potentiometer. Because the distance between Channel 6 and Channel 7 is a big jump in frequencies, some of these varactors aren't able to cover all VHF channels with different voltage applications. In this case, when buttons are pushed to change from high VHF to low VHF, a switch automatically is opened or closed to add or take out some of the coil of T1 in this case. This changing of the inductance again makes the varactor, X101, able to cover all channels, either 2 to 6 or 7 to 13.

We have now covered transistor tuners in general and those with varactor tuning. Now we will consider tube type tuners. Some early tube tuners were turret types with little plug-in coils that made contact by side connectors as the whole group of coils was rotated. Constant movement of the coils changed position of the tuning slugs in them and affected frequency adversely. Most tuners now keep the coils stationary and move wafer switches to contact the proper coils. This causes less movement and possibly detuning of coils. Figure 12-4, lower right, shows two basic types. The coils in series use all coils for low-frequency Channel 2 and just one coil for Channel 13, and more or less for channels in between. The parallel arrangement has a separate coil for each channel, and only one coil is connected at a time. The tuner in Figure 12-2 used a combination of both methods.

Let's analyze the schematic of Figure 12-4. The signal is picked up by the antenna and inductively (magnetically) fed to the secondary of L1. The signal on grid pin 2 of V1 controls the electron flow to the plate, pin 1. Coil L2 gets rid of interference to some degree. The amplified output goes in on the cathode of V1 at pin 8. While the grid affects flow from cathode to plate, it is the difference in voltage between cathode and grid that determines current (electron) flow to plate. In cases where the cathode has the incoming signal, the grid is tied to ground in some way, so as the cathode becomes less positive it is the equivalent of the grid being less negative. This is hard to understand, but it works. Since the signal would have to go through the 100,000-ohm resistor, R11, to get to the grid, the grid is signal-grounded through C7.

The amplified signal goes out plate pin 6 to top coil L4. The bottom coil of L4 is the oscillator. This oscillator has a feedback circuit from plate, pin 1, of the lower half of V2, down and just above the fine tuning capacitor to bottom of the coil marked A20. Then the signal goes up and through C12 to grid pin 6. C13, C14, the fine tuning capacitor, and the number of turns in the bottom part of L4, plus the setting of the slug in that coil determine the frequency of oscillation. C14 is adjustable with a screwdriver, usually on top the tuner. The fine tuning capacitor is just before the detent mechanism that locks the channel switch squarely on channels mechanically, in the upper-right-hand view. The outer shaft on the front of the tuner usually controls fine tuning.

Since the magnetism of both the RF and the oscillator are going across the center coil of L4, they mix there, and the combined signal goes to grid pin 5 of V2, the upper part of the tube. Of course, this tube is all one envelope and not cut in two as shown; the drawing is designed to clarify what happens. The mixed signals are amplified and go from plate pin 2 through a coil that rejects some interference, through C17, which blocks DC and keeps it out of the grid of the next stage, on to the IF first transformer. This transformer and

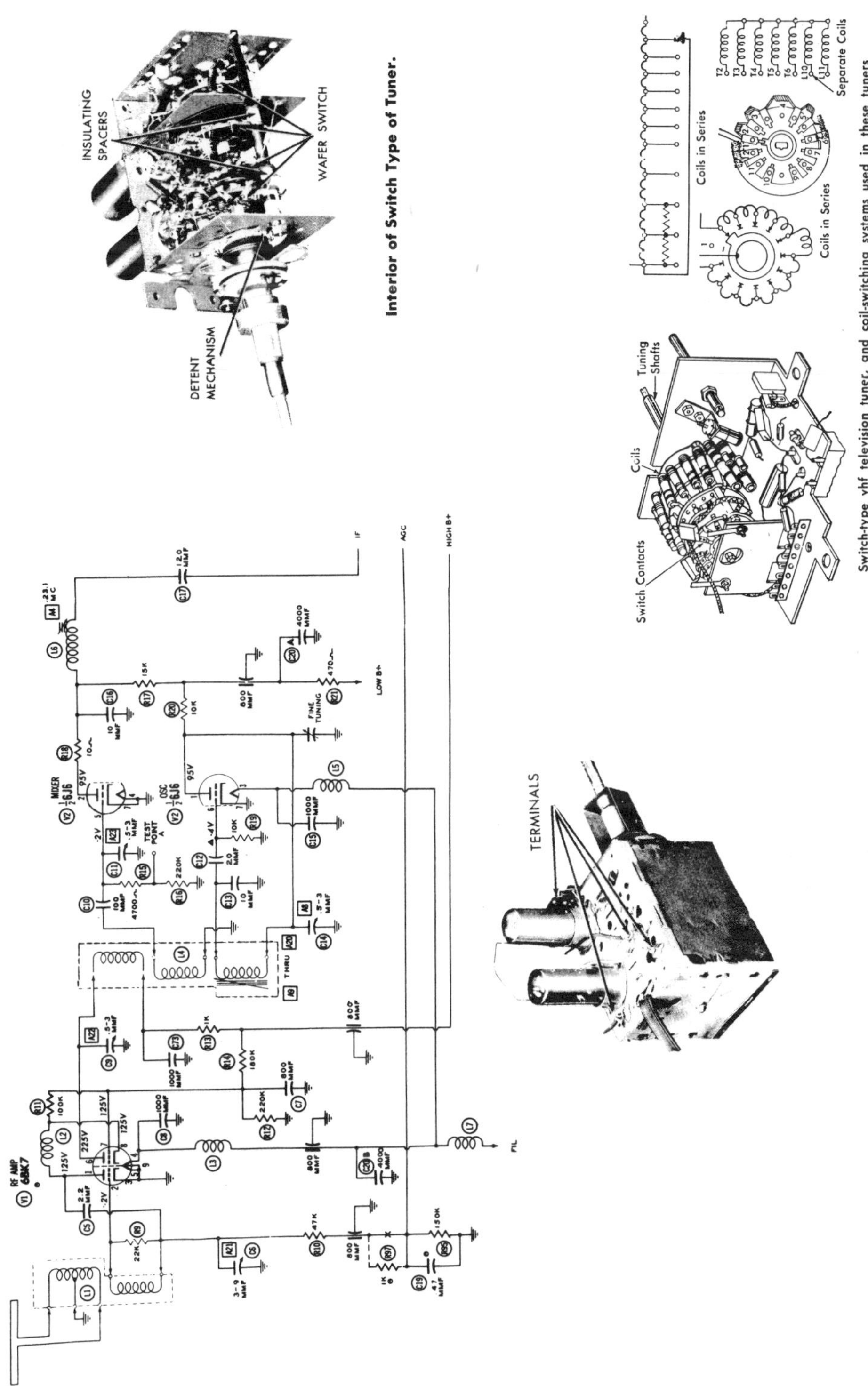

FIGURE 12-4
ROTARY WAFER SWITCH TUBE TYPE TUNER

others in the IF will accept the difference in frequencies between the local oscillator and the incoming signal from the station, but it keeps a 6-MHz bandwidth.

The automatic gain control biases the RF tube and some of the IF tubes on strong signals to prevent overloading. An overloaded signal might reverse the picture, making the white parts black and vice versa. It might cause the sync to tear the picture, and the picture could have so much contrast that there would be no pleasing in-between gray tones. Of course, a self-biasing arrangement like what R15 and R16 provide for the mixer could be used, but this could not anticipate extra strong signals. When you have checked

everything else in the RF, mixer, oscillator, and IF stages, measure the AGC voltage, in this case from pin 2 to ground of V1. Always remember that sets vary, and it wouldn't be pin 1 on every set; check with a tube manual or schematic of the set. In this case a minus 2 volts is desired.

Attach the red lead of voltmeter to the chassis and the black one to the grid with clips. Now adjust the AGC shaft adjuster at the back of the set. You should be able to vary the AGC bias a bit with this adjustment. If the adjustment has no effect, the AGC circuits are defective. Or if you have either a very strong bias or none, you may have AGC troubles. Naturally, strength of incoming signal will

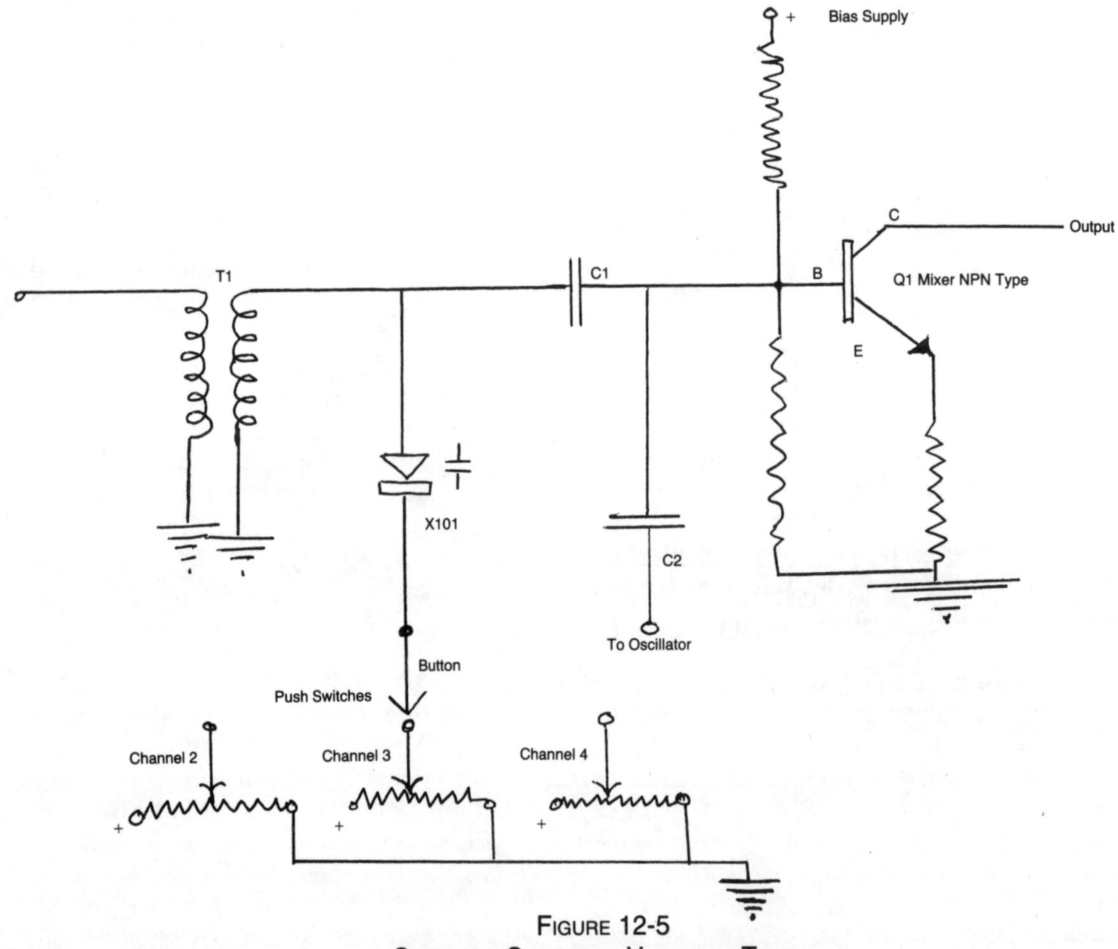

Figure 12-5

Simplified Schematic Varactor Tuning

change the reading if the AGC is working properly.

The AGC can be in working order but still you can have problems that appear to be AGC. In this case, you have a weak tube in the IF or RF. Even a little bias cuts it down so it can't amplify normally. This is tricky, so always measure AGC and if it is OK, check parts the AGC goes to.

Though it requires an extra set, you can check for oscillation in a TV by putting it beside another TV with an antenna. Flip from channel to channel of the bad TV until you interfere with the good one.

The tube set we are studying has a low and high B+ from power supply. These are both parts of the low-voltage power supply, but one is higher than the other. This has nothing to do with the very high voltage for the picture tube anode.

Any bending, moving, and so on of a tube shield, coil, or capacitor in a tuner will affect frequency. After replacing a tube or part you may need to touch up the slugs a little. Do this carefully. It is best to align a set only as specified by the manufacturer. Chapter 19 covers alignment.

In the case of Figure 12-4, 5 coils are changed every time the channel is changed. The wafer switch contacts get a lot of wear and in time will cease to make contact. Very careful bending can sometimes bring a channel back to life. Dirt also will prevent good contact, and special sprays can be used to clean contact points. Spray, rock the channel knob back and forth several times, and spray again. We have mentioned that the bottom part of L4 had a slug or core to tune. Usually there is one for each channel to put the oscillator on frequency for that particular channel. Normally you pull the tuner knobs (channel selector and fine tuning) off the front of the set. Inside, you will see a hole for each channel, each having a slot for a screwdriver. Use an insulated screwdriver, set the fine tuning for midrange, and set each channel for best picture with this screw setting. This moves the core slugs in and out of the oscillator coils. Tune in Channel 4 to set the Channel 4 slug, and so on.

Next to bad contact and poor adjustment of coils, weak and burned out tubes are the biggest trouble in tuner circuits. With solid state sets lightning damage can ruin transistors in tuners. In general with transistor TV sets, however, suspect capacitors first, then diodes and transistors, finally coils and resistors.

Here are some things to remember:

1. A TV tuner preselects the desired channel (RF amplifier), produces for heterodyne a local oscillator frequency (oscillator), and combines the channel and oscillator frequency and amplifies (mixer). The output goes to the IF amplifiers.

2. Wafer switches often change channels by combining or changing coil connections to oscillator, mixer, and RF amplifier.

3. Varactor tuners use a diode that changes its internal capacitance with voltage variations, in conjunction with a coil to tune channels.

4. If a TV gives trouble on one channel, the wafer contacts are probably dirty, causing poor connections.

5. Automatic gain control (AGC) is a varying voltage that biases the tube or transistor and controls the size of the signal.

6. Automatic Fine Tuning (AFT) is a varying voltage that changes the oscillator frequency slightly and keeps the signal tuned in.

Intermediate Frequency Stage

As explained earlier, a heterodyne system tunes more sharply than does any other channel selecting method. In addition, the intermediate frequency (IF) derived from it is capable of great amplification in the IF stage. The term *heterodyne* has to do with beating one frequency against another, and with television and its wide band requirements, it actually involves beating a whole band of adjacent frequencies against a higher oscillator frequency.

The channel selected and heterodyned in the tuner stage is fed to the IF stage as a band of frequencies almost 6 MHz wide, but on the frequencies of 41.25 to 47.25 MHz. Refer to Figures 12-6 and 12-8 throughout this lesson. The IF bandwidth carrier, with color, sound, video, and sync, is connected to the IF stage input by a tuned, specific length of shielded cable. Poor contact of this cable to ground at either end can seriously affect performance of the receiver. If the cable is broken internally, a specific replacement should be obtained from the manufacturer to assure tuning quality. This is true of modern sets, especially color ones.

C101 couples the IF signals to the first transistor, Q101. Tunable coils, L101A, L101B, L102A, L102B, as well as C102, C103, C104, C105, C106, and C107, tune to the intermediate frequency. Since the IF is 6 MHz wide they only tune to a portion of this bandwidth. Note that L102A is to be tuned to 47.25 MHz, or the upper end of the band. L103, L104, L105, L109, and L110 are all tunable. Small soft iron slugs that make the core of these coils are set with plastic screwdrivers by the repairman when needed, so that at different places along the way different frequencies between 41.25 and 47.25 MHz are tuned in. This accomplishes an overall tuning of the whole band. L105 specifies 41.25 MHz.

In Chapter 19 you will find directions for tuning these IF coil slugs. If such a coil needs replacement or if someone has tried to "fix" the set with a screwdriver, alignment is needed. A symptom in the set might be *ring*— several images or outlines of parts of the picture appearing close together.

Q101, the first IF amplifier, is protectively biased by R103. Protective bias works thus. The more current that goes through the transistor the more current has to go through R103. This develops a larger voltage across R103 as there is less voltage across the transistor, and series voltages must add up to supply voltage. The increased voltage makes more positive voltage at the emitter, and with an NPN transistor the more positive the emitter, the less current will flow through the transistor. This actually biases the transistor to work less. We need a forward bias to allow a decent current to flow in the emitter-to-base circuit and to allow an amplified current to go from emitter to collector. This bias is supplied by the AGC. It travels up from below the schematic through R102 to the base of Q101. The more positive the AGC voltage is, the more this transistor will conduct. Proper AGC keeps Q101 operating at just enough amplification to supply strong color, video, sync, and sound signals, without overdoing it and producing distortion or other problems.

Good level of signal is very important. Capacitor C112 is an emitter bypass capacitor. It keeps signal from developing on R103. True, the signal does not enter there, but as the collector sends out pulses of current in proportion to signals, the emitter will pulsate somewhat because they are to some extent a series circuit. Therefore, without C112, signal would develop across the resistor, R103, and bias Q101 incorrectly in pulses. Ring could result. C108 serves the same purpose in the AGC circuit feed to the IF.

The load of Q101 is developed across L103A and swamping resistor R105. A swamping resistor keeps the bandwidth broad. The composite signal is induced on L103B and coupled by C114 to the next amplifier, Q102. Q101 gets its collector voltage through R104 from a 24-volt supply. With this set sound signal is removed at the collector of Q102. In some sets it would travel through the entire IF system. Q102 and Q104 amplify the signals more at the IF, 41.25 to 47.25 MHz. But each separate signal holding its frequency such as video and sync at 45.75 MHz and color at 42.17 MHz, and the sound at 41.25 MHz. The sound leaves after Q102.

The biasing of Q102 and Q104 is a little different. They both have protective bias on the emitters, but the forward bias is not AGC. R116 and R117 combine to form a voltage divider. Current would travel up from ground

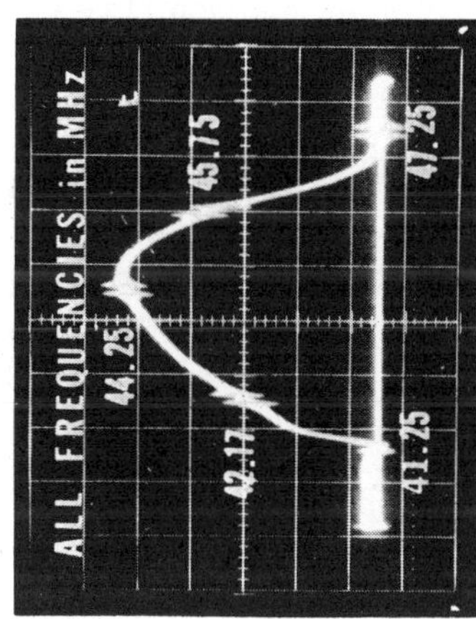

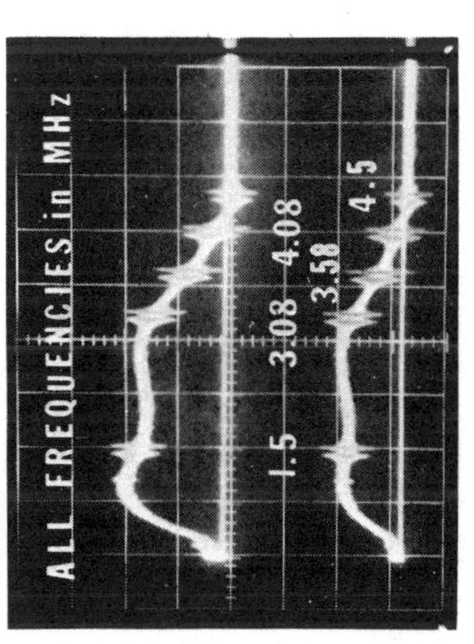

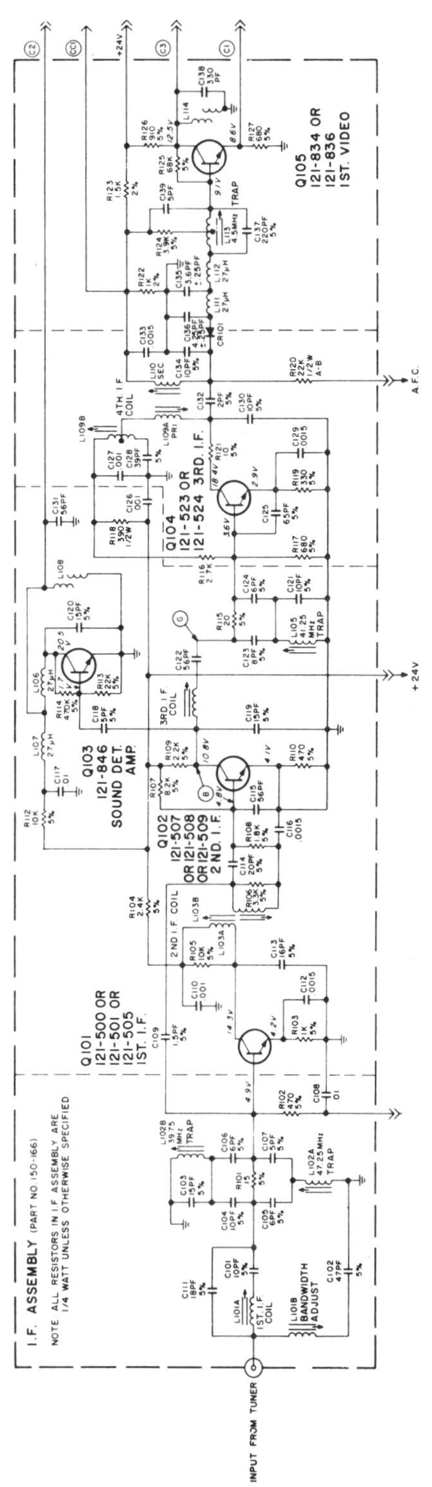

Courtesy Zenith Radio Corporation

FIGURE 12-6
INTERMEDIATE FREQUENCY STAGE AND DETECTOR

through R117 then through R116 to +24 volts. The base of Q104 is tied between them, and since R116 has far more resistance it will develop most of the voltage drop. Therefore only 3.6 volts is supplied to the base of Q104. The advantage of a voltage divider is that if the 24-volt supply should drop, so would the proportionate voltage at the base. If it should increase, the greater increase would be in a voltage drop across R116, so not too great voltage would develop at the base of the transistor. Naturally with a huge voltage like 200 volts applied, the transistor would be damaged anyway.

Some sets have a metal shield around the entire IF stage. Shields are necessary to prevent such oscillations as are given off by the horizontal oscillator from entering the IF, so be sure to replace the shield if there is one when you work on an IF system.

Besides producing ring, weak IF transistors can affect the overall performance of a receiver, and of course a dead transistor stops everything. A coupling capacitor, such as C101, C114, or C122, if open, would end all signals too. With tube receivers, the heaters sometimes inject AC into the composite signal by leakage from heater to cathode, producing wavy lines on the sides of the picture or bending the picture into waves. Poor filtering in the power supply could do the same thing to solid state IF circuits. Unshielded parts with shields left off or grounds not soldered to the chassis might let 60 Hz AC into the IF circuits.

Once in a while sync will be partly clipped in the IF stages while all else will get through properly. If a voltage divider bias system changed value—if heat caused the resistors to develop less resistance so larger voltage was on the base, for instance—the increased current flow might increase the gain of the transistor, and too great a gain could cause the sync pulses to be above the size the next transistor could handle. They would get clipped before they were separated. Too small a forward bias in the IF might not give the sync

pulses enough gain so that they could be separated from the video later on. Usually when there is sync trouble, though, it is in the sync circuits themselves. Poor setting of the AGC or defective AGC might cause too great or too small sync and a tearing picture.

Since IF and tuner stages share the composite or bandwidth channel it is often hard to know whether the IF or the tuner is at fault. As a general rule, if the trouble is on only one channel, suspect the tuner. Otherwise, if the trouble is with video, color, sync, and sound, it could be tuner or IF. This is not intended primarily as a servicing lesson, but it illustrates the importance of knowing what is normally expected of each stage in a TV receiver.

Negative picture can sometimes be an IF problem. What should be white will be black on the screen and vice versa. In this case look for a weak tube in the IF section. Open bypass capacitors to ground or a misadjusted AGC will sometimes cause a negative picture. In cities near the transmitter sometimes the signal is so strong that it will invert and make a negative picture. As you will recall, every time a signal is moved from grid to plate or base to collector, it is turned upside down, or inverted. The manufacturer of the TV has to plan so that from tuner to picture tube the picture will be inverted enough times to be of proper polarity on the picture tube. If the picture is positive at the picture tube it is put in on the cathode of the picture tube. If it is negative at the picture tube it is put in at the grid. With some color sets the video or brightness isn't put to the picture tube at all, but is mixed with the color signals earlier. Either way it is very critical that the picture be of the proper polarity for the set as manufactured.

Streaking in a picture can mean IF trouble, but it usually indicates video circuit trouble. A weak tube, lower than required voltage for plates and screen grids, and faulty capacitors, are some things that cause streaks. Sometimes a tube will become microphonic. Vibration will cause it to act differently. Jarring the speaker might cause a microphonic tube to change its physical space between grid and plate for every jar. Thus the set would amplify the noise of the speaker as jerks in picture and sync. Tapping a tube lightly with an eraser while it is on will often reveal whether it is microphonic. With transistors microphonic troubles usually stem from loose contacts to the leads of the transistor.

A typical tube IF system is shown in Figure 12-7. *c* is schematic, and *b* shows where some of the parts lie. The shaded area is the IF part of the chassis. Note that all tubes are the same type, 6CB6. Sometimes you can change them around and improve reception. If V7 were weak, it might not be able to handle the amplified signal in its spot, but would work as V5. This would indicate a need to replace the weak tube, but temporary changes help reveal the problem. These tubes have very little bias in this schematic. Note that V7 is supposed to measure 0 volts to ground. A 6CB6 can operate without negative grid bias if it doesn't have a large plate voltage.

All three tubes use a small cathode bias resistor. They are R26, R30, and the one marked C33, which should read R33. In addition, the AGC is placed on the grids of V5 and V6. L18, L19, and L20 are tuned for different parts of the IF bandwidth. This way the whole thing gets through. Note that this is an old-fashioned 20-MHz IF system. The screen grids (pin 6 of each tube) have as much voltage as the plates. Why don't the electrons go out there instead? The screen grids are a coil or mesh. Because of the holes, most of the electrons travel on to the solid plate. The screen grid bypass capacitors are very important. They are C35, C39, and C41. Some signal does go out the screen grid, though it is small. If it went to the power supply of 140 volts and interfered with every other part of the TV set, it would prevent the screen grid bypass capacitors from sending these signals to ground and eliminate them. An open screen grid bypass capacitor with tube equipment means more trouble than you would think. If you suspect one of being open, clamp

across it a good one of the same value and watch for an improvement.

If the bypass capacitor in this type circuit were shorted, all 140 volts would short to ground. This would likely cause real trouble in the power supply, and of course you'd have no plate or screen grid voltages. The cathode bypass capacitor, C42, keeps a signal from building up across the cathode resistor. Screen grids help attract electrons from cathode through control or signal grid to plate. They always have a high voltage on them, but not always will it be as high as the plate voltage. They also help prevent internal feedback from plate to grid and self-oscillation, which can occur if the input and output are tuned to the same frequency.

The grid at pin 7 is a supressor grid. It is grounded or sometimes connected within the tube to the cathode. Its negative polarity helps redirect electrons to the plate that hit it with such force that they bounce off. Point X is the heater voltage input. Note that even the heaters have bypass capacitors to ground, such as C36.

The IF signals come in from the tuner and are impressed as voltages on the control grid, pin 1 of V5. As these signals become more positive in nature the tube sends more electrons to the plate. This amplified signal develops across R27 and the primary of L18. Since it is tuned to 25.6 MHz this frequency goes inductively through the secondary to pin 1 of V6. The other IF (down to 21 MHz and up to 27 MHz) gets through with less strength, but it does get through. V6 amplifies the IF even more, but in its load L19, 23.3 MHz is tuned so the lower end of the band is emphasized. The swamping resistor, R32, helps lower the tuning ability of the coil, so low and mid frequencies between 21 and 27 MHz are both strong now. After V7 amplifies, L20 accepts 24.6 MHz part of the IF the strongest. All in all signals of about 23 to 25 MHz would be amplified the strongest and 21, 22, 26, 27 MHz would be amplified somewhat less. This makes for a response curve on an oscilloscope

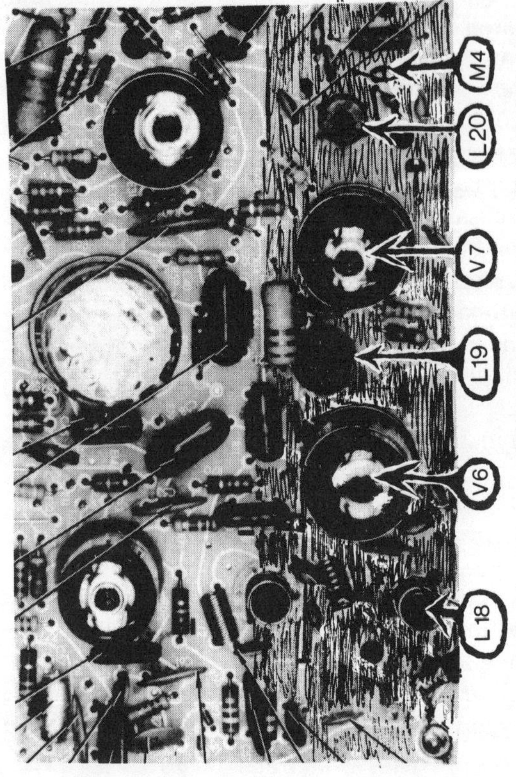

A. Tuner in Horizontally Mounted Chassis, Shield Removed

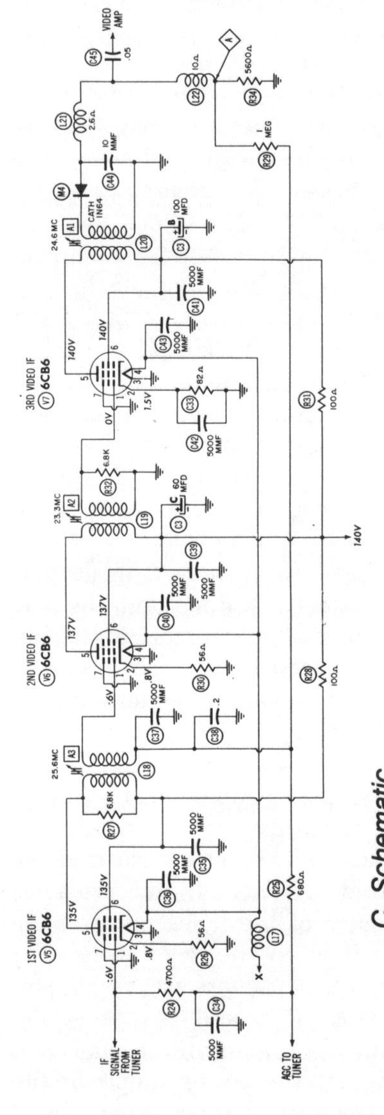

B. Typical Intermediate Frequency and Detector Parts Arrangement in a TV Receiver

C. Schematic

FIGURE 12-7
TUBE TYPE IF CIRCUITS

that resembles a mountain to some extent. Such waveforms appear in Chapter 19.

Remember the following about intermediate frequency circuits in TV:

1. The intermediate frequency stage of a color TV has a channel 6 MHz wide with color, sync and video, and sound frequencies within that bandwidth.

2. For older TVs the frequency range would be from 21.25 to 27.25 MHz and for newer sets from 41.25 to 47.25 MHz.

3. Stagger tuning of the IF coils or transformers (having one tuned to low, one to medium, and one to high frequencies) passes the entire bandwidth. Swamping resistors also increase bandwidth because they resist all frequencies.

4. The AGC biases one or more of the IF tubes or transistors as well as the RF amplifier in the tuner to control gain of the stages.

5. Too much IF gain can produce tearing pictures, negative pictures, severe contrast. Check the AGC setting.

6. Screen and cathode bypass capacitors with tube IF stages can cause all kinds of distortion if open.

7. Microphonic tubes or transistor connections can amplify speaker vibrations and produce distortion in the picture.

8. Shields around tubes or the entire IF are important and should be replaced after servicing.

Detectors, AGC, and AFT Development

The detector must cut the IF signal to a one-sided signal. Though radio transmits full positive and negative sides of the radio waves, TV transmissions are comprised of one side and a piece of the other side. Such transmission is called *vestigial sideband transmission*. Still, the partial signal must be detected. The sound can be taken off before the detector as it was in Figure 12-6. The sound is FM, remember, and it is the shifting frequency, not the amplitude of the signal, that carries the sound. Therefore it matters little whether it is a two-sided or one-sided signal. Everything but the middle will be clipped off anyway, and the frequency that is slightly off frequency will produce sound.

Video (the lightness or darkness on the screen), sync, and color signals must be detected. A diode such as M4 in Figure 12-7 or CR101 in Figure 12-6 is usually used. (In very early black-and-white receivers a tube such as a 6H6 diode was used.) You can check the diode with the power off and a high impedance ohmmeter. Measure both ways, and one resistance measurement should be high and the other very low. If you replace a detector diode, be sure you get the cathode in the proper way. L113 in Figure 12-6 is adjusted to get rid of any sound that might have traveled that far. If it gets into the picture, the picture will blink in sync with the sound. This 4.5 MHz trap sometimes needs adjustment to keep the sound out of the video and color.

The buildup of the AGC voltage for the set in Figure 12-6 is a circuit board all its own. Some AGC systems take a sample of detected signal, combine it with horizontal sync and call it "keyed AGC," and sometimes even delay it so the shift from weak to strong picture will be more gradual and supposedly more pleasing to the eye. Others just feed it back without amplification to the first IF and RF. Such a simple case is Figure 12-7. Part of the detected signal goes down through L22 and drops a voltage across R34 as it goes to ground. This voltage going through R34 is negative at top. The point where a current enters a resistor is where the electrons pile up, and that will be its negative end. This negative voltage is coupled through R29 to pin 1 of V6 via secondary of L18. Yes, R29 would use up some of the voltage. Only minus 0.6 volt bias is needed unless a very strong signal arrives. Also the AGC voltage is applied to V5 through R25 and R24.

The reasons for the extra resistors are: (1)

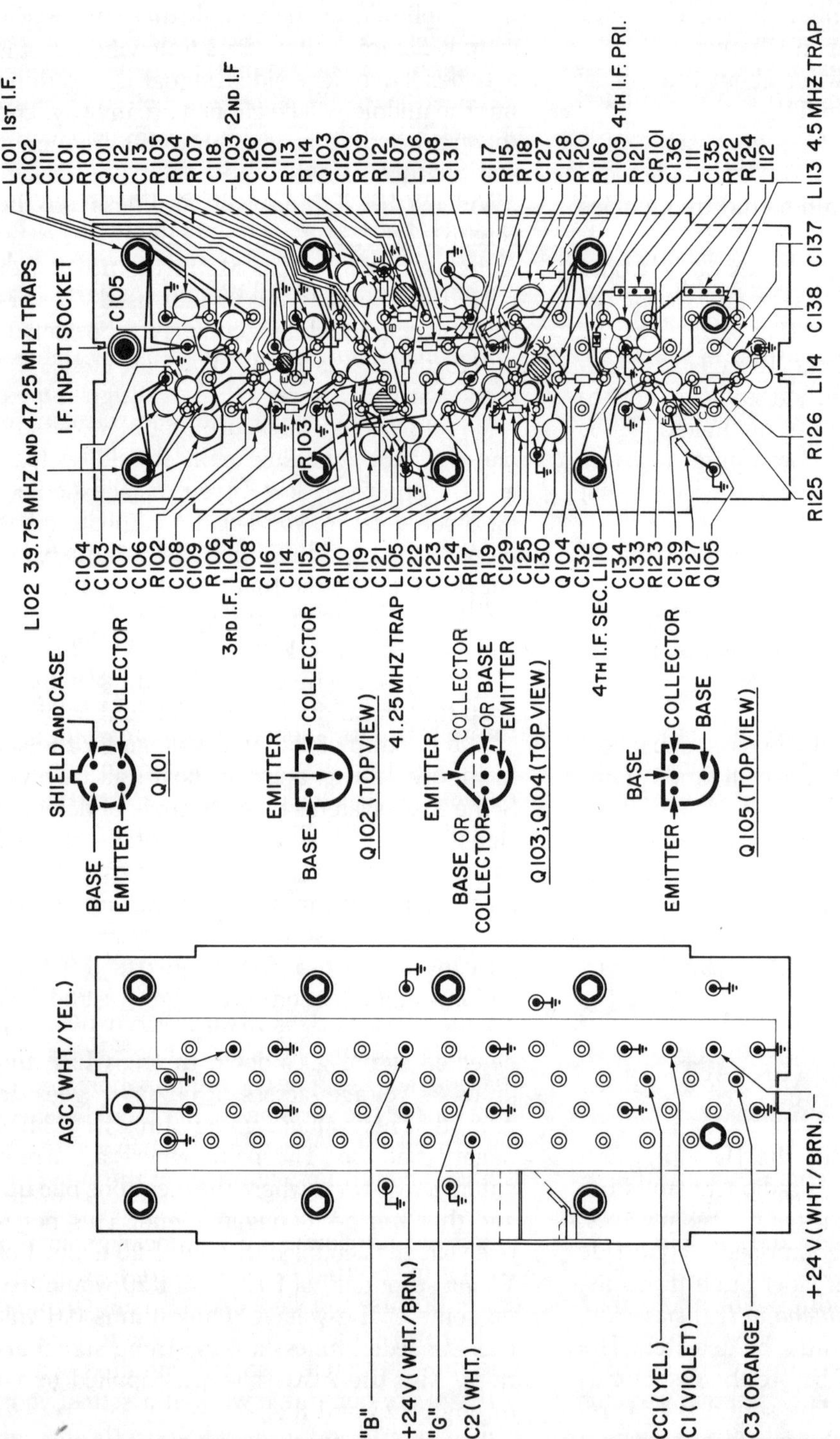

Courtesy Zenith Radio Corporation

Parts Location Diagram, IF Module 150-166.

FIGURE 12-8
TOP VIEW OF TRANSISTORS

without them the grids would be tied together with only a coil separating them, and oscillation could occur; signals might bypass from first grid to second and skip being amplified by V5; and (2) the second IF tube handles a larger signal than the first; after all, the first has already made the signal stronger and it does not need as much bias.

As the signal gets stronger, a stronger detected signal would build up across R34. This puts more negative bias on the tubes and cuts down their amplification, bringing the signal down to proper size. This is mainly a voltage that is supplied to the grids, though a small current must travel to get it there, but since the grids don't conduct by themselves, it is an end-of-the-street situation. Since very little current is fed back, the 1-megohm resistor, R29, doesn't use up a lot of voltage, nor does it change the polarity to positive. The larger current is from ground through M4, through L21 down through L22, through R34 to ground. Where did the energy come from? It came across L20 by magnetic force or inductance. The pulses that form across L22 and R34 are fed as signals across C45 to the video section of the TV.

The automatic fine tuning, AFT, called *automatic frequency control* (AFC) in other sets, is developed in Figure 12-6 across the secondary of L110. The voltages across a coil will vary depending on whether the signal fed to it is on frequency of the coil or not. L110 has to be set properly for this princple to work to advantage. After the AFC leaves the IF module, it would go to a transistor amplifier, then to a discriminator.

Look at Figure 14-1. This is not an AFT or AFC circuit, but a part of it will illustrate the idea. Note the two diodes and coil circuit. If the signal in the coil shifts from center frequency the diodes conduct. The more they conduct the more current travels from the center tap of the coil. If this were an AFT circuit this current would be fed to a varactor in the tuner that would shift its capacitance

and change the frequency of the tuned circuit slightly. So any time the carrier drifts at the TV station, a sample of IF signal as applied between parts of a tapped coil will cause diodes to conduct and produce a corrective current that will affect the varactor in the tuner to get the tuner to go to the frequency change. For this to work, of course, the drifts must be minor and the station must hold within reasonable bounds to its assigned frequency band. This system just saves you occasionally getting up and adjusting the fine tuning control as you watch a program. Many systems have a defeat switch so you can set the fine tuning by hand if you desire.

For laboratory work with this chapter, first do the previously stated antenna experiments. Next use a sweep generator, or your home-made signal generator as a signal injector. With the sweep generator you will have to set some audible modulation on, within it; the home-made one will produce its own signal, with harmonics enough to work channels 2 through 6. Disconnect the TV antenna lead in. Try the signal generator at the antenna terminal. You should get bars across the TV screen, as noted in Chapter 7. Use a layout and schematic of the TV and locate the IF input transistor or tube. Inject a signal at the grid or base. With a sweep generator you will need it to sweep the 40 MHz or 20 MHz range depending upon the age of the set. Older sets are 20 MHz IF type. With the home-made generator again you depend upon harmonic frequencies, but tune the regeneration control for a tone in the earphone. This should make a pattern on the TV screen. Such work is not refined enough for alignment of the set. This will be explained in Chapter 19. For now we merely want to learn how to inject signals into the set to see that the stages are in working order. Be sure to follow directions in Chapter 7. The capacitor before the probe to base or grid must be rated at several hundred volts.

If you are doing this work at a school your instructor will likely demonstrate the use of an oscilloscope in waveform analysis in IF stages and the detector.

At this point it would be good practice to align IF stages, but since TV IF demands precise alignment and should be attempted only after you have studied Chapter 19, I would suggest that you practice IF alignment on a pocket radio. Connect your lab equipment as a signal generator, Chapter 7. Place the home-made signal generator close to the pocket radio. Normally no wire or probe connection is necessary. The back must be removed from the pocket radio. Turn both on and tune the pocket radio for a spot on the dial without station reception. Tune the signal generator until a whistle appears in the speaker of the pocket radio. You may need to adjust the regeneration control, R5, to produce a tone.

You must now locate the IF transformers. Usually there are three in such a radio. There are usually two IF transistors, and there will be a transformer before, between, and after these transistors. Normally the transformer coils are within a small rectangular box of aluminum. There will be holes in the top of each, and there will be a core of powdered iron or ferrite material that has outside threads so it may be rotated upward or downward. You then use an insulated screwdriver to move these slugs in or out of the coils of wire. The more they align directly inside the respective coil, the lower the frequency it will resonate or tune to. Listen to the tone in the speaker and adjust for maximum volume.

This practice will aid you in adjusting TV IF stages in the future. TV IF alignment is much more complex. You need to sweep the IF stages with a full 6 MHz of signal, and you need to inject markers at specific frequencies so that each IF coil can be peaked at a certain specified frequency. The practice of tuning one coil to about 42 MHz and another to 44 MHz, and another to 46 MHz, so that all 6 MHz of signal get through is called "stagger tuning." Peaking a coil means tuning it so the output is maximum as will be seen on an oscilloscope.

QUIZ FOR CHAPTER 12

▶ 1. What is ring?

▶ 2. Describe negative picture.

▶ 3. What is a microphonic tube and what could it cause?

▶ 4. What is meant by stagger tuning? What section of a TV would have it? Why is it necessary?

▶ 5. What is the purpose of a swamping resistor?

▶ 6. What part of a TV develops AGC? What parts are controlled to a degree by it?

▶ 7. What part of a TV is controlled to some extent by AFT?

▶ 8. What kind of tuning might have no moving parts in the tuner itself?

▶ 9. Why are wafer tuners preferred to turret tuners?

▶ 10. In a grounded base transistor circuit the signal would be put in on the ______ .

▶ 11. In a grounded grid tube circuit the signal is put in at the ______ .

▶ 12. What item associated with a cathode bias resistor would prevent the signal from developing across the bias resistor?

▶ 13. What is the purpose of a suppressor grid in a tube?

▶ 14. Name two reasons for having a screen grid in a tube.

▶ 15. A folded dipole TV antenna would show ______ ohms between leads if the system were in good order.

▶ 16. Elements preceding the driven element of an antenna are called ______ and are (shorter, longer) than the driven element.

▶ 17. What antenna element prevents stations from opposite direction from interfering with the station desired?

▶ 18. ______-ohm twin lead antenna lead-in works with a folded dipole without transformers.

▶ 19. The bandwidth of modern IF stages is ______ to ______ MHz.

13. Sync and High-Voltage Circuits

PROJECT: Becoming Familiar with Voltages in Sync Circuits; Adjusting Hold, Height, Drive, and Linearity Controls

To BE certain that you understand how the sync circuits cause a picture to be traced on the screen of a picture tube, it would be helpful to review the last part of Chapter 9. To review how the beams form color pictures, re-read the last part of Chapter 10. You now know how the composite signal is tuned, amplified, and detected. Sound and color signals are separated by frequencies, as they are on frequencies 4.5 and 3.58 MHz different from the video and sync. Tuned circuits separate these signals, and the sound goes to its circuit, is detected by a discriminator or ratio detector (you haven't studied them yet), is amplified more, and goes to the speaker. The color goes to its circuits and later combines with the video to make a decent picture. This combining can occur in the picture tube or ahead of it. The color signals select the color and the amount of it, and the video determines the brightness. The video and sync are on the exact same frequency, so the sync cannot be separated from the video in a tuned circuit.

The sync moves the picture-tracing beams of red, green, and blue across and down the screen to trace out a complete picture. These sync signals do not return to the picture tube, but are connected to a yoke containing the vertical and horizontal coils. This deflection yoke fits on the outside neck of the picture tube. Magnetic pulses travel through the glass and bend the color beams to go to the right place when they strike the inside of the screen of the picture tube.

Here is what must happen to the sync signals before they can bend the beams and trace out the pictures: (1) the sync signals must be separated from the video; (2) the horizontal and vertical sync signals must be separated from each other; (3) the signals must be reshaped so they will move the beams either across the screen (horizontal sync signals) or down the screen (vertical sync signals); and (4) the signals must be amplified greatly, as it takes a healthy magnetic charge to bend these beams that represent the colors as they speed toward the screen.

During most of this chapter, study Figures 13-2 and 13-3, which show schematically the sync stages of a Quasar color TV. Note the oscilloscope waveforms at various places. Also refer to Figure 13-1 to see how the parts are actually laid out on the panel in the real TV. Though this diagram shows symbols, it also shows actual layout. The shaded areas are the copper-clad material that makes contact electrically between parts. In troubleshooting a circuit board like this, sometimes it helps to take it out and hold it up to a light so you can see where the connections really go. Printed circuits are very common today, and probably will be for years. They are fragile, however, and connections crack all too frequently.

Figures 9-1 and 9-2 show the sync pulses as the tallest part of the signal. With some sync separators this is the way the signal is separated. Remember that every time a signal goes through a tube or transistor the signal is inverted or turned upside down. In the case of the set we are studying the sync enters the sync separator with the video up and the sync

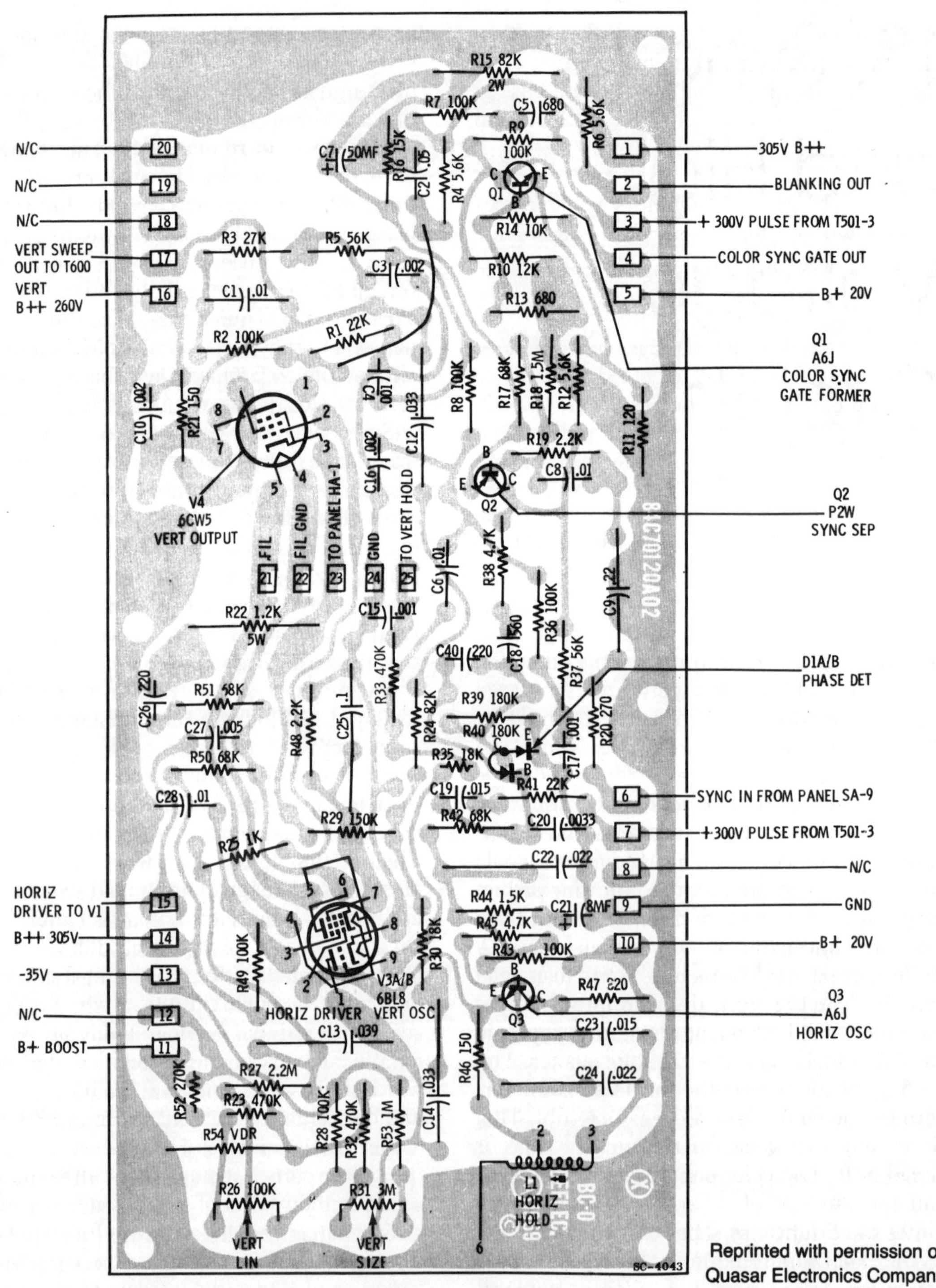

Reprinted with permission of
Quasar Electronics Company

FIGURE 13-1
SYNC CIRCUIT BOARD LAYOUT

as the lowest part of the signal. See W20 in the upper right-hand corner of Figure 13-2. This is the way the signal looks at the base of Q2. This sync signal came from the second video amplifier (shown in Figure 15-2). A sync and AGC take-off transistor, Q18, takes these signals. Above the emitter resistor goes a line down and to the left. This is the same line that enters at the left of Figure 13-2. The signal also went to the AGC circuit, but we will not concern ourselves with it now.

Note that at the base of Q2 is plenty of video and the sync. The aim is to get rid of the video and keep only the sync pulses. W21, the collector of Q2, shows only the sync pulses, so it must have eliminated the video. In case you are worried that all the video has been used up and none will be left for brightness on the screen of the picture tube look back at Figure 15-2 again. The second video amplifier, Q3, has a collector output to the sync take-off transistor. This we have traced to the sync separator. Note also though that there is another signal take-off point at the emitter of Q3. This goes to the contrast control and on to the demodulators. So the video is not lost altogether; only the part of it that goes to the sync circuits is eliminated. Would this video hurt anything if it continued? You bet it would! Strong video in the sync circuits would trigger the sync out of time, and the picture would tear, flip, and go crazy. Any time you turn the contrast or picture control of the TV up to normal and the picture starts tearing or flipping, you are putting video into the sync circuits. Usually the sync separator is at fault. You should be able to turn the contrast up sufficiently without the picture starting to flip. It could be that the sync separator transistor or tube is defective, but often the bias is wrong. Bias is very important in this sync separator.

Let's see how the separator got rid of the video that came to it. Note that Q2 is a PNP transistor. With a PNP transistor the more negative the signal, the more current will flow in the base-to-emitter circuit and open the

barrier more for the collector-to-emitter circuit. What is the most negative part of the incoming signal? The sync pulses. The bias of this transistor does not go to B− through some resistor or to the collector as it did with the radios you made. Those radios were made to amplify all the signal. This transistor is biased so that base and emitter are at the same voltage. With signal this is 36 volts; in fact, the base is even a half volt more positive. This keeps Q2 from amplifying. Only the very negative sync pulses are strong enough to change the 36.5 volts to less than 36 so a current will go from base to emitter.

Since only the very negative ends of the sync pulses trigger base current, they are all that allow the much larger collector-to-emitter current to flow. Nothing happens on video signals. This circuit need not have the entire sync pulse negative enough to make it work. Just so the most negative tips of sync pulses do it. Note that they are rectangular, so any part of them will properly synchronize the circuits. Because the transistor amplifies these sync pulses, they are up to 32 volts in height at the collector, as shown in W21. Though we see only two horizontal pulses, 262½ of them occur followed by a vertical pulse, and so on. The gate pulse former takes a sample of the horizontal sync to use in the color circuits to turn the burst signal on and off at exactly the right time. It does not have color information on it, nor does it do anything in the vertical or horizontal circuits. We will study it in connection with the color circuits.

The sync signal as separated is fed by a line at W21 down, to the right, and up for the vertical circuits through R37. It also goes to the horizontal circuits, entering at R38. Since the same signal goes to both vertical and horizontal circuits, some means of separating them at their point of entry is needed.

So far you should know the following:

1. Sync signals are used to trace the picture across and down the screen of the picture tube.

2. Before sync signals can be used they

must be separated from the video and from each other. They must be reshaped and amplified.

3. Separation from video takes place in a sync separator stage, which is biased so only the sync signals trigger amplification.

VERTICAL SYNC CIRCUITS

In Figure 13-2, R37, R36, C17, C16, and C42 form what is known as an *integrator circuit*. The shorter horizontal sync pulses are lost across the resistors and capacitors. The capacitors will charge and discharge with these short pulses, but will not send them to grid pin 9 of the vertical oscillator tube. Some sets use transistors throughout, while others use tubes or a combination of both. If transistors are used in the vertical and horizontal output stages, they have to be power transistors that can handle over 100 volts.

Horizontal sync pulses are so short in duration that they don't get through the integrator circuit. The longer vertical sync pulses are capable of charging all capacitors and still having signal for the grid of V3B or pin 9. The vertical pulses are now something like W24. Note that V4 is removed for this scope picture. Removing the output tube keeps the oscillator from oscillating, so we see only the pulses that have left the sync separator and passed the integrator. The vertical pulses are removed by the integrator. Note that the vertical pulses no longer have the equalizing serrations on them as shown in Chapter 9.

As mentioned earlier, the sync pulses must be reshaped before they will move the beams. A straight up-and-down pulse or even a rectangular one would not move a beam across the picture tube or up and down it. Such a pulse would hold the beam at one place on the screen. The pulses must change voltages very gradually from low to high or from high to low. The waveform must look like a slide. A very gradually increasing or decreasing voltage can produce magnetism in the yoke that will smoothly move the beams down the pic-

ture tube screen as the lines are being traced, and a quick change in the waveform will quickly move the beams to the top to trace out the next picture.

W39 is the desired waveform. The quick down movement moves the beam (three beams really) upward, and the upward slide moves the beam downward slowly. The less-than-perfect slide is necessary, as the coils of the deflection yoke store a little magnetic charge and the distorted slide produces perfect movement. This is called a *parabolic waveform*. It is what the oscillator should produce. You might think that the slide should go from up to down because that is the way the lines are moved, but it all depends on which way the coils of the yoke are wired.

Now let's look at the vertical oscillator. Since early in the experimental days of TV, sets have been made so that the vertical and horizontal oscillators work whether an incoming signal is present or not. For one thing, if they operate continuously they will be warmed up and keep on frequency better. For another thing, the horizontal sync makes the high voltage, and if it were not present when the set were on, a snap of the high voltage would result every time you changed channels. Also, it is convenient to have a raster (brightness on the screen) whether pictures are present or not. If a raster appears you know that the high voltage, sync, low voltage, and picture tube are working. Lack of picture indicates trouble with the tuner, IF, video, detector, or antenna. The raster makes for better troubleshooting.

The vertical and horizontal oscillators therefore must run without any signals coming to them. They must also be built so that an incoming signal will lock them in place and keep them in time with the sync coming from the TV station. Otherwise they could be too fast or slow and create a flipping picture or a picture with the middle split and the wrong half at each side. They must lock in with the input sync.

The tube V3B will not oscillate by itself. In W24 V4 was removed to show how the wave-

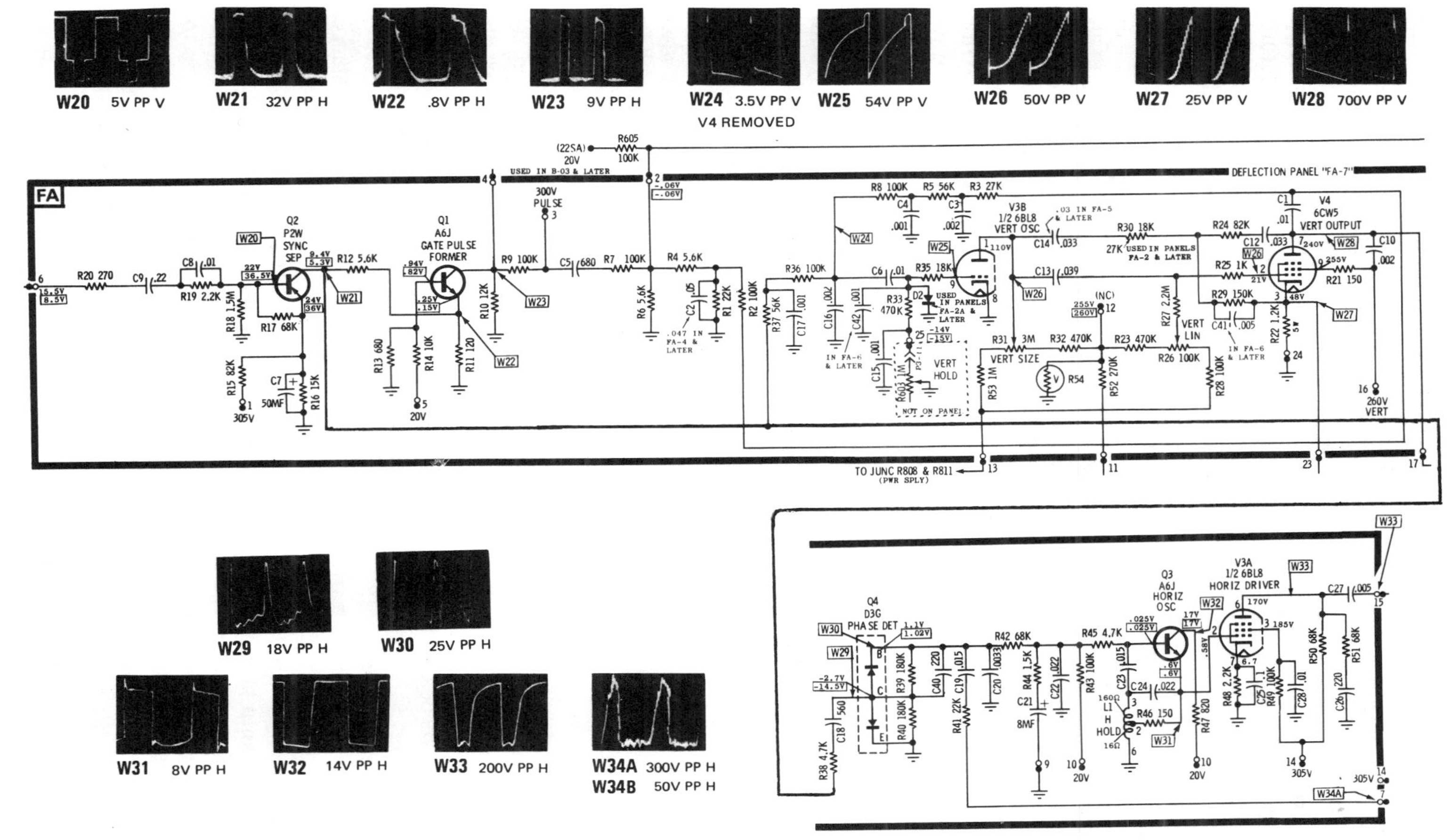

Reprinted with permission of Quasar Electronics Company

FIGURE 13-2
SYNC SEPARATOR, VERTICAL SYNC, HORIZONTAL
PHASE AND OSCILLATOR

form would look without oscillation. Now V4 is restored and oscillation occurs. The signal is amplified by V3B and goes to V4 by coupling capacitor C13. V4 amplifies the vertical sync and sends it back to V3B by the feedback path of C1, R3, R5, R8, and C6 again. This feedback produces the oscillations and really uses both tubes to get it. This can work with or without an input, but with input at grid of V3B, it opens or closes the ability of the oscillator to work, so it works in sync with the input sync. The oscillator must operate without input to trace a raster when no signal is present. W25 shows an unusual wave. The way the signal is fed back produces this curvy slide. No problem. By W28 at the plate of V4 the wave shows a nice smooth slide between spikes.

Learn to recognize the shape at W25, oscillator waveform. This is normal. The setting of R26 smooths the waveform out, and this linearity control is an important adjustment. In older sets, and in some modern ones, the linearity control will be at the back of the receiver, but in this case it is on the panel. Only the vertical hold control, with the dashes around it, is not on the panel. Customers have to adjust the vertical hold from time to time, so this control must be on the outside of the set. The linearity control determines whether objects or people on the screen will be in proper size relationship. Bad setting might produce a man with a huge head and a little body or make an auto all wheels. Its setting affects the bias to V4.

The size of the picture (in this case vertical size) is controlled on the panel here. In many TV sets this is a back-of-set control. Some sets term it "height." It is adjusted so the picture just fills the screen, and it controls plate voltage of the oscillator. Obviously, the more plate voltage, the more signal is developed, and the higher and lower the picture would be traced. For the picture to lock in with the incoming sync from the TV station the oscillations must be correct. If the bias of the oscillator is not right, this is impossible. The vertical hold control, R603, is set to keep a steady picture. If this control or R33 becomes old, or if C15 or the tube develops defects, it could become impossible to stop the vertical flipping. Replacing the proper component would restore a steady picture. Of course, too much or little sync signal input can cause the same symptom.

The output waveform, W28, has built up to 700 volts on the spikes that will move the beam from the bottom to the top of the screen. This large signal is put on the primary of the transformer, T600, Figure 13-3. It goes to 260 volts as the positive supply for the plate of V4. To the right are some windings that connect to the convergence panel, which takes some of the signal and by setting different potentiometers makes the different color beams travel at exactly the same rate down the picture tube. Otherwise the red might go faster than the blue, for example. Pins 3, 4, 5, 8, and 9 are connections to the convergence coils that fit just ahead of the yoke on the neck of the picture tube. Thus the yoke affects the movement and the convergence coils modify it to keep all three colors going at the same rate.

Below the primary of T600 is the output coil to the deflection yoke. The dashed lines are the yoke itself, which sits on the neck of the picture tube. Note that the horizontal coils are there also. Pictures in Chapter 10 show the arrangement of convergence coils and the deflection yoke. T500, the pincushion reactor, is adjusted so lines of picture near the top and bottom will be straight across, not bellied out.

To review:

1. Vertical sync stages separate the vertical sync by a group of capacitors and resistors called an *integrator*.

2. Vertical and horizontal sync oscillators run with or without an input signal. When there is a signal, the input sync locks the oscillator to its time.

3. Vertical sync needs a waveform with spikes to return the picture to the top and a falling or rising part in between to carry the picture down the screen.

4. The vertical hold is a customer control, usually a bias control of the oscillator.

5. Height of picture is controlled by affecting vertical sync signal size such as a plate voltage adjustment.

6. Linearity adjustment proportions the picture properly with the same number of trace lines in every inch of picture. Linearity is often accomplished by a bias to the output tube or transistor.

7. In color sets, not only does the vertical output transformer supply power to the deflection yoke, but a pincushion adjust keeps the sides and bottom of the picture from flaring. In addition, the convergence system uses some of the vertical (as well as horizontal) output to move all three color beams together.

HORIZONTAL SYNC CIRCUITS

Horizontal sync input accepts both vertical and horizontal sync signals as they are equalizing pulses during the vertical sync pulses to keep the horizontal oscillator running, thus the longer vertical sync pulses do no harm. Without equalizing pulses during vertical tracetime the horizontal sync might drift off frequency if cut off during vertical sync time. In addition, the high voltage would drop or quit during the vertical pulses, since it is made from the horizontal output. A blink would occur every time the picture trace was moved back to the top. So the horizontal sync is continuously running, normally on its own pulses and on the tiny equalizing pulses during vertical time. The equalizing pulses are too short to affect vertical sync.

Horizontal sync must not only be in frequency with the incoming signal as vertical sync is, but it must start in exact phase with the TV station for each line traced. If it does not, the picture could start in the middle of the screen instead of at the left side. The circuit that regulates this phasing is called *AFC* or *automatic frequency control*. That is the same circuit term that refers to keeping the general composite signal in the tuner on frequency. This appeared in Chapter 12. Try to keep the two separate in your mind; TV terms are confusing. The automatic frequency control of the horizontal oscillator really is more of a phase control. Here is how it works.

The input horizontal sync, which comes from the TV station, is applied at a set of two diodes in the phase detector. At R41 and C19 (Figure 13-2, lower part) a sample of the output waveform, taken from the output of the horizontal system at pin 3 of T501 (Figure 13-3), is added to the input wave at R41. If these two signals are exactly on top of each other, the diodes, being opposite in connection, offer no current to the circuit. If they are apart one diode will conduct more than the other and will change the bias on Q3, the horizontal oscillator. Changing bias affects the starting and stopping time of each pulse and thus gets the oscillator exactly in phase with the incoming signal. If the signal of the oscillator is late, one diode biases, and if it is early, the other. With color sets this phase of the horizontal sync is doubly important. The burst signal in the color circuits is controlled by the horizontal output system.

The feedback path of the horizontal oscillator, Q3, is a little different. First, we are dealing with a rather low frequency, 15,750 Hz. For that reason the tuning coil must have many, many turns. Usually these are of very fine wire so the coil will not have to be very large. L1, the horizontal hold, is the tuning and oscillator feedback coil. Don't forget to keep looking up the parts on Figure 13-1 also. L1 is at the lower right. On many sets horizontal hold is a back-of-set control for the customer to adjust.

Between base and ground is the whole coil that tunes in the horizontal frequency. Since this is an NPN transistor, current goes up from ground through the lower half of L1 and through R46 to get to the emitter, then on up to the collector when the base allows it. Of course, current must first travel through this path and out the base, down R45 and R43 to +20 volts. The phase detector also affects this bias, as you have seen. The current pulses through the lower half of L1 feed back energy to the upper half of L1, and this starts the oscillations. The emitter feedback to base can

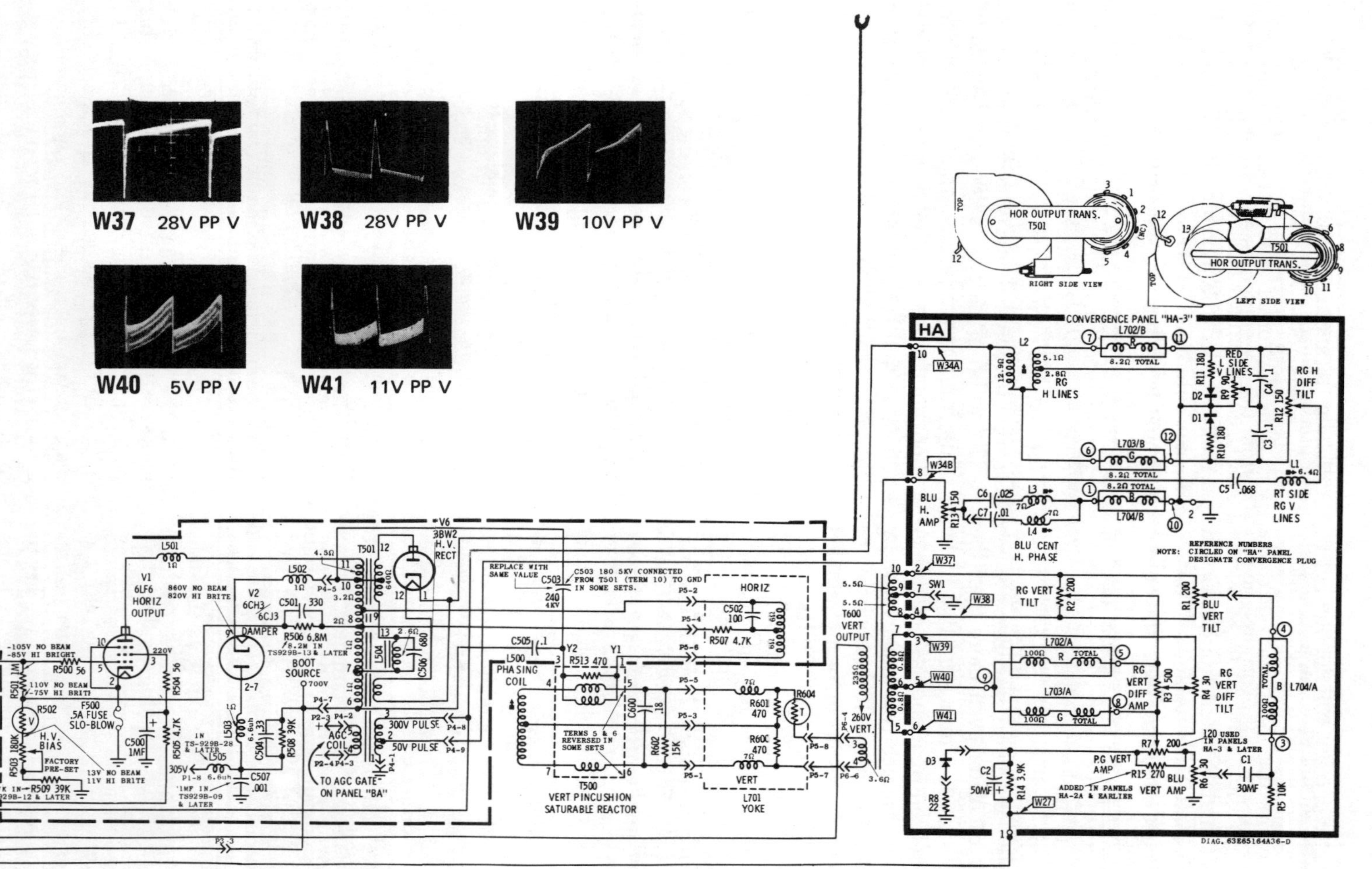

FIGURE 13-3
HORIZONTAL OUTPUT, HIGH VOLTAGE,
DEFLECTION, CONVERGENCE CIRCUITS

work as it does here, but it will not build up as large oscillations as would a collector-to-base feedback. An extra tube, V3A, the horizontal driver, builds up the signal to make up for this. Note that signals taken from diodes at W29 and W30 show some waviness. The signal is trying to get on phase. W31, the output of the oscillator, has a good signal, but it is only 8 volts. The input was 18 volts, so voltage has been lost to make oscillations with an emitter-base type oscillator.

Adjusting the slug of L1 changes the frequency of the oscillator even more than the phase detector can. It is important that L1 be set just as much on frequency as possible. A bad setting can produce two images side by side, four images, or more. If the adjustment is just a little off, the result can be a scrambled picture or one that doesn't start at the left side. W32 shows an amplified waveform, but it doesn't do anything, as the output is taken at W31. The horizontal driver tube greatly enlarges the horizontal pulses. Note that at W33 they are 200 volts in size. The driver tube is just a good amplifier with appropriate cathode bias, screen voltage and bypass, and so on.

This large signal is now put to the horizontal output tube through C27 and R500 (Figure 13-3). From here on we cannot look at waveforms, because the signal is too strong for a scope. In fact, measuring the DC voltage at the plate cap of V1, the horizontal output tube, can only be done with a special high-voltage probe; otherwise you would get shocks and damage your voltmeter. Locate the horizontal output tube in the lower right of the set shown in Figures 10-4 and 10-5. Most output tubes have a cap for the plate and a clip and lead that go to the horizontal output transformer. A few types, such as portables with compactron tubes, don't have the plate cap. The horizontal output transformer is also called a *flyback transformer* or a *high-voltage transformer*. A drawing of one is shown above the convergence schematic in Figure 13-3. Schematically, T501 is it. These transformers serve several purposes. They have many turns of very fine wire and change an already high voltage to about 17,000 volts for black-and-white sets and over 20,000 volts for color.

The horizontal sync pulses actually work the entire length of T501, but the part that really works the horizontal defection and moves the lines of picture across the screen would be at tap No. 9 over to the top of the yoke. The midpoint of the yoke goes back to tap No. 8 on T501. The bottom of the horizontal deflection coil system has two paths. Both are coupled by capacitors after going through part of the pincushion correction mechanism. The width of picture is controlled by L504. Adjusting its slug affects the length of each horizontal pulse. This coil is on top of the horizontal output transformer in the upper right-hand drawing above the convergence panel.

Note that the same point that gave back a reference wave to compare with the input also sends energy to the top of the convergence panel. Not only must the red, green, and blue guns keep together going down the screen, but they must trace across on the same line at the same instant. The convergence sends currents through pins 1, 6, 7, 10, 11, and 12 to the convergence coils on the picture tube neck, to keep the beams together. We will learn how to set the convergence in Chapter 18.

Things to remember about horizontal sync circuits are:

1. The horizontal oscillator operates even during vertical sync time by use of equalizing pulses.

2. The horizontal oscillator must not only be on frequency, but must be in perfect phase with the broadcast sync.

3. The horizontal oscillator coil is often the horizontal hold control also. It is set by a slug adjustment.

4. The width of the picture is often set by a coil in close proximity to the horizontal or flyback transformer.

5. The horizontal output tube usually has a plate cap on top with high voltages.

6. The flyback transformer takes the horizontal sync pulses and sends them to the horizontal deflection coils of the yoke, but it also

makes high voltage for the picture tube, and it sends signals to aid the convergence coils in keeping the three color beams close together.

TROUBLES IN VERTICAL AND HORIZONTAL CIRCUITS

If the picture on a set is folded over and the control setting won't fix it, the trouble is usually a bad output tube in the vertical circuits, although it could be an open cathode bypass capacitor, an open grid resistor, or a resistor of changed value. Another vertical trouble is a thin line across the screen while the rest of the picture tube is dark. Check the tubes, the voltages to plates, and the capacitors between stages. A picture that is much too short probably has a weak tube or low voltages on the plates. In solid state circuits, look for weak transistors, leaking capacitors, and bad resistors.

The coils in the deflection system also can go bad. If both vertical coils go out, only a thin line appears across the screen, probably at the middle. The resistors across the deflection can also burn out. If one coil is broken, a partial picture will form, usually taller on one side than the other. This is called *keystone effect*. Often the entire yoke assembly rather than a coil must be replaced, depending on how the manufacturer puts up parts.

Don't forget the vertical output transformer is a source of trouble. You may have decent voltage on the plate, and since that voltage has to go through the primary of the output transformer, you'd assume all is well. Just a few shorted turns in this transformer could upset the height and linearity greatly. Of course, it is sensible to check and substitute tubes, capacitors, and resistors first. They are cheaper and more likely to be the trouble. A distorted picture with a wave in it usually indicates a heater-to-cathode leak in the tube.

In horizontal systems, a scrambled picture means that the frequency is wrong or that the phase is wrong and the frequency about right. First try setting the horizontal hold control. If

that doesn't do it, check the AFC tube if there is one. If the controls are nervous and get out of adjustment after a while, chances are the set has a resistor or capacitor that heats up and changes value with the heat. When the set is first on everything is fine, then it gets the jitters. Sometimes you can heat different parts with a soldering gun when the set is first on and discover the bad part. The diodes of the phase detector are possible trouble spots with a jittery picture. A microphonic tube can upset a horizontal oscillator stage greatly. If tapping the chassis causes the picture to scramble, there may be a microphonic tube or a loose connection.

A hashed-up picture—the "Christmas tree effect"—means an oscillator way off frequency. If the picture won't fill the screen from right to left, chances are the trouble is in the driver or output. The screen grid of the horizontal output tube usually can be measured. For the plate cap, either you need a high-voltage probe or, in a darkened room, you can hold an insulated screwdriver near the metal and a purplish fire will go from the cap to the screwdriver. This corona discharge indicates the presence of some plate voltage and indicates that high voltage is being made.

A high-voltage probe has a ratio. If it were 100 to 1, you could use the 500-volt DC range to measure up to 50,000 volts with the probe to meter. As with vertical stages, a keystone effect can occur—a picture that is very short at the top of the screen and full at the bottom, looking like a wedge, or normal at the top and forming a point at the bottom, depending on which deflection coil is bad. A picture with lines that are a little wavy but otherwise in place indicates an overactive AFC tube or parts, often caused by a microphonic tube or parts that change value.

HIGH-VOLTAGE CIRCUIT

Every horizontal output sync pulse is composed of a slanted line that will put increasing or decreasing magnetism in the deflection

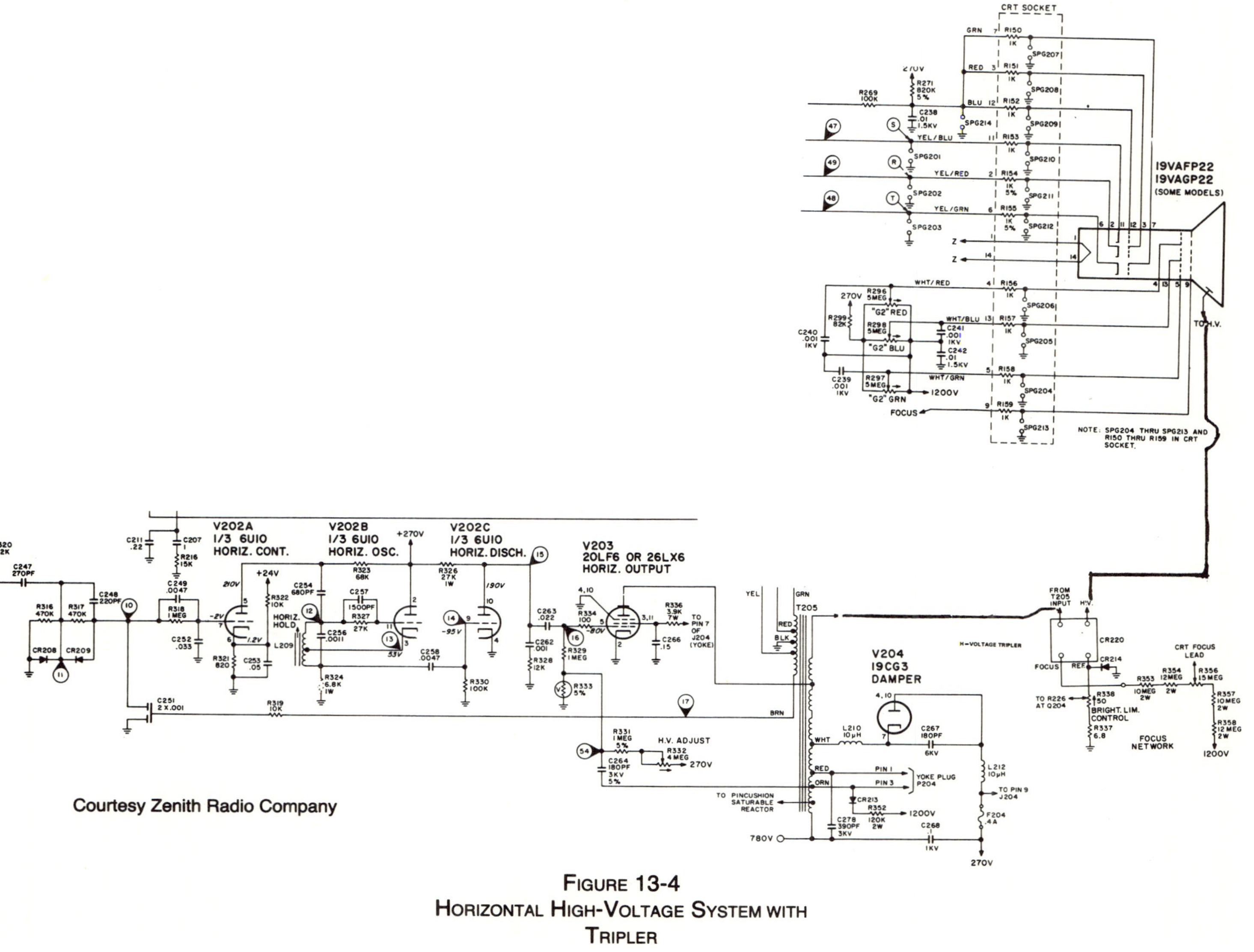

Courtesy Zenith Radio Company

Figure 13-4
Horizontal High-Voltage System with
Tripler

and move the beams across the screen, and a spike that quickly returns the beams to the left side for the next line trace. These quickly occurring spikes trigger tremendous surges of energy through the horizontal output transformer. Since 15,750 of them occur a second, they put a lot of voltage into a secondary. These spikes, and the rest of the pulses travel through the primary of T501. Although they aren't drawn on the diagram, between pins 12 and 13 of this transformer secondary are many, many turns of very thin wire. Far more turns occur in this secondary than from pin 6 to pin 11. Note in the drawings at the upper right above the HA panel that the left-hand member of each transformer is much fatter than the right. This is the secondary with all the thousands of turns of wire. Since there are so many more turns in the secondary from 12 to 13 than in the primary, a voltage step-up from possibly 500 or 1,000 volts at the most through the primary to over 20,000 volts in this secondary is achieved. Since this is AC (DC would not pass a transformer), it must be rectified or converted to DC.

V6 does this. Heater pins 1 and 12 are connected to the horizontal output transformer with a few turns to get energy to heat the cathode. On every positive cycle there is a positive charge on the plate cap of this rectifier tube. During this half cycle of AC, current flows through the picture tube from cathodes in beams of color-making electrons to the picture screen. The electrons bounce off and collect on the sides of the picture tube near the screen. An internal path connects them to the high-voltage anode, where a heavy lead is connected, allowing current to travel down and to the cathode of the rectifier. The current goes as electrons across the rectifier and down the secondary of T501 through other components and to the low-voltage power supply.

On negative charges on the rectifier plate cap, the circuit is broken so the high-voltage anode in the picture tube is always kept very positive. This very high positive voltage is needed to attract the three color beams of electrons, because the distance from the neck to the screen of the picture tube is great and because the aperture mask makes it all the harder for the beams to get through to the screen. High-voltage corona discharge is easy to show at the rectifier plate cap and at the high-voltage anode at the picture tube side. It is extremely easy to get a very unpleasant, but fairly harmless, shock by just getting near these terminals—you don't even have to touch them to get the shock. Normally these shocks are not harmful, as the current is very light, even though the voltage is great.

In many sets the high-voltage system is in a cage. In others the flyback transformer is in a cage and the horizontal output, high-voltage rectifier, and damper tubes are just outside with the plate caps going in. This high voltage has a way of attracting dust. Once enough collects, the very high voltage will travel on the sticky dust and part will arc to ground. It is normal for the flyback transformer, such as T501, to hum at the 15,750-Hz frequency, but if it starts to fry, something is wrong. The arc to ground will be a frying sound, and in a dark room the purple corona will show.

The damper tube, or in the case of completely solid state sets, the diode, rectifies some extra voltage that is used in certain parts of the set. With color picture tubes usually a very high focus voltage is needed to apply to the last set of grids for the three beams before they are converged and deflected. This high-voltage focus grid will be noted on Figures 15-1, 16-2, and for the set whose power supply we are studying, 16-1, the part marked "Boot B++." This voltage is a little high for a standard low-voltage power supply to make, unless the manufacturer spends more money making heavy-duty parts. Instead, the damper produces this high voltage for the focus grids. Though the illustrations show it and other grid connections along the neck of the picture tube, they are all connected at the socket at the rear of the tube. Drawing them properly would produce congestion of wires that would be hard to follow. Only the very high-voltage anode is connected to the side of the picture tube.

Because high voltages are being developed in T501, the flyback or horizontal output transformer, tapping up a distance on this transformer can produce a high enough voltage for our purpose. Every time the polarity is right through the transformer, negative at top and positive at bottom of primary of T501, V2, the damper tube, conducts, cathode above to plate, below. A voltage is developed across R508 with the boot source on the positive end. This is then connected to the boot at pin 9 on the picture tube. You will need to consult Figures 13-3 and 16-1 to see it. If a set doesn't have the focus or boot voltage, the damper tube or rectifier diode should be suspected first.

We have talked about the horizontal pulses developing the great voltages necessary for the picture tube anode. Before these voltages can be developed, a positive voltage must exist at the plate of V1, the horizontal output tube. This positive voltage is supplied by the low-voltage power supply. See Figure 11-3. Upper right shows the B++ 305-volt supply. This point with socket 1, pin 8 has (in Figure 13-3 now) plug 1 with pin 8 connected to it. Through L505, R508, the entire flyback primary, L501, we have backtracked the positive voltage source to make V1 operate. Once it is operating, V1 (with the flyback transformer) makes the very high voltage for the picture tube; with the aid of the damper it makes the rather high voltage for the focus grids of the picture tube; and it also develops the proper horizontal waveforms to pull each line of picture across the screen. The high-voltage bias on the output tube, V1, is its grid bias. This is a critical adjustment, R503, and the manufacturer's instructions for each set should be diligently followed.

As mentioned, high voltage is very likely to

short to ground around the tube sockets and other places. Cleaning the high-voltage tubes and heavy-voltage plate wires, tube cap connectors, and such often stops arc-over and its problems. Sprays are made that will help insulate these terminals so they won't make an easy path for high voltage to ground. Before you handle any plate cap, have the set unplugged. Holding well back on the handle of a plastic-handled screwdriver, short the plate caps to ground. You may get a snap long after the set is off; stored charges are common. When you can no longer hear the snap as you touch to ground, chances are it is safe to touch these parts.

The high anode lead to the picture tube is especially likely to give shocks. Some picture tubes store huge charges after the power is removed. You need a clip, wire, and insulated screwdriver. Connect the clip to a good ground and secure the other end of the wire to the metal of the screwdriver. Hold the handle way back and touch to the anode point. Even with the best of care you will sometimes get a tingle or two. After it quits sparking, hold the connection a while longer, as some recharge will take place if you don't. Then you can touch the high-voltage cable safely. *Always* discharge the high-voltage anode before replacing a picture tube. The high vacuum on the tube makes its glass quite dangerous in case of breakage, and getting a shock from a hand-held picture tube can make you drop it quickly.

Tubes in the high-voltage stage are likely to go bad. With solid state systems the high-voltage diode that replaces V6 is subject to so much surge that it can fail. It is usually a long device and hangs between the flyback transformer and the anode cap as part of the high-voltage wire. It does not need a heater supply. A horizontal output transistor can certainly be bad. A solid state diode that replaces V2 as damper is often bad. The flyback transformer is very subject to breakdown, and an exact replacement is usually necessary. A fuse in the horizontal output tube circuit such as F500 often blows. Replace it, but if it blows soon again, problems are indicated. Too high a setting of the bias, R503, might cause the fuse to blow often.

You can check out the flyback transformer with an ohmmeter, though this is not a perfect test. Note that low ohm readings are indicated for most parts of the primary. The high-voltage secondary calls for 640 ohms with this set. A bad flyback transformer will often smell bad, show charring, and smoke when the set is on. A charred resistor in these circuits should be replaced. If the fault occurs again or a tube keeps blowing, chances are a capacitor is shorted. Some spark gap terminals in high-voltage systems build up dirt and short. A clean high-voltage system is often a good high-voltage system. Clean it with care, as the wires of the flyback are very fine and delicate.

Some sets have a voltage regulator tube in addition to those we have talked about. It conducts when voltages get too high and keeps the high voltages at a proper level. If the high-voltage rectifier tube keeps blowing, try a little less heater coupling. If it has three loops to the flyback transformer, cut to two or place a series resistor type wire to drop the heater input to a good level. It is normal not to see a light in the heater of this tube. Naturally if the horizontal frequency is way off at the oscillator, no high voltage will be made. A bad AFC system could knock off high voltage too. And of course if there is no B+ from the power supply for the horizontal output tube, there will be no high voltage.

Another type of high-voltage system that is fast becoming standard is shown in Figure 13-4. Most of the circuit is similar to that we have studied. Note at point 17 the feedback of horizontal waveform to compare with the input and keep the phase correct. Horizontal discharge is the same as the driver. Instead of the top of the flyback T205 going to a plate cap of a rectifier tube, it goes to a tripler circuit. This box raises the high voltage greatly. From it, the higher voltage is supplied, posi-

tive side of course, to the high-voltage anode connection on the side of the picture tube. Such a tripler must have rectifiers to change the AC that must work its transformer to DC. Sometimes the extra part will be a quadrupler instead of a tripler. Usually this voltage increaser with rectifiers will be very close to flyback transformer in function.

As laboratory work, use a schematic and layout of a TV and connect a high impedance voltmeter between collector or plate of the horizontal oscillator and ground. Make sure it is the oscillator, not the horizontal output! Vary the horizontal hold or horizontal frequency adjustment a little and note the voltage. Now change the meter to the collector or plate of the vertical oscillator and adjust the height and hold or linearity controls one at a time. What effect does this have on voltage? Which makes the most noticeable change?

If available, connect the oscilloscope to the output of the vertical amplifier tube or transistor. This would be a plate or collector connection and should be isolated through a capacitor at the probe point. Note the changes of the waveform pattern when the vertical controls are changed. Sweep frequency for the scope, in case you haven't figured it out, is 60 Hz.

To observe horizontal waveforms do *not* under any circumstances connect the scope to the horizontal output because the high voltage is partly at that point. Instead look on the schematic and find a horizontal test point and connect the scope there. Then adjust horizontal controls slightly to note the effect on both the TV and the scope screens. Note on Figure 13-2, V3A plate is such a point and is designated W33. You must use the schematic of the set you are using, however, to locate such a point.

Here are a few important things to remember about high-voltage systems:

1. High voltage is a by-product of the horizontal sync system.

2. High voltage is for the anode of the picture tube so the color-producing beams will be attracted to the screen.

3. High voltage is made by a flyback transformer that is also the horizontal sync output transformer. It makes use of the many pulses of horizontal sync, with a step-up transformer to generate great voltage.

4. The high voltage must be rectified.

5. A damper tube or diode rectifies more voltages from the horizontal sync for other circuit parts. With a color set these usually are the focus grids of the picture tube.

Quiz for Chapter 13

▶ 1. Explain how sync signals are removed from the video even though both are on the same frequency.

▶ 2. What circuit rejects horizontal sync pulses but accepts vertical pulses at the beginning of the vertical stage?

▶ 3. What kind of pulses keep the horizontal stage oscillating on frequency and in phase during vertical sync pulse time?

▶ 4. Explain the difference between height and vertical linearity.

▶ 5. What is the purpose of a phase detector in horizontal systems and how does it work?

▶ 6. The horizontal system assures that each line of picture will start at the proper time, but what other reasons are there for keeping the horizontal system on frequency and in phase?

▶ 7. What kind of system can be used to measure voltages in a high-voltage system such as horizontal output plate or picture tube anode?

▶ 8. Explain how to avoid shocks when working with a high-voltage system if you have to remove tubes or the high-voltage wire.

▶ 9. What is the function of the damper tube or diode?

▶ 10. The horizontal hold control coil is often also the _______ for that stage.

▶ 11. What other stage or stages besides the high-voltage circuit itself must be operating properly for high voltage to be present?

▶ 12. If setting the contrast control of the set on high produces tearing of the picture, what stage is likely at fault?

14. Sound Systems

PROJECT: Using a Signal Tracer to Trace Sound from Volume Control toward Speaker

WHEN THE sound leaves the video IF system it is still frequency modulation. The carrier is being bunched and spread in accordance with the sound produced at the TV station. At the input point to the sound stage you could not just place a detector and amplifier and hear anything. A detector cuts off half of each cycle or alternation of frequency of AC, but it would not recognize changes in frequency that represent the sound. It is necessary to remove this sound by a different process. Then the sound must be amplified considerably to drive a speaker or speakers with comfortable room volume. This chapter will cover FM detectors and audio amplifiers that make the sound stronger.

When the sound enters the sound system, it is at the IF frequency set for it, usually 41.25 MHz. In some sets this frequency is amplified with more IF stages, and in some it is fed directly to the sound FM type detector. See Figure 14-2. D3, the detector, merely detects the 41.25 MHz and does not change the FM to AM. The signal then enters the integrated circuit, IC1. Several necessary parts must be added at the pins of this integrated circuit (IC) and are shown below it. The first triangle, "IF AMP/LIM," amplifies the IF frequency of the sound, and then clips off all amplitude. In an AM radio this would be a disaster. With FM, as you know, nothing important rides on the size of the carrier. All the voice and music is on the slight shifts of the frequency of the carrier waves. Clipping the amplitude off gets rid of static, manmade noise, and so on, that ride in on the carrier, perhaps from the next house or apartment. This system is not 100 percent effective, as a passing truck may inject ignition noise into a TV.

One way the limiter works is by taking a large signal and biasing a transistor or tube so that only part of the signal can get through, then, because the signal gets inverted, sending it through another heavily biased tube or transistor that loses part of the other half. This leaves only the middle with the carrier intact. This carrier with shifts of frequency to represent sound is fed to the FM detector, which shows very little circuit.

You can see what happens inside much better in Figure 14-1. T602 and T603 are tuned to the center of the intermediate frequency. As long as the TV receives only the carrier (time between sounds) with no bunchings or spreadings, there is equal current flowing in the diodes, R608 and R609. But when modulation comes in to the transformers it produces a bunching or spreading of the carrier, and that is a change of the frequency, slight to be sure, but a momentary change. Since the transformers are tuned for the center carrier, any change of frequency that is placed on them produces a small voltage. This voltage is developed because the diodes are not completely balanced, with a frequency on the coil that is not correct. As the diode currents try to equalize, the voltage difference is placed across R607, and it comes in pulses in accordance with the music or voice that is transmitted. These pulses are put on the volume control, R630, and then an audio amplifier makes these audio pulses loud enough for the speaker.

Figure 14-3 is another sound system that uses an integrated circuit for the FM detector, and Figure 14-4 shows the actual module. 1101 is the quadrature coil, which acts somewhat like the transformer T603 in Figure 14-1. T1102 is the input from the IF circuit

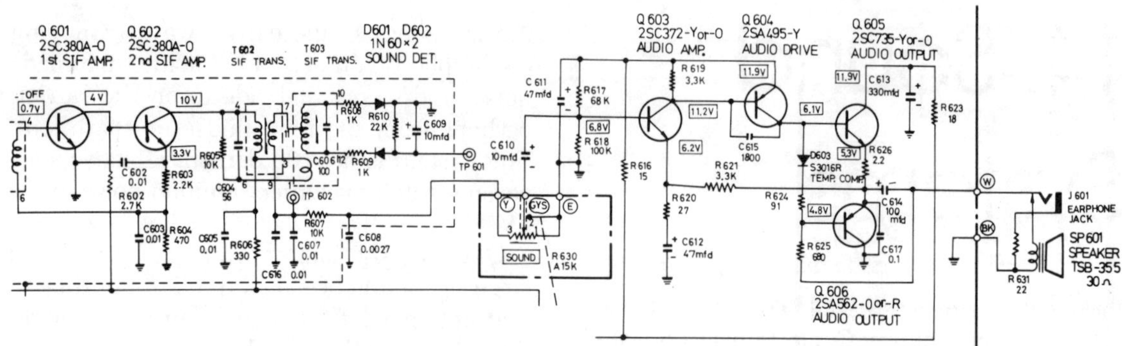

FIGURE 14-1

RATIO DETECTOR AND AUDIO FREQUENCY

AMPLIFIERS

board. This circuit is a continuation of Figure 12-6. Q103 in Figure 12-6 is the sound detector and amplifier. This part of the sound circuit on the IF board explains how the circuit in Figure 14-3 seems not to have enough parts. The connection to Figure 14-3 is at C1112. Note the shielding on the high side of the volume control, R211, to prevent stray pickup. The output of IC1101 at pin 12 is directly coupled to Q202, the audio output transistor. In this FM detector circuit we have an integrated circuit, IC, that is mounted on a module. That is a part of a part mounted on a part. IC plug-in units are convenient for testing by substitution.

Now that the FM has been changed to pulses of audio, it must be amplified before it works a speaker. We refer to this as *audio frequency* or *AF*, because it is in the hearable range. The most simple amplifier is shown in Figure 14-3. A single amplifying transistor, Q202, makes the AF stronger and couples it through a transformer to the speaker. Q202 is a power transistor. It can handle a large current, and it can amplify greatly. Usually such a transistor is mounted on a metal frame that allows heat dissipation through the frame. Often the collector is the shell of a power transistor. For that reason the frame to which it attaches is live and not at ground. When taking voltage readings on such a transistor,

attach the ground probe of the meter to a sure ground in the set such as the metal shield of the tuner. Note the collector voltage, 135 volts. The applied voltage to the circuit is 270, but R209 drops the difference.

The audio output transformer has a high-impedance primary and low-impedance secondary, matching the transistor collector to an 8-ohm or 4-ohm speaker. The primary leads are usually blue to collector and red to power supply, and the speaker leads often are just enameled wire. Some outputs have no color leads, but tabs that solder into a printed circuit board. R206 and R207 are protective biasing resistors. C204 is a bypass. The forward bias comes from pin 12 of IC1101, and the signal travels this same path. R204 is a temperature compensating resistor that changes resistance with temperature as a means of keeping a proper voltage on the collector of Q202.

Now let's examine a sound system without an output transformer. This kind of system is called *output transformerless* or *OTL*. Figure 14-2 is such a system. The whole idea is to get rid of the output transformer. Since power transistors draw a fair amount of current, and current is needed to drive a speaker, it is possible to put the speaker directly to the collector of a power transistor and the positive voltage (with NPN transistors of course) to

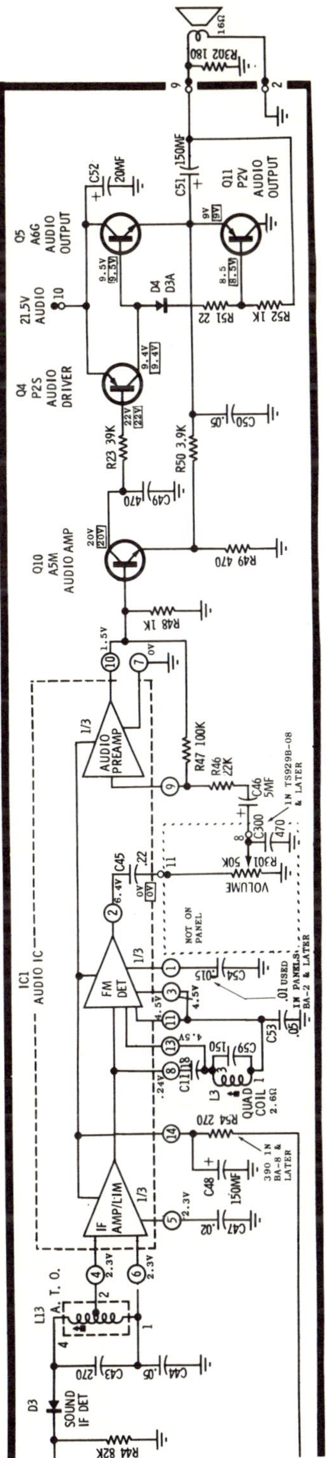

Reprinted with permission of Quasar Electronics Company

FIGURE 14-2
IC DETECTOR (FM) AND OUTPUT
TRANSFORMERLESS AF AMPLIFIERS

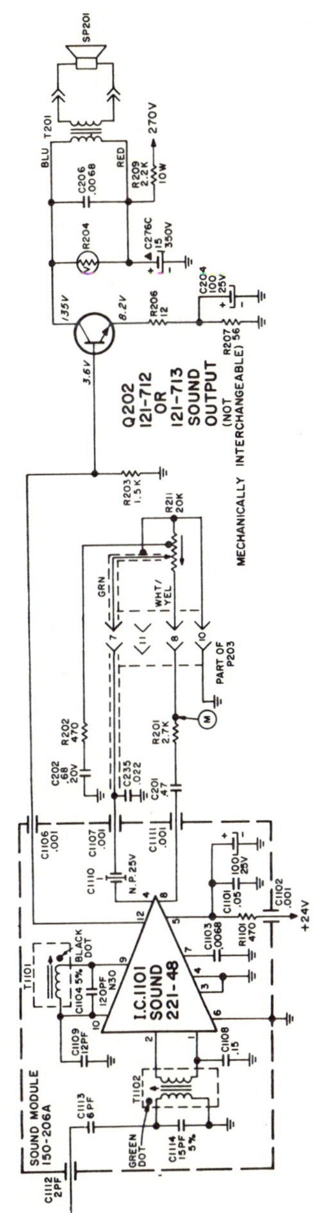

Courtesy Zenith Radio Corporation

FIGURE 14-3
IC QUAD FM DETECTOR AND TRANSFORMER
OUTPUT

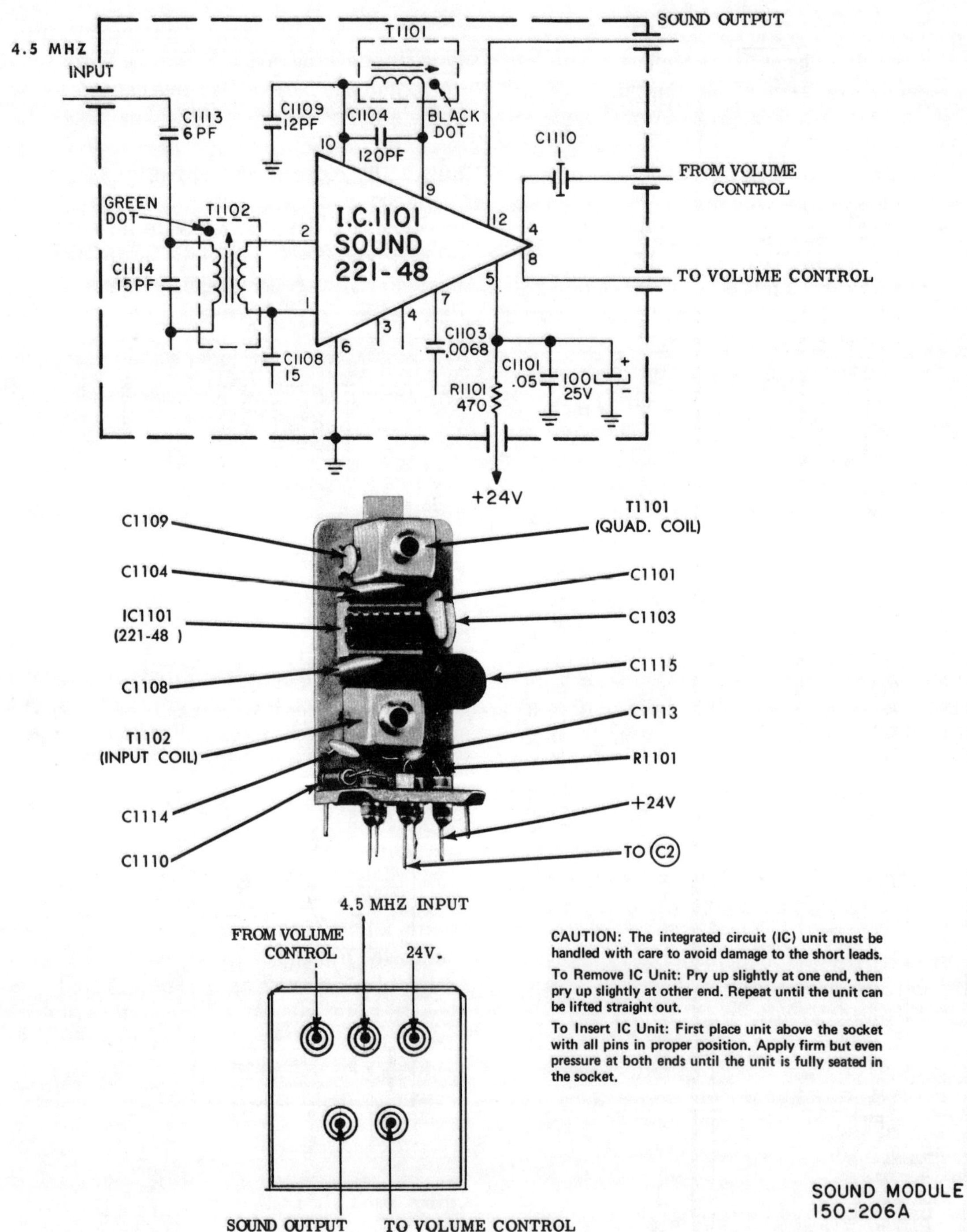

Courtesy Zenith Radio Corporation

FIGURE 14-4

the opposite side of the speaker. This does not work too well, at least not as well as the use of an output transformer. The output needs a bigger load than a few turns of voice coil wire offer.

The solution is to have two power transistors, one above the other, and to have the voltage applied to the top one. The top transistor thus acts somewhat like a load for the circuit, but since it draws a fair current, it is a reasonably good (though not perfect) impedance match for the speaker. In this case Q5 and Q11 are the voltage divider circuit. The speaker is at midpoint. Note that a 16-ohm speaker, not the normal 4–8-ohm speaker, is used. Using a higher-ohm speaker with more turns of wire in the voice coil helps match impedance.

Note that Q5 is an NPN transistor and Q11 is a PNP type. The way they are situated, current could flow up from ground collector to emitter of Q11 (that is normal current flow for a PNP) on up and from emitter to collector of Q5, and out the 21.5-volt positive power supply point. The biasing of these transistors prevents any large amount of continuous current. First there is a DC path from ground, through the speaker, through R52 to the base of Q11. This puts a negative bias on the base, and since negative on the base of PNP makes it conduct, Q11 is ready to go, but the bias on Q5 is wrong, so it is not ready to conduct. The setup is at rest position.

A pulse of positive signal at the base of Q10 causes Q10 to conduct considerably in pulses. This current goes through R23 and out the base-to-emitter circuit of Q4. Since Q4 is a PNP type, current can easily travel from base to emitter. This base-to-emitter current causes Q4 to conduct greatly from collector to emitter and out the 21.5-volt source. By chain reaction, current is drawn from Q5 base, as an NPN transistor has an easy path for current to flow from emitter to base. This current opens the collector of Q5, and big pulse of current flows through the output transistors Q11 and Q5, causing a pulse to be drawn across C51.

Recharging C51 requires a large current to flow through the speaker. Now a fairly small current at the base of Q10 has caused a large current to go through the speaker. This is not a continuous current, obviously, but is in pulses that correspond with the voice to be produced in the speaker. R49 to a minor degree keeps a signal of opposite phase on the Q11 circuit. Such a system as this can be made to give very deep tones with certain settings of the biasing network.

Let's review:

1. The FM is detected by a frequency-sensitive tuned circuit that develops abnormal voltages when off-frequency signals arrive, and through diode conduction a current is taken to a volume control or other load and to an amplifier.

2. Some sets have a limiter to clip off the amplitude of the FM.

3. Output transformerless (OTL) audio systems have transistors in series and take the output for the speaker off just one transistor.

4. Output transformers in more conventional circuits convert the high-impedance output of a transistor or tube to the low 8-ohm impedance of a speaker.

Tracing Sound

Servicing a dead sound system, when the rest of the receiver is working normally, is a matter of repairing the FM or the AM. Incidentally, FM radios work just the same way as the sound part of a TV. The FM radio will have the tuning mixer, oscillator, and IF stages, then an FM detector, then audio. The frequency of radio FM is 88 to 108 MHz, or between Channels 6 and 7 of TV. The IF for FM radio is not 6 MHz wide. There are two basic forms of FM detectors—the ratio detector that you saw in Figure 14-1, and the Foster Seely discriminator that turns the diodes the same way. If both diodes face left or right, a limiter is needed before the discriminator to help rid the signal of noise.

With TV or FM radio, you must first determine whether the trouble is in the audio or in the IF or FM detector. A good quick way to tell is to set your radio signal generator for a tone in the earphone and inject a signal on the center pin of the volume control with the probe. If you hear the tone in the speaker, you can start with troubles in the IF or ratio detector, because from the volume control on back you have an operating set. If you can't get the tone there, try farther toward the speaker. The trouble has to be between volume control and speaker.

In the IF stage you would need a sweep generator and an audio tone to modulate it to test from stage to stage. With many sets, though, if you get the tone to the speaker at the volume control, you can just pull out the IC and replace it. If that fixes it, the IC was bad. If not, suspect the parts that surround it, such as C1110, T1101, and C1155 (Figure 14-4).

An integrated circuit, IC, usually has many transistors in it and perhaps some diodes and resistors. You cannot tell what is inside. Certain manufacturers of IC components publish manuals showing internal parts of their ICs. In general, you don't have to know all the parts to service such units if you understand the proper function the IC should perform.

As suggested lab work try the tone injection to the volume control of your TV using the home-made signal generator. Then change the signal generator back to an audio signal tracer as done in Chapter one and trace through the TV audio circuits. Again use the TV schematic for this work.

QUIZ FOR CHAPTER 14

▶ 1. Name the two normal types of FM detectors. Which one needs a limiter ahead of it?

▶ 2. What does a limiter do? How does it do it?

▶ 3. Why are at least two output transistors used in an OTL audio system?

▶ 4. What easy-to-find part of a sound system is a logical place to put in a signal first to decide which half of the circuit is bad? What kind of signal should you inject here?

▶ 5. What components often make up an IC? What advantage does it have when one is servicing a receiver?

▶ 6. How is the voice carried from the TV station on the carrier?

▶ 7. What kind of cable is often used from FM detector to volume control to avoid hum and stray noise pickup?

15. Color Circuits

PROJECT: Identifying and Testing Color Circuits

COLOR CIRCUIT GENERAL ANALYSIS BY BLOCK DIAGRAM

THE COLOR signal is actually a combination of two signals—the burst and the chroma. After the two are combined in separate color demodulators the three primary colors result in signal form. The picture tube beams strike the proper phosphor dots to make the colors. Figure 15-1 shows each stage of a color TV in block form. In each block would be the transistors, resistors, capacitors, tuning coils, and so on as needed. The arrow shows which way the signal information goes. Dashed lines show control voltages or currents that show how one circuit controls another to make it function properly. The solid lines are signals that will affect the picture tube or speaker.

You have studied the tuner, IF, detector, sync separator, vertical sync, horizontal sync, and high-voltage circuits, and you have learned about the low-voltage power supply, which is not shown but would have an arrow to all circuits, as it supplies voltage to all. We will now consider the color circuits. Someplace in the video amplifiers the color signals are removed and put into the chroma bandpass amplifier and the burst amplifier. The color signals are 3.58 MHz above the video frequency so they can be separated by tuned circuits.

The signal to the chroma bandpass amplifier comes from the TV station during the video trace. A control voltage from the horizontal sync system, through the color killer, turns the chroma bandpass amplifier on when the sync is not operating. In this way it gets only the signal of color that is broadcast during each line, and the amplifier is turned off during sync time so the burst signal will not be amplified.

The burst amplifier is just the opposite; it is turned on during sync time so the burst signal gets amplified, but it is turned off during line tracing time so none of the bandpass signal gets amplified. We have now separated all the color signals by frequency and further separated the burst from the chroma or bandpass color. This separation is necessary, as it is the difference in phase between them that will make up the different colors. This seems very complex, and it is.

The color killer has another function. When a program is being broadcast only in black and white (and this is rare today), no burst signal is transmitted, so lack of burst causes the color killer to cut off the chroma bandpass amplifier altogether, so no stray colors are shown on the screen.

The two signals—burst and chroma—are mixed in the demodulator. But the burst occurs only during horizontal sync time, so it would be a spurt, quit, spurt, quit signal, and the chroma could not be compared with it. The burst has to be amplified, phased correctly, and it triggers an oscillator that runs continuously. The burst, when on, keeps the oscillator in perfect rhythm with the TV station burst samples. This oscillator usually has a crystal that oscillates on 3.58 MHz. These two signals—one at burst phase and another line by line in phases proportionate to the different colors—enter the demodulator, where either diodes or tubes or transistors or a combination amplify the difference in phase. The small part of a cycle these signals are out of phase with each other represents the colors and creates different voltages to represent them.

This voltage difference signal is fed to color amps that make the color voltages or signals

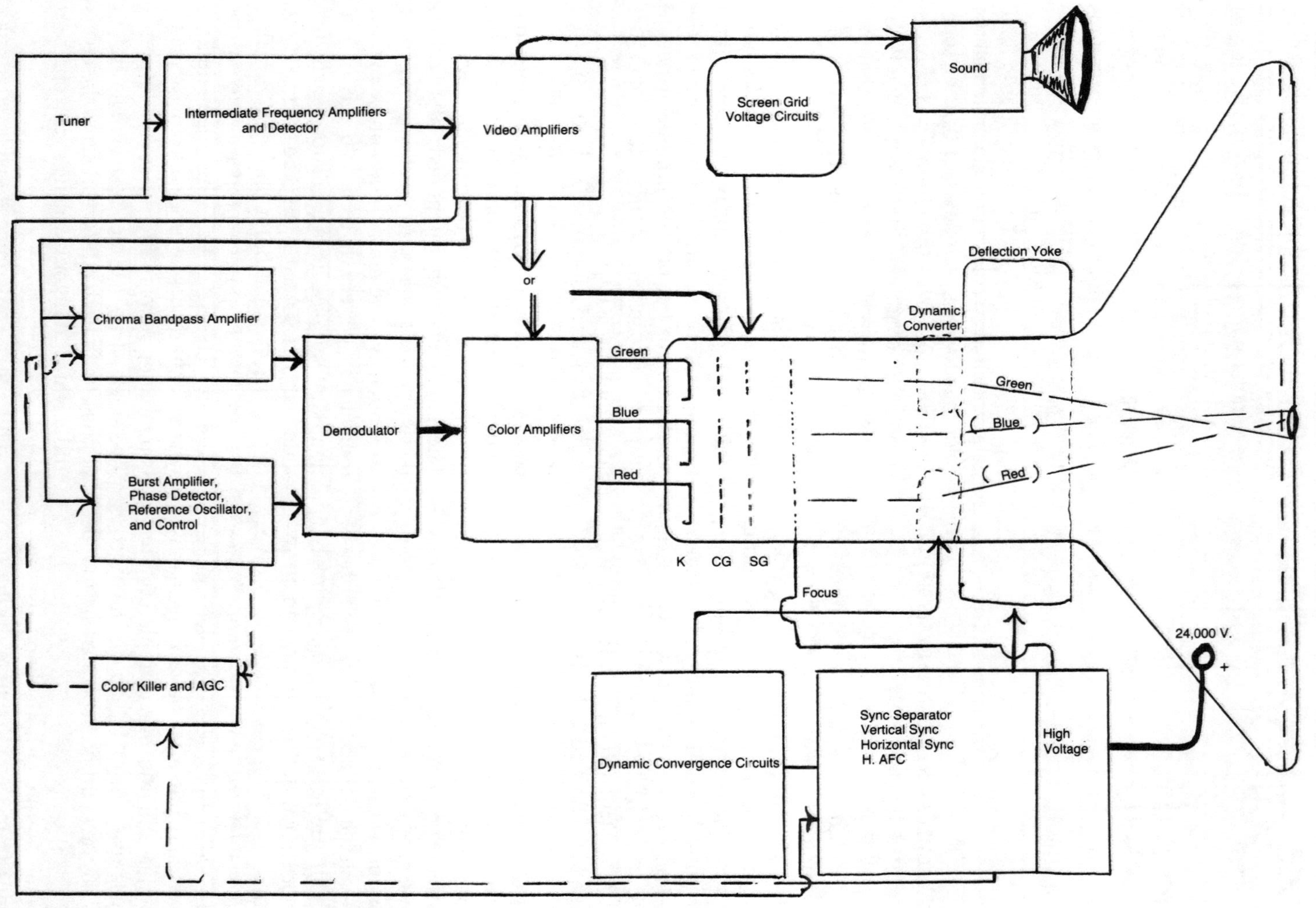

FIGURE 15-1
BLOCK DIAGRAM, COLOR TV

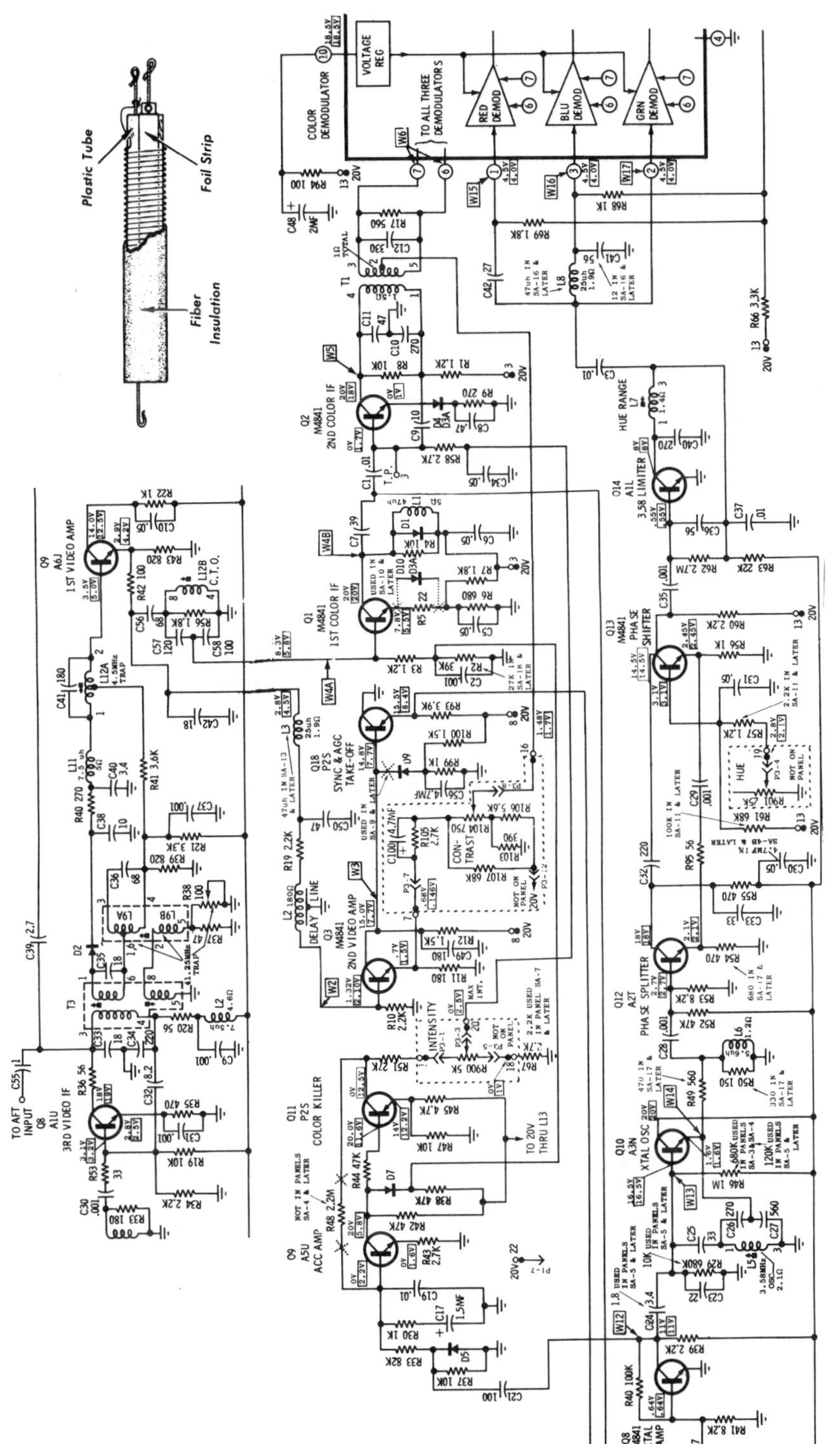

Reprinted with permission of Quasar Electronics Company

FIGURE 15-2

CHROMA BANDPASS (COLOR IF), COLOR KILLER, COLOR SYNC, DEMODULATORS

large enough to go to the picture tube. As shown, these signals can be put in on the cathodes, but they can be put in on the control grids instead, depending on the manufacturer.

Now the video or brightness signal can be put in the picture tube directly on a grid, or video can be put on the cathodes and the color on a grid, or the video can be put in at the color amplifiers before the picture tube. The video determines how much lumination or light a color will have, but it does not determine the tint or type of color.

The screen grid voltages are set for the proper intensity so the proper amount of each color will appear on the screen. Screen grid voltages have no signals. The dynamic convergence gets energy from the sync signals, as you have learned, and it helps keep each color beam close to the next one.

The chroma bandpass amplifier circuit is sometimes called *color IF,* a poor choice of words because it creates confusion with the real IF circuits of a TV, and it is not an intermediate frequency anyway. It is the 3.58 MHz that the color has been on all along. The set in Figures 15-2, 15-3, and 15-4 is referred to with the color IF terminology, and so are other sets. You must get used to this.

The "color IF" really is a set of amplifiers that pass and amplify the 3.58-MHz signals of color that take place during trace time of the picture.

CHROMA BANDPASS OR COLOR IF AMPLIFIERS

We will be using Figures 15-2, 15-3, and 15-4 for some time now so keep the places. There are two outputs to the first video amplifier, Q9. One is to the second video amplifier, Q3, with the delay line in between. The delay line keeps the video from getting to the picture tube or color amplifiers ahead of the colors. Note the cutaway view of a delay line at the upper right in Figure 15-2. The video is amplified and adjusted by the contrast control, then sent to the center tap of the

secondary of T1. In this case the video is put in ahead of the demodulators. This is a rather early insertion. Often video is added to the color amplifiers or in the picture tube.

The other signal that leaves the first video amplifier is the chroma signal. L12B combined with C57 and C58 are a tuned circuit sensitive to 3.58 MHz. Actually, the circuit tunes a band from 3.1 to 4.1 MHz. The resistor, R56, broadens the band. The signal bandwidth is then applied to the base of Q1, the first color IF. Remember, the color IF is also called the chroma bandpass amplifier. In other circuits that term will be used. The bias for Q1 is set by the ACC (automatic color control) amplifier at left, Q9. L1, D1, and R4 are the tuned circuit for the bandwidth, so it travels to Q2 through coupling capacitors C7 and C1. Part of the signal, as a control only, is fed at a tie point between C7 and C1 to the color sync gate (Figure 15-3) to keep the color sync off during chroma signal time. Otherwise the sync could lock in with chroma signal phase and defeat the whole plan.

The second color IF transistor, Q2, has a cut-off bias system connected with the color killer to disable the color IF during black-and-white programming. In that condition the color circuits do not affect the demodulators or outputs to the picture tube. The output of the chroma amps or color IF amps is taken at T1, which is tuned to the chroma 3.58-MHz frequency in its primary, but accepts the video in secondary along with the chroma. The combination is fed to each of the demodulators at pins 6 and 7. The diode in the emitter circuit of Q2 is placed so that normal currents that go up into an NPN transistor can go through, but should spikes get into the circuit of wrong direction they would not get to ground.

AUTOMATIC COLOR CONTROL AND COLOR KILLER

The major control over the amplification of the color IF stages is exerted by the automatic color control (ACC) circuits. ACC not only

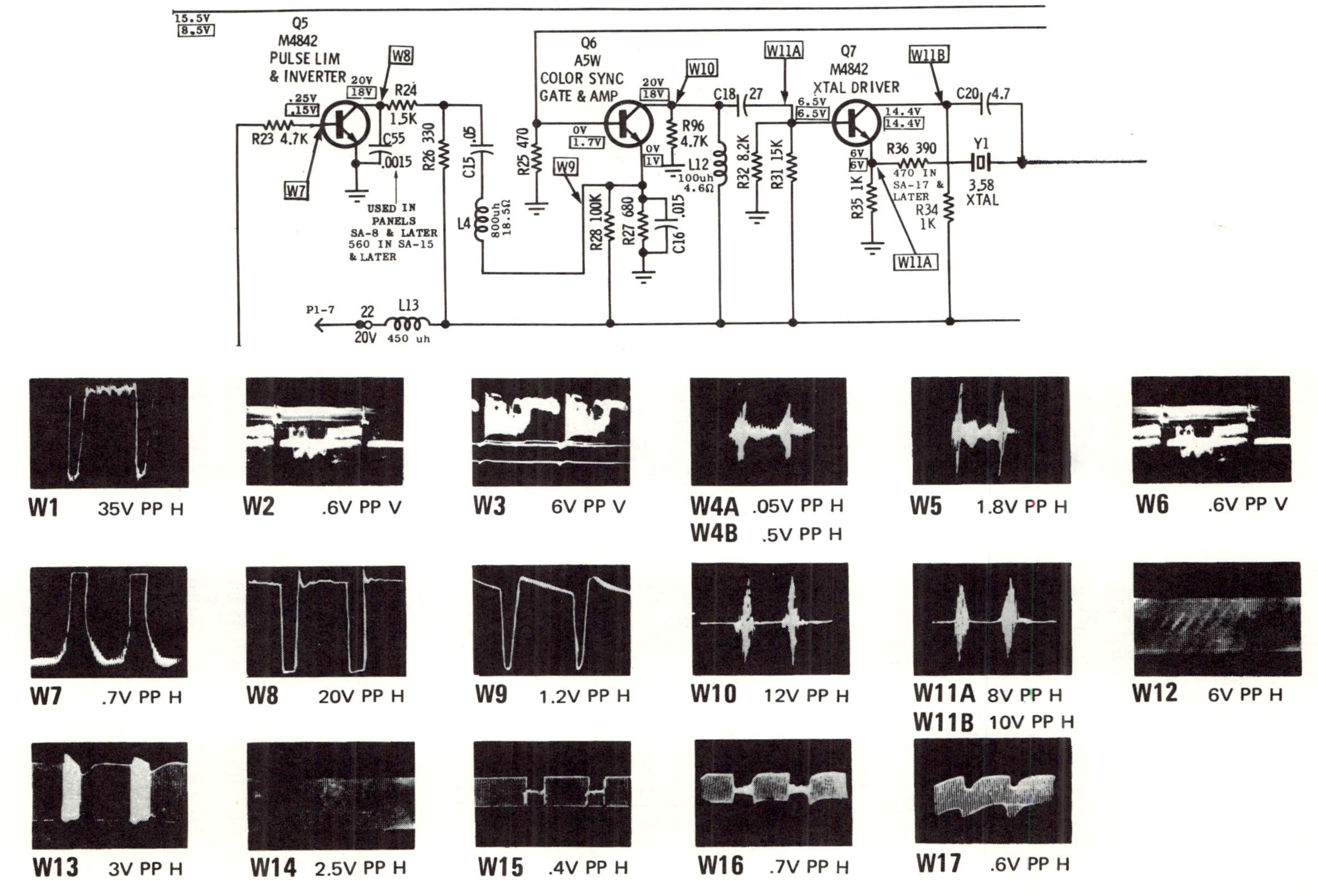

FIGURE 15-3

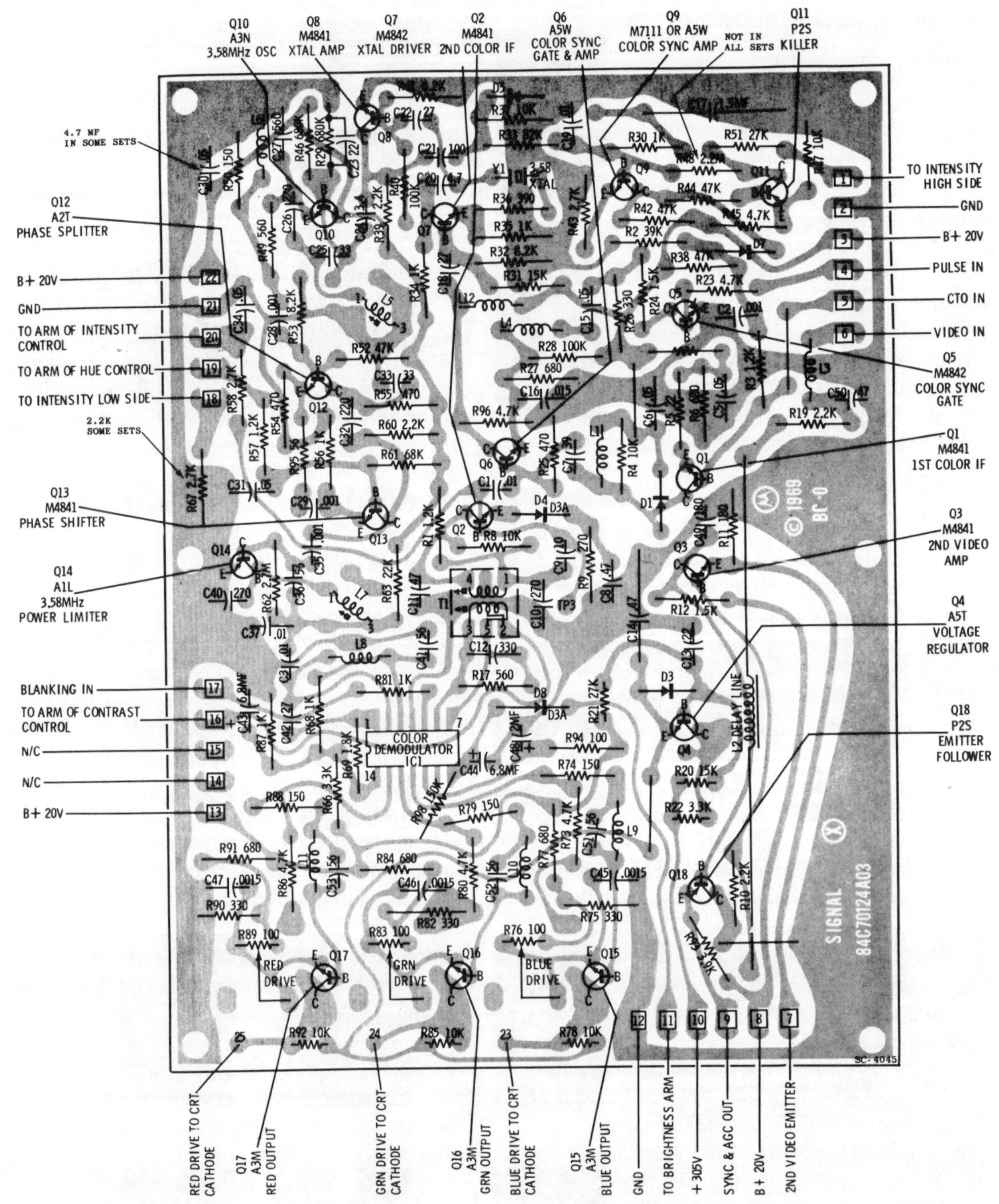

Reprinted with permission of Quasar Electronics Company

Figure 15-4
Color-Video Panel "SA" - Circuit Side

controls the gain of the color IF, but it also cuts the color IF stage or stages off during black-and-white transmissions. In the case of the Quasar set in Figure 15-2, the ACC transistor, Q9, controls the bias of the color killer, Q11. In turn, the control of Q9 is affected by its base bias, supplied by the crystal (XTAL) amplifier Q8. The signal that enters the base of Q9 is a result of the burst signal. The strength of the burst determines the conduction of Q9, and therefore the operation of Q11.

Though transistors are current amplifying devices, their operation can also be illustrated with voltages. In the figures you are studying, the voltages listed in the rectangles are with signal present, and the voltages outside the rectangles, above them, are voltages with no signal present. So, for zero voltage on the base of Q9, a full 20 volts of applied voltage occurs at the collector. This is a condition of a transistor at rest when the collector voltage is the same as the applied voltage, to the collector from the power supply. The applied voltage is through R42. There is no voltage drop across R42. The reason no voltage drop occurs across R42 is that there is no current flow with Q9 having no forward bias. A full 20 volts can be measured at the base of Q11.

Since Q11 is a PNP type transistor, a heavy positive base cuts off any current flow through it. Under this condition the base bias current path for Q2 is stopped at the collector of Q11. Let's explore this path. From ground up through R9, D4 in emitter of Q2, on to base of Q2. Since Q2 is an NPN transistor, current travels easily from emitter to base to form forward bias if a positive source is supplied farther on. Continuing, current goes through R58, part of R900 (the intensity control), depending on setting, to the collector of Q11.

Now the circuit is broken by the heavy base positive bias on Q11. Remember that current can only go through the barrier of a collector to base when a good forward bias from base-to-emitter current flow is developed in a PNP transistor. This is not the case now, because the supply current for forward biasing tie base

of Q2 is broken, Q2 will become inoperable. Note that the upper voltage is 20, the applied voltage. In this condition the color IF system is disabled, because there is no burst to operate Q9. This would be the condition of a black-and-white picture.

Now let's assume that the burst signal is present and that we have 2.2 volts on the base of Q9. Now Q9 is biased to allow current flow, so the voltage goes down to 5.8, because with current flow a voltage drop occurs across R42 and R43, so when current flows in a transistor the more current flow, the lower the voltage will measure at the collector. This in turn lowers the voltage at the base of Q11 to 11.6 volts. The voltage is still positive, but since the emitter voltage is now 12.2 volts, a current can flow from 11.6 to 12.2, as there is a negative-to-positive potential of 0.6 volts. This current going to emitter opens the PNP transistor and the path from emitter of Q2 now is complete through Q11 to R45 and positive voltage to power supply. Now with current flow from emitter to base in an NPN transistor, Q2 in this case, Q2 is biased to work and will amplify the chroma signals. Note that now the collector of Q2 has 18 volts instead of the 20 applied, so current is flowing.

All this process is necessary to work the color killer when no color program is being transmitted. When no color is transmitted, no burst signal is transmitted, and the lack of burst starts the whole turnoff sequence just described. The beginning of burst opened Q2 to operate. Now it might occur to you that burst appears only on the back porch of the sync, so during video time when the chroma signal is present why is the color IF not cut off? Because capacitors charge up and keep the color killer from cutting the second color IF amplifier off during short intervals.

The other control circuit to the color IF or chroma amplifier circuits is a bias that is set on Q1. It is not a complete cut-off, but a control device to build up the chroma signal on weak station color and cut it down on strong transmissions of color. The theory is that on weak color inputs the burst signal would also be weak. So the burst keys the amount of bias on Q1. This is similar to AGC in the video IF and tuner stages.

To get a forward bias for Q1, let's trace the necessary path of current. It goes from ground up through R6 and R5, emitter to base (it's an NPN) down R3, and across to a voltage divider center point between R38 and R44. D7 allows current to pass easily upward. Since this circuit is not to cut off Q1 completely but to limit its amplification, the current path to +20 volts is always there through R38, producing some forward bias at all times. Under this condition, if the other path is blocked through R44 and Q11, the voltage at the base of Q1 would be 8.3 or fairly high, and Q1 would work hardest. True, it would produce the most amplification with no burst signal at all, but since Q2 will block such a signal under those conditions, no harm is done. The advantage is that on very weak burst signal, Q1 will do its best to build up the chroma, which is likely to be weak too.

Now let's assume that a strong burst is present so that Q11 conducts after receiving forward bias from Q9. The second path for the Q1 bias current is opened so the voltage at the base of Q1 drops to 5.8 volts. Note that the emitter voltage changes also. A positive voltage is supplied between R5 and R6. The overall effect is that under weak signal conditions the difference between base and emitter voltages is 8.3 minus 7.8 or 0.5 volts that would induce a larger current than 5.8 minus 5.5 or 0.3 volts difference. So Q1 conducts more during weak signals.

The intensity control, sometimes labeled *color killer,* can be adjusted to the point that it will kill the color on weak braodcasts. It must be set just below this range. The setting of this control, R900, must be high enough to cut out color noise spots, but not high enough to cut out weak signals. With no burst signal it will cut out the color amplifiers anyway, but the setting determines whether it will cut out for weak burst too. Note that the bias of Q2 must be positive, but applying more to ground

or less to ground would affect how positive the bias can be. Remember that the color killer cuts off the color IF or color bandpass amplifier, while the ACC controls the strength of the chroma signal.

BURST AND REFERENCE OSCILLATOR CIRCUITS

D5 detects the burst signal with associated resistors to produce a voltage rather than burst signals. C17 is a long-time constant capacitor that charges the detected voltage and prevents the control system from turning the color IF off between bursts.

Now we will consider the burst signal. Figure 15-3 shows the beginning of it, and it continues in Figure 15-2. The long line to the right at the top of the schematic in 15-3 connects to the 27 capacitor, lower left, and to the base of Q8. As you study the circuits, refer to the waveforms at the various points as shown in lower Figure 15-3.

W7 shows horizontal sync pulses with burst supplied by the gate pulse former, which you saw in Figure 13-2. The collector of Q5, the pulse limiter and inverter, does just what it says. It turns the signal upside down at the collector, and it limits to only sync, with burst of course, as shown in W8. At W9 C15 and L4 have changed the shape a bit, but the basic horizontal sync pulses with burst signals are still there. This signal is placed on the emitter of Q6. The control signal from the color IF is placed on the base of Q6, which is called a "gate" because the base is "opened" only during burst time. The rest of the time the "gate" is closed, so only the burst signal gets by Q6. Note the waveform at W10—no more horizontal pulses, just the burst. This 3.58-MHz burst signal is sometimes called "color sync," as if we didn't have enough confusing terms. True, it does synchronize the crystal oscillator to operate on frequency and in phase with itself, but remember that this color sync is *not* the sync that moves the beam across the picture tube screen. The crystal driver enlarges the burst some, from W11A with 8 volts to

W11B with 10 volts. The crystal driver does not amplify a lot but feeds a crystal the burst frequency often enough to make the crystal operate continuously at 3.58 MHz and in phase with the incoming burst signal.

We how have a steady reference signal, bursts not only on the back porch of the horizontal sync, but all the time. W12 shows this steady signal; it is the waveform at the collector of Q8, the crystal amplifier. This steady signal will be fed to the crystal oscillator to trigger it into oscillation exactly in phase and exactly at 3.58 MHz.

Q10, the oscillator, has a feedback path that might not be apparent at first. The signal is built up in L5 and the capacitors C26 and C27. The signal is amplified in the normal way, base to collector, but the collector goes directly to B + without a return path. The input through the emitter, though, is through L5 and the capacitors—that is, the signal input is there; the direct current input is through L6. Since what flows to collector must first flow through the emitter, the signal is fed back by the emitter's tie to center tap between C26 and C27. The output of Q8 is capacitively coupled to the base of Q10 by C24, triggering the oscillator into action at exactly the right phase and frequency.

At W13 we see a signal with bunchings corresponding to horizontal sync pulses. These are removed at the emitter follower of Q10 (W14 waveform), but appear again somewhat diminished at W15, W16, and W17. Since these bunchings represent time between traces, they will do no harm now that a continuous 3.58-MHz oscillator signal is operating.

We have learned that the crystal oscillator locks in exactly on the 3.58-MHz frequency of the burst as transmitted by the TV station. It not only locks in on frequency but also must be exactly in phase. It must start each wave at exactly the same time it starts at the TV station. Then the difference in the chroma signals will produce the colors.

The set has now achieved this oscillator signal, but because of distance from the TV station, interference, the fact that no two TV picture tubes are precisely alike, and other factors, it is necessary to be able to set the signal a fraction of a second different from its present phase to vary the colors a bit to get pleasing flesh tones and other colors. Changing the chroma signals would be very difficult, and they are constantly varying in phase, too. On the other hand, the reference signal that the oscillator has made can be changed, because it is constant, and shifting it up or back a bit will change all the colors a little. The hue or color control (usually on the front of the set) does this. R901 is this control with this Quasar set. To change the phase of a signal it is necessary either to delay the signal slightly or to accelerate its wave. A signal at an emitter follower, at R54 and emitter Q12, is out of phase with a collector signal, at collector Q12. This is a natural situation. Both of these signals that are out of phase are applied to Q13, the phase shifter. The phase of the color sync on the collectors has been amplified by Q12 and cannot be amplified further by Q13 as it does not go to emitter or base so that it could be amplified. On the other hand, the phase of the signal from emitter to emitter can be amplified by Q13, as it enters at the emitter.

Now we have a common base or grounded base transistor in Q13. The base is signal grounded by C31, but its bias is controlled by a voltage divider of R61 and R901. R57 controls the bias to a lesser degree. Since R901 is variable, adjustment of the control on the front of the TV sets the forward bias of Q13. Changing the bias determines how much Q13 can amplify, thus the emitter phased signal can be strong or weak. The output of Q13 then will be a combination of the phase of both and not two signals.

The hue control usually is set for the best flesh tones, and all other colors have to do. Viewers expect people to be their proper colors on the screen. The phase shift is now adjusted for most pleasing picture and the limiter, Q14, merely amplifies the sync signal a little, and the hue range control, L7, is set for the general 3.58-MHz range so that the proper phase may be had by adjustment of the hue control.

Color Demodulators

Now that the two color signals have been produced, they must be combined to produce voltage variations for each of the three primary colors. The demodulators do this. Some sets have two demodulators, usually labeled "X" and "Z". The X demodulator would develop the red signal and the Z demodulator the blue. The green signal would be produced by combining portions of these in a matrixing system.

Other systems use a different demodulator for each of the three colors. A demodulator is a type of detector system that is sensitive to the difference in voltages of two signals that are out of phase. In other words, the more closely the two signals are in phase, the greater will be the voltage output. The color demodulators also are sensitive to amplitude, like video detectors. Thus they not only detect the difference in phase and convert it to voltage difference pulses of new signal, but they also detect the strength of the chroma put into the circuit. The term *demodulator* is logical enough. The information on a signal is called "modulation," as you know, so if the information is taken off the carrier, the modulation is removed.

Figure 15-5 is a color circuit board schematic with a demodulator system that shows all its parts. Later we will deal with those that have integrated circuits that hide individual parts in the demodulator. The outputs at the right of Figure 15-5 go directly to the right inputs of Figure 16-2, which is part of the same circuit board. On Figure 16-2 the top output is the red signal, the middle one is the blue signal, and the bottom one is the green signal. These are labeled R-Y, B-Y, and G-Y, as the Y stands for video, and these signals don't have the video mixed in. In the RCA set we are studying, the video, or Y, signal is added at the cathodes of the picture tube, and the color signals are put in on the first grids of the picture tube.

Much of this circuit is similar to that of the Quasar set we have been studying. It would be well to look over most of the circuits of color and compare them with those of the other set. The tint control, Figure 15-5, is comparable to the hue control. The output of the reference oscillator signal, the constant signal kept on frequency and phase by the burst, is at C of the transformer above Q711 and Q712. In this set the reference oscillator is called the *CW oscillator*.

The output reference oscillator signal is applied to three sets of diodes that are in turn tied to a secondary of T303. All three sets of diodes get the chroma signal from the bandpass amplifier in the same phase. Note that the top of the transformer (T303) goes to the top of each set of diodes (CR717, CR719, and CR722). The bottom of the transformer is connected to CR718, CR720, and CR721. Each would detect the same signals the same way, except that current through the middle pair would go the opposite way from that through the top and bottom set of diodes. It is the reference oscillator signal that will be converted to three phases.

It would have been possible to have left the oscillator phase the same for all three and to have taken the diode feeds from chroma off at different parts of the secondary, or to have used different secondaries to change the phase. In this case, however, the phase of the reference oscillator is changed. The green detection is left as is and put in through R727. For the blue, L710 shifts the frequency enough for blue detection. For the red, L711 and C759 get another shift in phase to produce red signals.

Coils and capacitors, as you have learned, change the phase of a signal because a coil has a delay action on higher frequencies and a capacitor has a delay action on lower frequencies. A slight retardation is all that is necessary, for if it were too long an improper color would result, and the delay might affect the next wave.

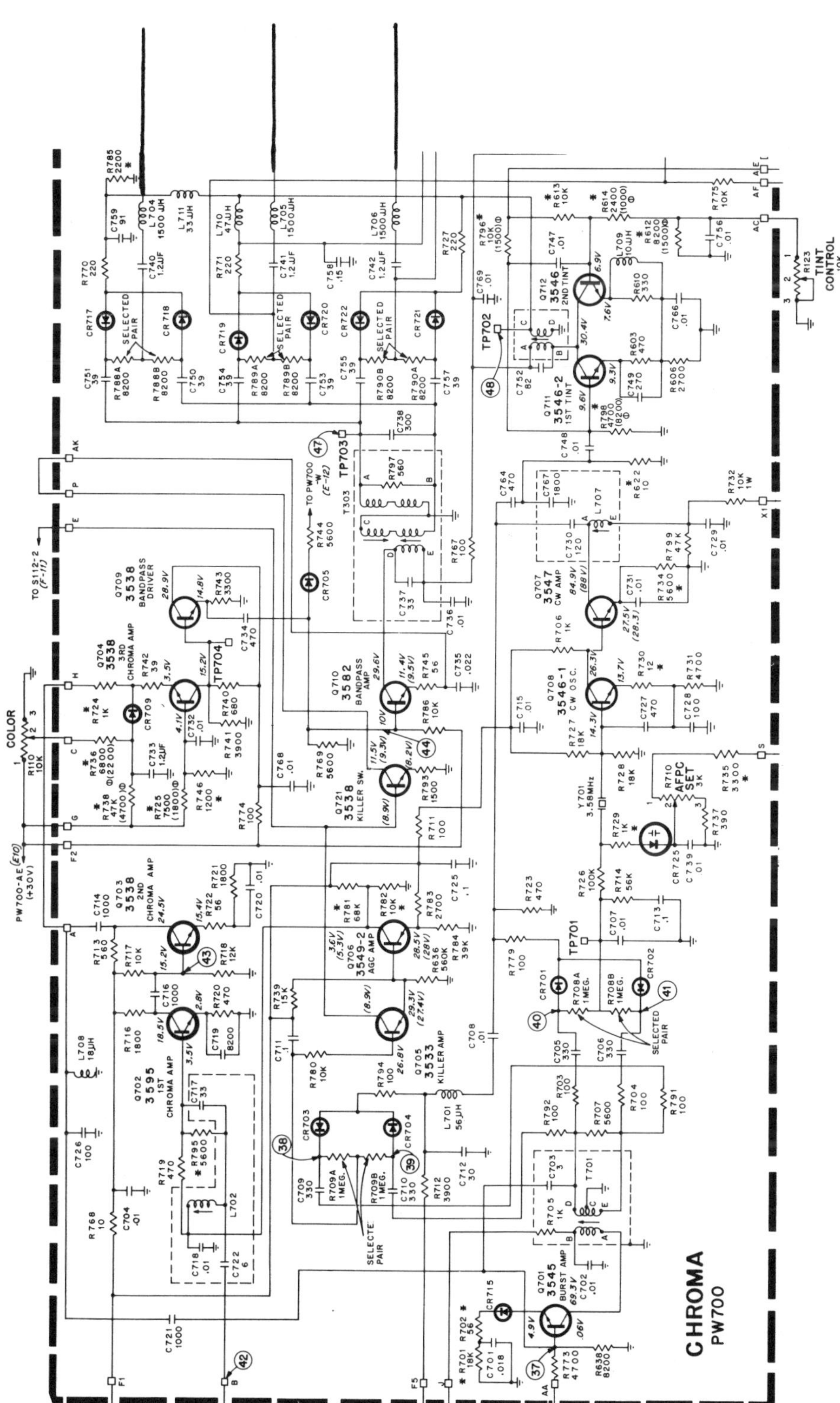

Courtesy RCA

FIGURE 15-5
COLOR CIRCUIT USING SEPARATE DIODE
DEMODULATORS (THIS CIRCUIT CONTINUES AS
FIGURE 16-2)

Here is an example. The oscillator is sending the signal up to all three sets of diode-associated circuits. The red turned on by L711 and C759 is of the phase to enter at the proper phase for red from the chroma. In other words, the oscillator signal is getting there at exactly the right time, and it would get to the other two circuits a little earlier or later. The broadcast signal right now happens to send out a portion of picture that should appear bright red on the screen. This phase signal gets to the top circuit as well as to the other two. The reference oscillator will not be entering the lower two circuits at this split second. To get the red circuits to amplify a current must be flowing from emitter to base of Q713 (Figure 16-2), an NPN transistor. For this to happen current must go from ground up R753, the emitter load resistor, through emitter to base of Q713, then out the left to Figure 15-5. This will be a pulse current and not a bias; the bias is provided with R747 and R748.

The pulses go through L704 and C740 to a pair of resistors, R788a and R788b. The transistor emitter-to-base path is not furnishing a current pulse. Something must provide it, and it must be of positive polarity. For red information a pulse of current would come from the transformer T303, go up through C751, through R770 and down to the reference oscillator transformer at C to D and ground.

The upper resistor, R788a, develops a voltage negative at the top and positive at the bottom, bringing current across from the R-Y driver emitter to base, and Q713 (Figure 16-2) is set for current flow. The reference signal has to be in a downward current flow for this to happen. When it is not, as at opposite phase time, the current through diodes will go only through CR717 to CR718 and back to the chroma output transformer, T303. This current flow balances voltages across R788a and R788b so little or no positive potential opens up Q713 and very little red signal will be developed.

Remember, it takes the proper phase of both chroma and reference signal to cause the

driver to conduct. One thing to bear in mind is that much of the time a blend of colors is required, so when the red is partly in phase, the blue might be too, so both could work to some extent, instead of one being absolutely off and the other on at full intensity.

Since integrated circuit (IC) color systems are here and before long may appear in most sets, it might be well to look at a Zenith color system composed of three integrated circuits and necessary exterior components—Figure 15-6. IC901 is the chroma amplifier or bandpass amplifier, the same as color IF. The input at T13 from the video is coupled to the IC by C901, and a tap on L901, through C903 and into the IC. The color killer is in there, too—note connections A9, A11, A15, and T9.

The output of chroma signal is coupled to the demodulators by C913 and L902. The video is added after the demodulation. See Figure 16-3, which shows the same set. The video goes through the emitters of the color outputs. The outputs of the demodulators in IC902 go to the bases of the color output transistors, as shown in Figure 16-3.

Now back to Figure 15-6. R359 is the tint control that selects the proper color for skin tones. It adjusts the phase of the oscillator and would compare to R901 in Figure 15-2. CR1001 in Figure 15-6 is the 3.58-MHz crystal. It is not inside the IC, as it sometimes needs changing and that frequency demands a larger crystal than could fit in a small IC. It connects to pins 6 and 7 of IC1001.

The oscillator system outside the IC but on the "Sub Gen" module gets two separate phase reference signals to feed to the demodulators by coupling the blue-sensitive oscillator signal at the top of L1003 and the red at the bottom of L1003. The signals across a coil are always out of phase from top to bottom. With two signals instead of three input to the demodulator from the oscillator, this system has a good chance of being an X and Z demodulator type, and the green is derived from the two in the demodulator IC902.

One nice thing about such a system is that if colors are drifting through the picture the

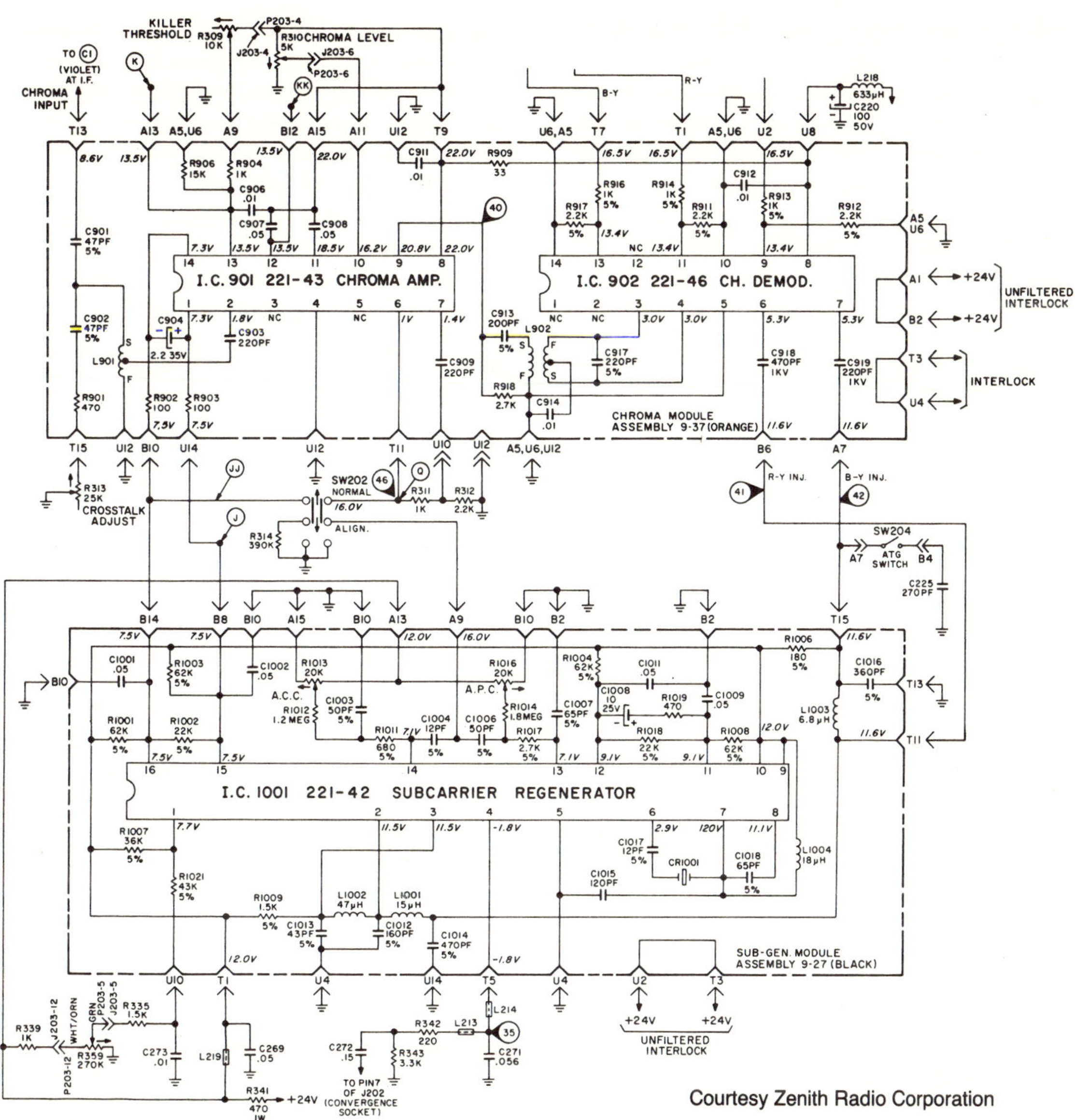

Courtesy Zenith Radio Corporation

FIGURE 15-6
ALL MODULE, IC COLOR SYSTEM

repairman can substitute another IC1001 and see whether it eliminates the trouble. If not, he can replace the entire module assembly, 9-27. If that doesn't work, he should suspect such parts as R359. In time, of course, the module parts must be tested and the module must be repaired. If the IC is bad, then replacing it is the repair. The main disadvantage of this system is the need for substitute IC and module boards. In addition, you can't tell the exact working of parts in this schematic.

The Quasar TV, Figure 15-2 also is composed of IC demodulators. The reference signal is phased for each one by C42 for the red, L8 for the blue, and nothing for the green. Every set has a slightly different system, and wise is the repairman who doesn't try to be an engineer and only replaces parts rather than trying to rebuild a circuit to his liking. At T1 the chroma connects to all demodulator ICs, but note that the video is center tapped into the transformer, T1, from the contrast control. So chroma and video input at pins 6 and 7 of each demodulator. The output of each demodulator goes to an IC amplifier of its own color, not shown, and on to the output transistors for each color, as shown in Figure 16-1.

TROUBLE-SHOOTING COLOR PROBLEMS

Certainly other parts of a TV can make it have poor color. Weak IF stages could still allow sync and sound to be fairly normal but cut out much color. One should always suspect other circuits as well as the color ones when color is bad. As to the color circuits themselves, color troubles usually fall into two categories: (1) colors moving through the picture and (2) incorrect or weak colors.

Colors drifting aimlessly through the picture are a problem of the reference oscillator, burst amplifier, phase setter, 3.58-MHz crystal, and related circuits. Since this system often has a gate controlled by other circuits, the other circuits could be bad also. A slight horizontal sync phase error can certainly affect the reference oscillator circuits. One must be able to distinguish between colors drifting through a picture and a picture in which the entire images are drifting. If the black-and-white part is also moving, the trouble is with the sync that you studied in Chapter 13. If the sync is holding but just colors are moving, then the trouble is with the reference oscillator or other circuits connected to it.

Before assuming that the reference signal circuits are bad, try adjusting the following controls—fine tuning, horizontal hold, color killer, AGC, and noise gate control (if any). If this doesn't clear up the problem then trouble-shoot the color reference oscillator circuits, burst, and so on. Capacitors in the oscillator circuit are possible bad items.

The other category of problems with color

circuits—weak color, no color, or wrong color—indicate bandpass trouble. The problem can be in the bandpass amplifiers, the ACC that controls their gain, or the color killer. Let's consider these separately.

1. *No color or weak color.* Suspect a bad tube or transistor; improper ACC bias on a tube or transistor; poor contact of a tube to a socket or a socket to a printed circuit board, or a transistor making poor contact; low plate or screen grid voltage; excessive cathode bias; defective bypass capacitor; open coupling capacitor; or bad color or color killer controls.

2. *Smeared colors or poor color fit.* The cause could be a gassy tube, an open bypass capacitor, or misaligned slugs in transformers.

3. *Ring in picture.* If a ring occurs in black-and-white pictures, troubles are likely in the IF section; if the ring is in color only, tuned stages of the bandpass amplifier should be aligned.

4. *Incorrect colors.* Look for improper hue control setting or defect, a possible reference oscillator circuit problem, a misaligned transformer, or a defective capacitor in the tuning transformer circuits. If coaxial cables are used between stages, check to see if whether the outer conductor has proper ground.

5. *Intermittent color reception.* Check the antenna and connectors to the set. Color requires better antenna pickup than black-and-white. In color circuits look for places where heat can make parts vary after the set is hot. Transistors, especially, conduct more when warm.

When the problem is no color at all, first check the color killer circuits, bandpass amplifier, burst amplifier, and oscillator. If the colors are incorrect and the hue control will not rectify the trouble, try the burst amplifier stage, and the reference oscillator. And occasionally the 3.58 crystal goes bad. If one of the primary colors (red, green, or blue) is missing, chances are one section of the picture tube is out. If not, the matrix system may be bad (we'll get it later) or the demodulator system should be checked. For instance, if the picture lacks red, check Q17 (Figure 16-1) and the red demodulator IC (Figure 15-2, or with the RCA set Figure 15-5). Check CR717 and CR718 along with R770, C759, R788a, R788b, C740, L704—in short, everything in the red part of the demodulator. Chances are, however, that the trouble will be with the picture tube.

Matrixing is the process of demodulating two signals and combining them in such an order that three emerge. The X and Z systems do this. Some older tube sets have two tubes for color demodulation and then feed to three with a combination of resistors and transformers.

You have a lot to remember about color circuits; here is some of it:

1. The TV station broadcasts two color signals. One is a reference burst signal that occurs at the end of horizontal sync time at precisely 3.58 MHz. The other is a 3.58-MHz signal that occurs during trace time and shifts phase from the burst signal to the degree of the exact color needed.

2. Though both signals are at 3.58 MHz, they don't interfere with each other because the burst occurs between line trace and the chroma occurs during trace time.

3. The signals are separated in the receiver by gates turning each one's amplifier on and off just as the proper signal occurs.

4. Tuned coils select the 3.58-MHz frequency.

5. The burst signal works a crystal oscillator that keeps going in phase even between burst intervals.

6. The crystal signal works another oscillator that makes the signal a reference. In the output the reference oscillator signal can be shifted in phase enough to get good colors. This is a customer control.

7. The bandpass chroma signal is amplified, cut off by the color killer if no color is being transmitted, limited by the automatic color control, and fed to the demodulators.

8. The reference signal is fed to the de-

modulators and can be phased properly for each color demodulator by coils and capacitors, or the chroma can be phased instead.

9. The combination of strong signals in phase with the reference signal makes an output that drives an amplifier of a particular color.

10. Usually more than one demodulator is working for each part of the picture tube picture.

Connect a color bar generator, if available, to the input of the tuner. Note the color bars on the TV screen with respect to hue and placement. List on a paper. Now carefully lower the red drive control of the TV until reds disappear from the screen. Note the pattern of colors that now appear, and compare with before. List the order and appearance. Now advance the drive back for as close to previous setting as you can get it. Do the same with the blue drive and note the colors. Then do so for the green drive control after restoring the blue. This list you have made will aid you in detecting when a color is missing—you will get an idea of how all colors are affected. This lab work can be done without aid of a bar generator, but it will be a little harder to know how all colors have been affected by removal of a primary color.

You can adapt your homemade signal generator, Chapter 7, to produce color disturbances on a TV set. Modify the tuning coil, L1, and feedback coil, L2, by removing turns so that each coil has 6. Connect the leads from the signal generator to the antenna terminals on the color TV.

With both TV and home-made signal generator working you should be able to tune color signals to appear on the TV screen as you adjust the frequency of the signal generator through 3.58 MHz. Try setting the color killer of the TV to minimize this pattern. Remove one lead from the generator. Now the color killer of the TV should be able to eliminate the color disturbance made by the generator. This demonstration represents static and its effect on color reception.

QUIZ FOR CHAPTER 15

▶ 1. What is the smallest number of demodulators that could be used in a color TV system?

▶ 2. If a picture never has blue in it and if the picture tube and blue adjustments for the picture tube are good, what other circuit components or stages should be checked?

▶ 3. A color killer control set too high would produce what effect with weak signals? If set too low what might be noted?

▶ 4. Why is a phase shifting circuit needed for the reference oscillations?

▶ 5. If the colors are scrambled and the outlines of the picture are likewise scrambled, what stage of the TV should be checked?

▶ 6. Why should a set have a color killer?

▶ 7. What device in the video circuits prevents the video from arriving at the mixing point before the color?

▶ 8. At what two places is the video mixed with the color?

▶ 9. What are the three primary colors for TV?

▶ 10. What components rephase the reference oscillator signal for the different demodulators?

▶ 11. For IC and component board circuits, explain which part to check first, how to check it, which part to check next, and finally what to do if nothing so far fixes the set.

▶ 12. Why should antennas, tuners, and IF stages be checked when there is a color problem?

▶ 13. How can improper horizontal sync affect color reception?

16. Video and Matrixing Circuits

PROJECT: Adjusting Controls

YOU HAVE learned that the video is the light or black-and-white signal. In monochrome, or black-and-white reception, the video is all there is. With color reception the video adds the outlines of the picture and adds brightness as needed. The video amplifiers strengthen the video and often one or more of them amplify the color before it is sent to the color circuits.

In Figure 15-2 the first video amplifier also amplifies color and sync signals. The video goes to the second video amplifier, Q3, through the delay line. Sync also goes through this circuit. The sync is removed from the collector of Q3, and the video is taken on from the emitter circuit. The contrast control determines how much video will be sent on. If set low, a dimly outlined picture will result. If set high, a lot of black and also very bright parts will be in the picture. A setting somewhere in between produces a normal picture. From the contrast control, R104, the video is taken to the output transformer, T1, of the color chroma circuits, and very early mixing of video with color results. You have studied this, but repetition never hurts.

With Figure 16-1 we can continue. Since the video is already mixed or matrixed with the color signals, it all goes in on the cathodes in the picture tube. The grids can be adjusted for either 38 or 75 volts, whichever works best. No signal goes in there.

Now let's see how the color signals get to the picture tube. Each color has its own amplifier. The tuned circuits to each base of Q17, Q15, and Q16 keep all but the color signals out. Each transistor has emitter bias. There are fixed resistors, R90, R75, and R82 plus an adjustable resistor for each color: R89, which sets the red bias and therefore the drive of the transistor; R76, which does so for the blue; and R83 for the green. The adjustment determines just how much each can amplify. Therefore, signal strength is controlled, and since cathode current for each beam depends on this, the amount of each color on the screen is controlled. A fairly large current flows here; it flows from ground and (for blue) through R75, R76, base to collector of Q15, through R13, R200, middle cathode, as an electron beam to the picture tube viewing screen, to anode and back to high-voltage rectifier, and so on. These transistors are power types and, for transistors, take very large collector voltages. The voltage regulator controls this from a 305-volt power supply.

The setting of these drive controls determines the background of a certain color, and they are set to avoid tinted backgrounds. If a green haze always appeared in the picture, the green drive control, R83, would be adjusted just enough to remove it. All three drive controls (sometimes called *background controls*) should be set just shy of tinting the picture with their own primary color.

The screen grids in the picture tube, labeled G2, are also set by variable resistors. In this case, R204, R205, and R206 are the screen controls, set for proper intensity for each color. They peak each color as much as needed. Since the positive voltage on them attracts more electrons than otherwise would be the case, they increase the beam of their particular color. Usually screen controls are adjusted by flipping a service switch in the

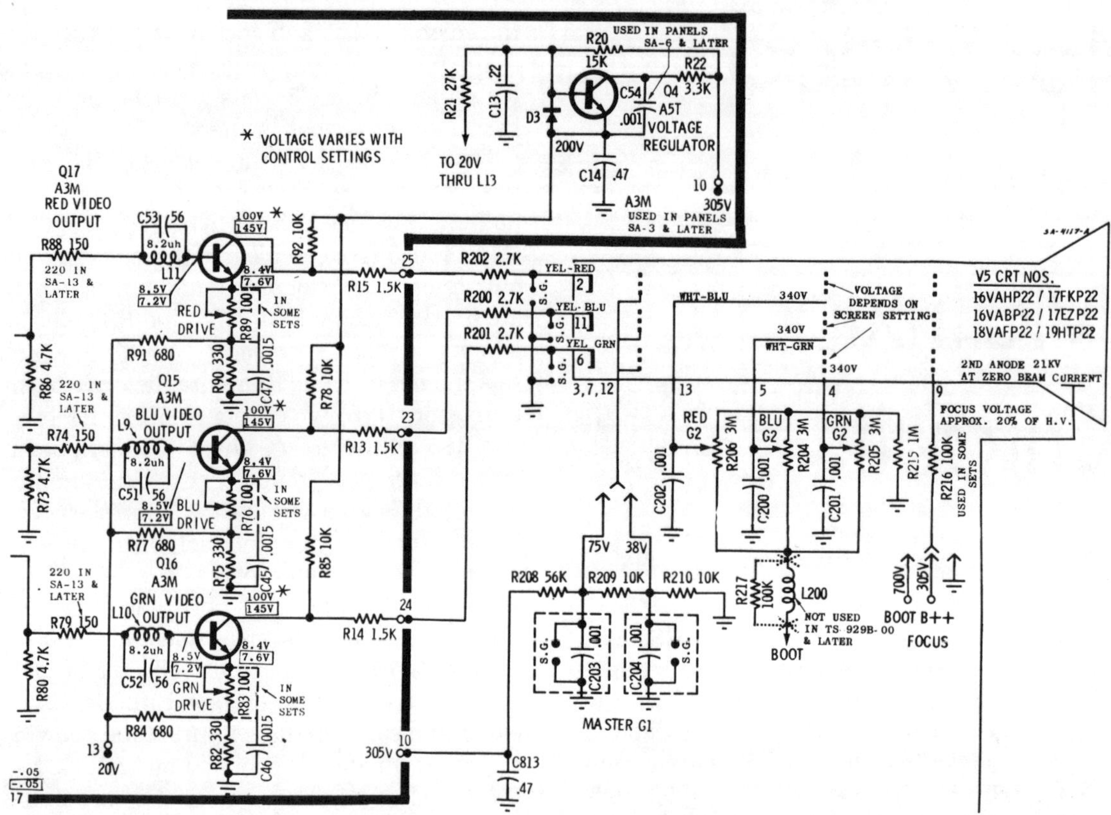

Reprinted with permission of Quasar Electronics Company

Figure 16-1
Chroma Video Outputs and Picture Tube

set. A thin line appears for each color, and one at a time they are run up until a thin line just appears on the picture tube in that color. (More on this in Chapter 18.) Since the color guns don't weaken at the same rate, the drive and screen controls must be adjusted from time to time. A gun in the picture tube consists of cathode, its associated control grid, screen grid, and accelerator or focus grid. Some repairmen always refer to the green gun, red gun, and so on.

The Zenith system you have been studying combines the video with the color in the color output amplifiers (see Figure 16-3). The second video amplifier, Q205, puts the signal to Q206, the third video amplifier, through the delay line. The output of this amplifier is into the emitters of the color output amplifiers. Current must go through ground, collector to emitter of Q206 (a PNP), and then through each of the color outputs. The video signal gets into them by direct line with small bias resistors, R272, R273, and R274. The chroma demodulator IC902 in Figure 15-6 delivers color signals to the bases of the output amplifiers. At the collectors the video and proper color are combined. The gain controls R276, R281, and R283 are the drive or background controls you saw in the other set. They are adjusted just shy of tinting the picture. The outputs go to the cathodes of the picture tube. See Figure 13-4. They come in at No. 47, 49,

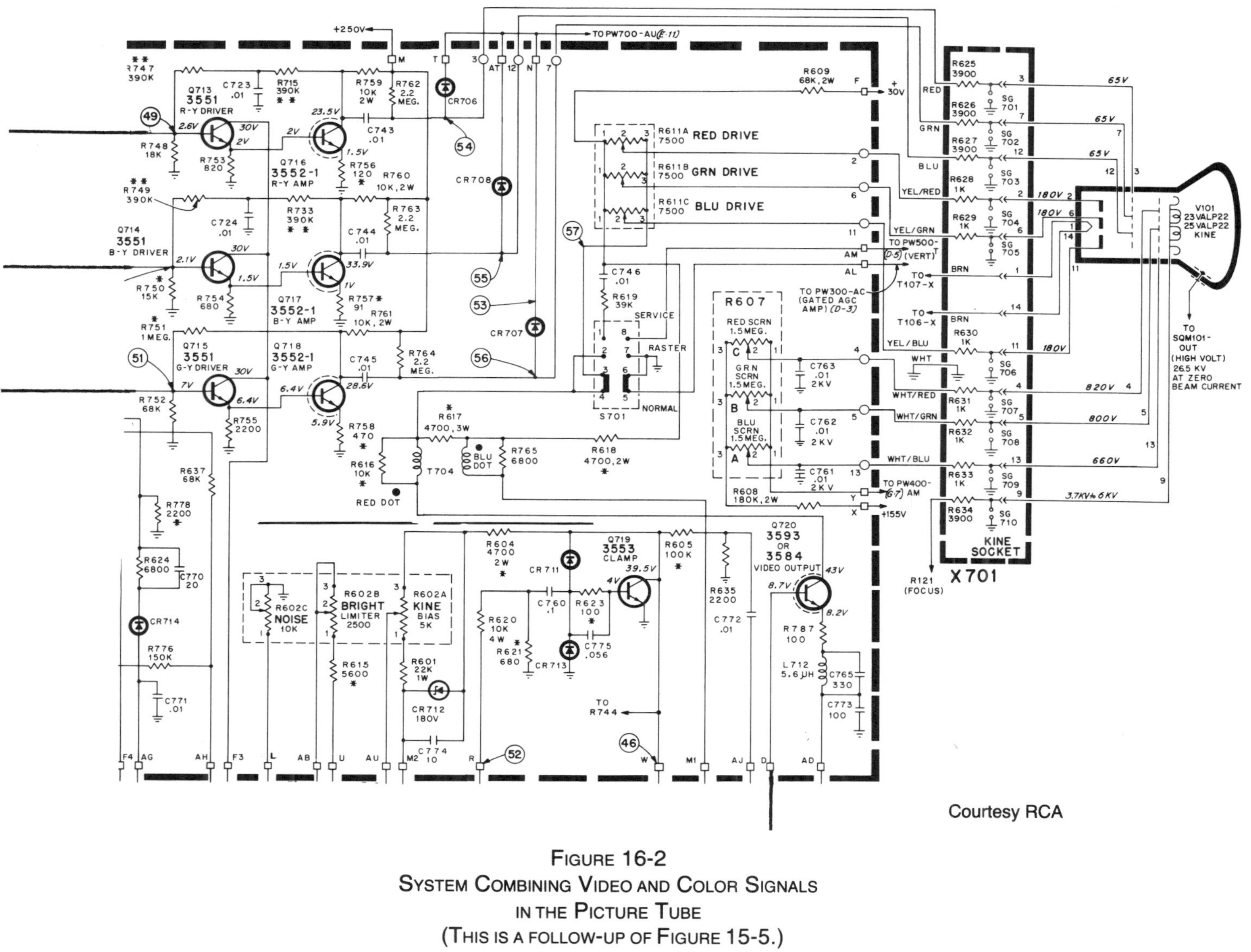

Courtesy RCA

FIGURE 16-2
SYSTEM COMBINING VIDEO AND COLOR SIGNALS
IN THE PICTURE TUBE
(THIS IS A FOLLOW-UP OF FIGURE 15-5.)

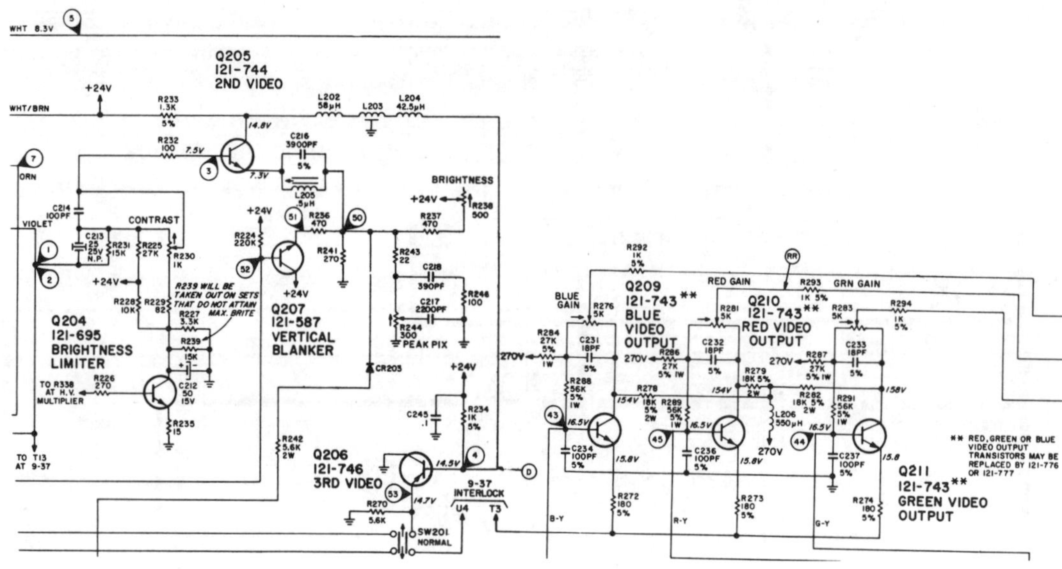

Courtesy Zenith Radio Corporation

FIGURE 16-3
VIDEO AMPLIFIERS AND CHROMA-VIDEO MIXERS

and 48. The screen grids are set as before for proper peaking of each color by R296, R298, and R297.

You have seen sets that combine the video and color before demodulation, in the last color amplifiers, and now let's consider a set that combines them in the picture tube (Figure 16-2). The video output transistor, Q720, sends its signal up to three color drive controls. The color signals are not here at all, but by adjusting the amount of video for each cathode, the amount of color is also controlled. So drive is merely setting the video input for each cathode of the picture tube. Now let's see how the color signals get to their appointed places. Earlier we discussed a current that traveled out the left of the schematic to allow a red signal to continue. As current pulses go left, they must go through R753 (for red), and it would develop a voltage positive at top. This (+) side of R753 is connected to the

base of Q716. This puts positive voltage on the base, and since Q716 is also an NPN transistor, current goes emitter to base, and a larger current goes through R756, emitter to collector, through R759, the load resistor, and to power supply. The signal is coupled by C743 through R625 and to the red grid of the picture tube. The blue and green systems work similarly. The screen grids are set by a system labeled R607 A, B, and C. As a laboratory exercise, slightly move each screen grid and drive control and rate the effect on the picture.

Important things to remember with video circuits and matrixing are:

1. The video can be combined with the color either before or in the picture tube.

2. Video makes the picture light or dark in any part and gives the most distinct outline.

3. Setting the drive controls sets the background of the color for the picture tube.

4. Setting the screen grids controls the peaking of each color.

QUIZ FOR CHAPTER 16

▶ 1. What other signals besides video does the video amplifier system sometimes carry partway through?

▶ 2. Describe in detail how these signals are removed from the video amplifiers.

▶ 3. Why do screen grids need to be set from time to time?

▶ 4. Why do drive or background controls need setting occasionally?

▶ 5. Tell how each of these controls is set by looking at the picture tube.

17. Repair Procedures

PROJECT: Finding Simulated Faults

ALTHOUGH THIS entire book is based on the assumption that you will be repairing television sets, much of it has been concerned with teaching about the operation of the various stages. This emphasis is necessary, as you must understand how a device works before you can repair it sensibly. In the lessons to date, however, considerable repair information has been given, and this chapter ties the ideas together to present TV repair as a logical sequence of events. The best way to service a TV set is first to note its faults or symptoms. Then you should figure out logically which stage or stages could be causing this trouble. Finally, you narrow the problem to a stage, then to a part of a stage, and then search for the one or more components that are at fault.

USE OF PICTURE AND SOUND ABNORMALITIES TO DIAGNOSE TROUBLE STAGES

In many cases the TV picture itself tells you which stage is defective. The condition of the picture, the sound, frying noises, odors—these often clue you to the source of trouble. Figures 17-1 and 17-2 show typical defects as shown on the TV screen, and Table 17A lists symptoms and their probable causes. Let's analyze them.

The entirely dark screen at least tells you that the picture tube is bad or something that produces its voltage is bad. First be sure that the set is turned on, that power is available at the wall, and that the circuit breaker is connected. These obvious things should not be overlooked. Turn up the brightness control and be sure it hasn't just been turned very low. Now look for a light in the neck of the picture tube. If it is absent, either the heater is getting no voltage, or the connection to the heater is bad, or the heater is broken internally. The latter case requires replacement of the picture tube. If the TV is a tube set and the other tubes are not lit either, try measuring a voltage here and there; if you don't get any, the low-voltage power supply must be at fault.

If the screen is black but the sound is working, then the tuner and IF must be working to carry the sound, so the trouble is in the high voltage, provided a heater light is visible in the picture tube and the picture tube is not defective. You can check the picture tube with a tester for that purpose; a schematic for such a tester appears in Appendix B.

Now suppose the set has a raster but no picture appears even with the best of adjustments. You have two choices (see Table 17A). If the sound is poor or absent the trouble is likely in the tuner, antenna, IF, AGC, or the first video amplifier if the sound goes that far before being removed. Absence of sound tells you to search in circuits that are common to sound and picture. On the other hand, if the sound is good, then the later video stages or the picture tube must be bad to block picture information.

A weak picture (weak on video as well as colors) suggests antenna, tuner, IF, or AGC set with too much cut-off; and if the sound is good probably the video. If the sound is poor suspect the antenna, tuner, and so on. If the picture and sound are bad on only one channel, be sure the TV station is putting out a good signal by confirming with another set. If only the set under repair is getting weak picture and sound on that channel, suspect the tuner wafer switches or tuner coils.

Poor to no sound while everything else is good obviously means a poor discriminator or ratio detector or a bad audio amplifier or

TABLE 17A

PICTURE SYMPTOM SERVICING

Monochrome Servicing

Sound	Picture	Stage Bad
1. None	Screen black, no raster	Low-voltage power supply
2. OK	Screen black, no raster	High-voltage, horizontal system, picture tube
3. Poor to None	Raster but no picture	Tuner, IF, video amplifiers, antenna system, AGC
4. OK	Raster but no picture	Video or picture tube
5. Poor to None	Weak picture	Tuner, IF video, antenna, AGC
6. Poor One Channel	Weak one channel	Tuner
7. Poor to None	Good	Discriminator, audio amplifiers, speaker
8. Buzz	Tearing or negative picture	AGC, bad IF stage, tuner
9. Buzz	One or two bars across picture or waves in picture	Poor filtering in low-voltage power supply, cathode-to-heater short in a tube somewhere
10. Usually Normal	Flipping	Vertical oscillator, hold control
11. Usually Normal	Foldover up from bottom	Vertical trouble, usually cathode bypass capacitor or bad tube
12. Normally OK	Wider at top or bottom, or wider at left or right side, keystoning	Deflection coils or yoke; replacement of entire yoke normally necessary
13. OK or Noise	Ring	Unwanted oscillation in tuner, IF video circuits
14. OK to Noise	Shows Christmas tree display	Horizontal oscillator or horizontal AFC
15. OK to Noise	Picture tears or has several pictures side by side, overlapping	Horizontal oscillator or AFC
16. Usually OK	Picture too narrow side to side	Horizontal amplifier, occasionally deflects coils
17. Usually OK	Picture too short	Vertical amplifier, occasionally deflection yoke
18. OK to Distorted	Bars form in sync with words being said	4.5 MHZ sound trap
19. Sizzling but not from Speaker	Lines in picture, purple high-voltage seen in set	Arcover; clean high-voltage cage

speaker. Sometimes the tuned circuit that separates sound from the video needs resetting.

TVs are subject to two buzz problems with different effects in the picture. A buzz caused by leakage of 60-Hz AC into the signals will often make a pattern on the screen like that shown in the center left picture of Figure 17-1. Many power supplies have two diodes, each one detecting half the AC. This changes the 60 Hz to 120 Hz ripple even though it changes to DC. Recall that the capacitors and choke remove this ripple as much as they can. If they are defective, 120 Hz would be in the circuits, and since the vertical frequency is 60 Hz and the AC pulses are in the DC at 120 Hz, two lines will appear on the screen as shown. If there were leakage from a heater to cathode in a tube this would be 60 Hz and would cause one gigantic bar across the screen.

The other type of buzz is not accompanied by bars across the picture. It may have a tearing picture such as the one at center right in Figure 17-2 or a negative picture such as Figure 17-1, center right. It is true that the sync is affected by the tearing picture, but a problem earlier in the set is making the sync behave badly.

Learn to tell the difference between a buzz with this problem and a high-pitched whine that might be heard with a twisted picture

Too Narrow & Distorted:
Horizontal Amplifier

Too Shallow:
Vertical Amplifier

Tilted Picture:
Deflection Yoke Crooked

Sound Bars:
4.5 M.C. Trap Circuit

Only White-Line Raster:
Tuner-IF or Video or
Antenna

Entirely Dark:
Low- or High- Voltage Power
Supply or Horizontal Sync

Negative Picture, AGC Bad:
Too Strong a Signal (unusual), Defective
Tube in IF or Tuner, Defective Parts
Same Places

Smeared Picture:
Bad Capacitor or Resistor in IF Stage or
Bad Peaking Coil in IF or Video Stage

60-cycle AC:
Some Tube Cathode-to-Heater Short or
Poor Power Supply Filters

Complete Loss of Sync:
Trouble in Sync Separator

Ring:
Unwanted Oscillation in IF or Video Stages

60-cycle AC Interference

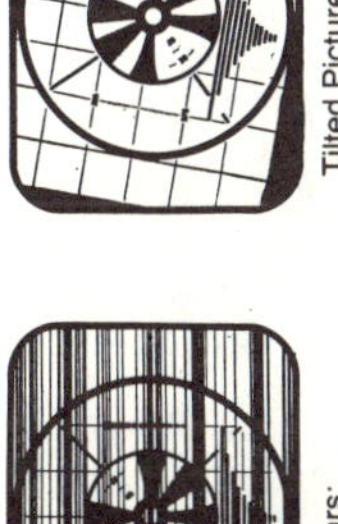

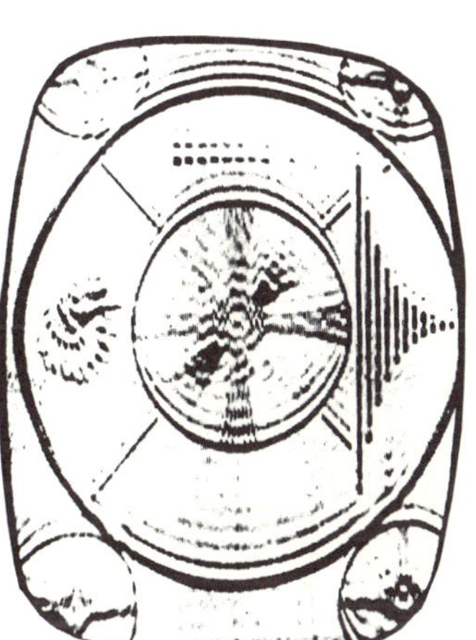

FIGURE 17-1

Foldover:
Vertical Sync, Often Cathode Bypass Capacitor
Loss of Horizontal Sync:
Horizontal Oscillator or Horizontal AFC
Loss of Horizontal Sync:
Oscillator Trouble Likely
Flipover:
Vertical Oscillator
Christmas Tree:
Horizontal Sync
Tearing:
Horizontal Sync Stage
Keystoning:
Vertical Deflection Coils Shorted
Keystone:
Horizontal Deflection Coils on Picture Tube Shorted
Horizontal Phase Off:
Horizontal Phase Circuit or Oscillator

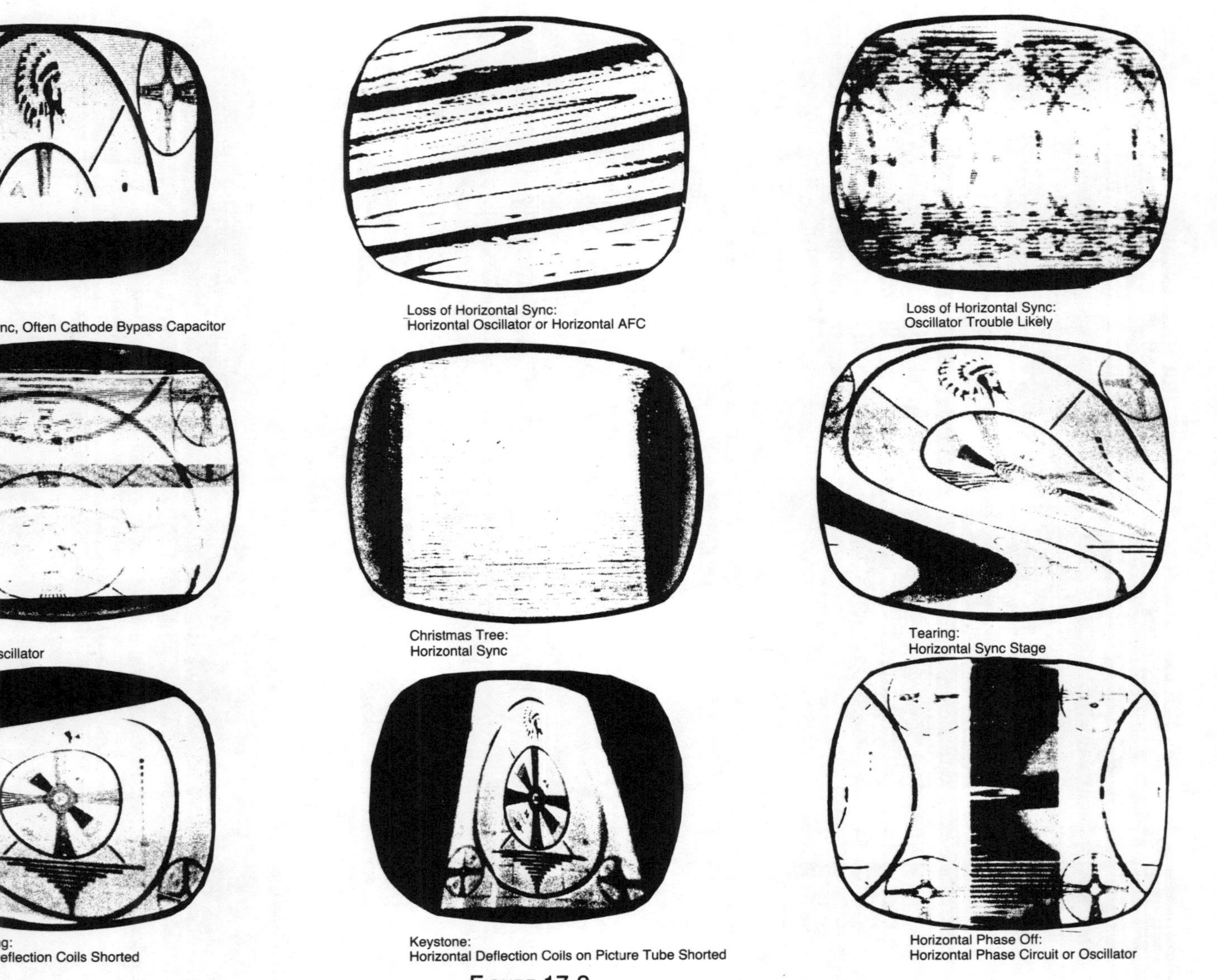

FIGURE 17-2

when the horizontal sync is at fault. The buzz and negative or tearing picture indicate too much signal. The station might be transmitting that way, but normally the AGC is set too high or is defective. The large signal is too much for the IF stages, and either some of it is clipped off or the video is getting into the sync stages because it is so tall. This would make the sync work even on video signals and, of course, out of time. The video and maybe the color are being detected as frequency variations; hence the buzz in the sound. A weak IF stage might not be able to handle a normal signal, or the tuner may not be working properly.

SYNC PROBLEMS

Several problems relate to sync. If the picture drifts both horizontally and vertically or can't be centered in either part, the trouble is in the sync separator, or possibly, though rarely, the IF or video could be sending bad sync to the sync separator. Figure 17-1, lower right, shows this condition.

Vertical flip sometimes just indicates the need for adjustment of the hold control, but if the setting is quite nervous and changes from time to time in normal viewing in a decent viewing area, the vertical oscillator is usually at fault. The exception is if turning the contrast control up makes the vertical sync roll on the screen; then usually the sync separator is letting video through and is not biased properly. A folded over picture (Figure 17-2, upper left) or a picture that won't fill the screen usually means a vertical output problem. In tube sets often one glass envelope houses both tubes, the oscillator and the output. With transistors it is more of a chore to find the problem. It certainly may be other than a tube problem; a cathode bypass capacitor in the output stage may be defective.

Horizontal stage problems often show in symptoms such as that illustrated in Figure 17-2, top center. If the horizontal oscillator is off frequency a little it will produce lines as

shown. If it is off frequency a lot several images will appear, as shown at the upper right, or the Christmas tree effect may appear (center), accompanied by a whine in the speaker. Horizontal tearing is the beginning of the condition shown at top center. The improper phase as shown at lower right is usually the fault of the phase control diodes or tube or transistor that controls the horizontal frequency just before the horizontal oscillator. In the other troubles the phase control rather than the oscillator itself may be bad.

A picture in which lines are not traced the full length of the screen—that is, a short picture horizontally—is often caused by a bad horizontal output circuit. This may be accompanied by picture dinmess, as the high voltages may be affected too. Sometimes setting the high-voltage bias, horizontal drive, or horizontal output bias will correct this condition.

A keystoning picture, wider at one point than another, indicates a deflection yoke problem—either the coils or a bad resistor or capacitor connected in or near the deflection yoke.

A picture with ring, Figure 17-1, bottom center, can be caused by misadjusted IF coils that are producing oscillation in the IF stages, or it can mean bad components, usually in IF but possibly in the tuner or video. Do not worry about this condition if it can be easily tuned out with the fine tuner. In some areas and with some sets this ring will appear with certain settings of the fine tuning.

A tilted picture is corrected by loosening the yoke locking nuts and rotating the picture to straighten it out. With a color set be sure you don't move the yoke forward or backward, or purity and convergence will need to be set. Blinking bars in the picture in time with the sound mean that sound is getting into the video circuits. A 4.5-MHz trap must be adjusted in the early video or at the detector to stop this.

A too shallow picture can look as it does in Figure 17-1 or can be just a thin line across the screen. Usually this means a completely inoperative vertical sync stage. Turn the brightness down first so that the line won't create a permanent problem on the screen. Check to make sure the normal-service switch has not been flipped to service position by mistake. If it is normal then the vertical sync is at fault.

The 60-Hz AC interference may just twist the picture and not buzz the set (lower left, Figure 17-1). In this case the trouble is more likely to be, but is not always, in the video or sync circuits, past the place of sound takeoff. Sometimes this AC interference will form a perfect 60-Hz sine wave across the screen, just as you would see on an oscilloscope. Heater-to-cathode leakage of some tube is likely to cause the 60-Hz ripple. Figure 17-1, center, shows a smeared picture. Colors may or may not smear with the video. As explained under the illustration, this is usually a problem involving the IF, usually a coupling problem between transistors or tubes in the stage such as a peaking coil, the IF coils that tune to a certain portion of the bandwidth. L102A or B is such a coil, Figure 12-6. A defective coupling capacitor can also cause smear in the picture.

COLOR PROBLEMS

Color symptoms in the picture are covered by Table 17B. Observe carefully to see that a color symptom is truly a color trouble and not a general problem with the video. For instance, trouble 6 (whole picture travels) involves both the color and the video. We have covered that. On the other hand, trouble 5, colors moving through the picture while the outlines and black are not scrambled, more likely indicates a color circuit problem.

Many color problems are really troubles in the IF, tuner, or another component. If the color circuits are at fault *only* the color will be incorrect or will occur at the wrong time. Remember, the color travels through tuner, IF, and part of the video circuits. If servicing the color circuits reveals nothing, try these other places. Jittery or misaligned colors can also be caused by ripple of AC in the circuits. An antenna usually needs better orientation toward the station for color than for black-and-white. Trucks moving through also create more interference in color than in black-and-white. And distance from station is more critical for color.

Use Table 17B as follows. First find the symptom as shown on the picture tube. Then make the adjustment indicated to see whether that will clear up the problem. Then look in the right-hand column for instructions relative to whether the control adjustment helped or not and what is possibly the problem. Of course, if the adjustment stopped the problem and it does not recur, then adjustment was all that was needed.

When one color is weak or out, suspect the picture tube gun of that color. The trouble also could be the color output amplifier or its setting; or part of the demodulator might be at fault; or somehow the bandpass might be set to cut out one color. But the most likely culprit is a bad picture tube gun. Normally the red gun deteriorates first and gives weaker emissions at the cathode. When the combination of colors is all wrong in the picture, try turning up the screen grid controls one at a time and see that you have all three primary colors. Then reset them as explained in Chapter 18 for a good white highlight of the picture.

Testing the picture tube is possible with a tester-rejuvenator, as explained in Appendix B. You test one set of cathode, grid, and screen grid at a time, or one gun at a time. If it tests low, you can try to rejuvenate it. Sometimes a gun will have internal shorts. I remember an advanced student I once taught who had a small portable color set. He tried rejuvenating the red gun to no avail. Finally he started tapping on the neck with the rejuvenating voltage on, and red came to life. Somewhere in that tiny little neck of picture tube was a minute short that he jarred and voltage-shocked loose. He was already taking service calls, and the customer was glad to get the set fixed reasonably, since it seemed to have a bad picture tube. If you try this, tap with care.

TABLE 17B

COLOR TV TROUBLESHOOTING

Symptom	Adjust	Likely Fault
1. Weak color	Channel selector	If it helps, tuner contacts
	Fine tuning	If only good at very end of control, check alignment of tuner and IF and maybe replace a tube
	Color control	If it helps, fine; if must be advanced to maximum, weak video, IF, or tuner stage; if it won't help, trouble in bandpass amplifiers, picture tube, or high-voltage
2. One primary color missing	Screen grids and drives of picture tube circuits	If no help, either picture tube has one dead beam or troubles in matrix or bandpass amplifiers or demodulators; see text
3. All colors wrong color	Hue control	If not easily adjusted, trouble in automatic color control or reference oscillator circuits
4. Colors too bright or dim	Color control	ACC or color killer
5. Colors travel through picture	Horizontal hold	If it solves problem, maybe horizontal trouble or just needs adjustment
	Hue control	If OK with slight adjustment, no problem; otherwise reference oscillator
6. Whole picture travels	Horizontal or vertical hold	Horizontal or vertical trouble, not a color problem
7. Colors slop over their outlines	Fine tuning	Usually to no avail; trouble either in
	Hue, color	bandpass amplifiers or set needs converging; also ring possibilities in color bandpass or video, IF tuner
8. No color	Advance color control	If nothing then
	Advance color killer to give less killing	If nothing, is there luminence on screen? If so, then color killer or bandpass amps or demodulators; if no light on screen, suspect high-voltage
9. Streaks of color through screen	Color killer	If no help, troubleshoot color killer circuits
10. Background tint of color to picture	Drive control	If no good, try degaussing (see Chapter 18); maybe bad picture tube

General weakness of colors, if they are correct in hue, is often a chroma bandpass problem, though it could be a weak IF that manages to get enough sync and picture through so that only the color is not amplified enough. The color control mentioned in fault No. 1 is the intensity control, not the hue control.

If the colors are wrong in hue (green-faced people, orange grass, that sort of thing), try the hue control. If it won't bring them in properly, the hue range control in the color circuits should be adjusted. If this won't help, the reference oscillator is out of phase. It would be on frequency and out of phase and staying the way it is. If it were drifting in and out of phase the colors would vary on the screen, being correct at times and incorrect at others, or colors would drift through the picture. Check out the reference oscillator circuits for some component that is keeping the oscillator out of phase, or check the automatic color control circuit; it might be holding the phase exactly wrong instead of exactly right.

If colors drift through the picture the reference oscillator circuits are at fault. It could be that the burst is not arriving; the 3.58-MHz crystal might be defective; the burst oscillator might be unstable; or the color sync gate might be opening and closing late or early. The phase shifter may have a bad component, or a coil in the demodulator might have changed value. This last, however, should cause incorrect color rather than drifting color.

Poor color fit, No. 7, is often hard to service, as so many things can cause it. Improper setting of the color killer can cause no color, static in color, or colors that are too vivid or too dim. Always try adjusting the color killer when you have color trouble. The automatic color control associated with it can also cause such problems (see also troubles 8 and 9). Problem No. 10 is covered in Chapter 18.

To summarize, weak color usually indicates bandpass amplifier trouble; no color means either color killer trouble or bandpass amplifier trouble; incorrect or moving colors are usually a reference signal or burst signal problem; and a missing color is either the picture tube or an output color amplifier trouble.

FINDING THE FAULTY PART

Finding the troublesome stage is easier than finding the faulty component. In some cases one of several stages could be at fault. In other troubles you know the bad stage, but since it has several tubes or transistors, you need to know which particular part of the stage is causing the trouble. It would be a huge task, however, if you knew you had an IF problem to check out every capacitor, transistor, resistor, and coil in that system plus all connections for poor contact. You need a way to pin down the fault closer than just knowing the stage.

One method you have studied is the use of a signal generator. With a set that produces no picture you can start injecting a signal at the video stages and watch for an indication on the screen. If you get it, you backtrack toward the antenna until the signal quits. This is a most efficient method for finding a weak or inoperative stage. It has less value for distortion problems. For instance, if a tube is injecting 60 Hz AC by leakage from heater to cathode, the tube will still amplify in some cases. You could put signals in all the way back to the antenna but you could not tell where the AC was getting in.

For checking distortion in signals or for signals that are incorrect in shape or size, or are entirely wrong, the oscilloscope is a very useful instrument. It will show on its screen the pattern of the wave you are checking. You can see whether it is distorted, too tall, too short, or absent. Though on-the-air signals, especially the video, are constantly changing in size and other charactistics, some troubleshooting can be done with a scope and an on-the-air signal. Other times you should inject a steady pattern at the antenna terminals with a crosshatch or bar generator. Usually these are set to operate on one specific channel, and you must set the tuner for it. Needless to say, it pays to read over the manufacturer's operation manual for any piece of test equipment.

Suppose you suspect a color trouble. With Figures 15-3 and 15-2, you could start with the scope set for vertical sweep rate, 60 Hz, and see whether you could get a pattern similar to W2 at the base of Q3 or at the delay line, whichever one is easier to find. If it looked good, with the vertical size of the scope pattern showing about 0.6 volts, you would proceed to look for W3, the collector signal of Q3. The signal should be inverted and much larger; the waveform shown says it would be 6 volts. It may be necessary to cut down on the vertical gain of the scope to get it all in the picture. This is good, as it shows the transistor amplified the signal. You could then look for the chroma signals at W4 and W5. This set calls them "color IF."

Note the H on the scope samples. You would need to set the scope sweep for 15,750 Hz. The appearance of the waveforms would vary depending on what color is being received. To get a steady color, a rainbow pattern could be put in by the generator, or a station could be observed at ID time when the picture is still. The main thing is that the chroma signal should be larger at W5 than at W4. If it is not, you may have a lazy amplifier or not enough voltage for the collector of Q2 (R8 may be changing in value from 10,000 ohms to 20,000 ohms, for example.)

Then you could look at the combination of chroma and video at W6. If you get a good pattern (change to vertical sweep rate) video and chroma are probably OK. Then you would start checking the reference signal. A quick way to test it out would be to look at W15, W16, and W17 at the 1, 3, and 2 pins of the demodulator. Using horizontal rate of sweep, you would see whether the waveforms somewhat resemble those shown. If they do, then the trouble must be on toward the picture tube. On the other hand, suppose you get waveforms at W15 and W17 but none at W16. The reference oscillator, called "XTAL Osc" here, is working, or W15 and W17 would be out too. That leaves just three components to check unless there is a short in the blue demodulator. Check R68, C41, and L8, because one of them must be bad.

If everything is down—that is, if you get nothing at W15, W16, or W17—start backing up until you find the waveform. If you get a steady reference with horizontal sweep on the scope as shown at W14, then start in with a voltmeter and hunt for troubles with the phase splitter, hue control or phase shifter, or limiter, hue range coil C3, and other parts. If nothing is amiss after checking voltages, unplug the set and check with an ohmmeter each part that is in that circuit between Q10 and the demodulator. Of course, if you found no waveform or an incorrect one at W14, you would back up until you found something.

An oscilloscope is also very useful in checking out sync troubles. See Figure 13-2. Sometimes the sync tears and jitters so much that you can't tell whether the vertical or the horizontal or both are at fault. In such a case, you would look at the input to the sync separator, W20—in this case the base of it. The video probably won't be standing as shown but will

be constantly falling and rising with the picture changes. The sync should be steady, though. Naturally, the sync pulses are the low parts. Since this is vertical rate, the horizontal sync is compressed too much to show.

At W21 the video is removed, as you learned before. If it is not on your scope trace, you have found the trouble—the sync is being triggered by video at the wrong time. If everything is OK here, you can eliminate the sync separator as the problem and go to the vertical oscillator. Check its output at W26, the plate of V3B. Then check the horizontal oscillator output at W32, and see which one is nervous. By skipping a few waveforms you can tell if the trouble is between there and the one you checked before and then check the intermediate ones if it appears to be. This is the quick way to do it.

Always pay attention to the sweep rate, whether vertical at 60 Hz or horizontal at 15,750 Hz. Older scopes may be set up in cycles rather than hertz, but they mean the same thing. Always pay attention to the size of the waveform too. With many scopes you can calibrate them, with an inch being equal to so many volts. The instruction book with the scope will help. In some cases you can troubleshoot without using a voltmeter with such calibration.

Quick scope troubleshooting may be done by checking the waveform at the video detector output. Set the scope for a 3-volt signal to fit in the circle of the viewing screen. Use 30-Hz sweep. A pattern like that shown in Figure 9-3*a* would indicate that tuner, IF, AGC, and detector are OK, if you check on all good channels. Then the trouble is likely in video, color, and so on. Change to 7,875 Hz and see whether the waveform like that in Figure 9-3*b* forms. Now you can check vertical sync. If both are OK at the detector, but you have bad sync at the picture tube, such as tearing, flipping, multiple signals and the like, the trouble must be in the sync circuits and not in the IF. This is really worth knowing.

Placing the scope at the detector is something like injecting a hum at the volume control on a radio. You know which half of the set contains the trouble rather quickly. (Naturally, if the picture clues tell you the logical stage, this is a waste of time.)

Once you have the stage narrowed down, and the scope or signal generator has narrowed the trouble to a particular tube or transistor and its surrounding parts, start the voltage readings as recommended by the manufacturer on his schematic. Any large variation should suggest a possible part at fault. Usually the ohmmeter is the final testing apparatus. Then finally you will try a substitute part to substantiate your diagnosis of the faulty item.

Some shops have a substitution box containing capacitors of many values and a switch to change from one to another. Two leads with alligator clips come from the box so a substitute capacitor can be tried before you go to the trouble of removing the old one. In a few cases there will be a warning in the schematic not to try parts this way, but often this technique can be used. If the capacitor under suspicion is open, substitution is a valid check. If it is shorted, substitution will prove little. You would need to desolder and lift out of the circuit one end of a capacitor you suspect to be shorted. Then the substitution box would be connected to the circuit both ways, not to the suspect capacitor. Always replace parts with one of equal voltage or wattage rating and of the same ohms or microfarads.

Repairmen have a tendency to use general replacement transistors when the exact replacement is hard to find. Often an ad will say that a given part replaces 50 listed transistors. In radio circuits usually such replacement is safe, and it may be with amplifiers and the like, but with high-frequency circuits it is best to replace with the manufacturer's own part. It may take longer to get and may cost more, but it is the best bet, especially when you replace transistors. Resistors don't usually have to be any particular brand, and capacitors are about the same. Tubes usually differ very little from brand to brand, but use the exact number tube called for.

In replacing flyback transformers in older sets you may have to use a universal replacement. Make absolutely sure it is a close match. It must have the same type primary and secondaries. Some are auto transfomers with only one coil and many taps. Others have a primary and several secondaries. Get a match. Don't start rebuilding the entire high-voltage circuit.

Power transformers may also be hard to find for older set replacements. If you must use a universal, be sure it has the right heater voltage or voltages. Some have 5 volts for the heater of the rectifier tube and 6.3 for everything else. Many though, have 6.3 volts for all. Match up secondary voltages to what the set calls for.

For many picture tubes, a rebuilt tube is the only available replacement. Explain this to the customer, and tell him that the guns, phosphor dots, coating, and so on all are new. Only the glass carton or container is used. When it is necessary to change a picture tube, follow the manufacturer's recommendations to the letter, not only as far as safety is concerned but also for the break-in and setup of controls once it is installed. A picture tube usually will require a complete purity, height, and size adjustment, and convergence. (See Chapter 18.) Handle a picture tube only after it is completely discharged of electricity. Hold it by the body—never by the neck! Wearing protective goggles as you install the tube to save your eyes in case of breakage is not a bad idea.

Important points to remember are:

1. Use picture tube and sound clues to tell which stage of a TV is bad.

2. Use a signal generator or oscilloscope to trace troubles down to one tube or transistor and the surrounding area.

3. Search for defective components with voltmeter, ohmmeter, and substitute suspected parts.

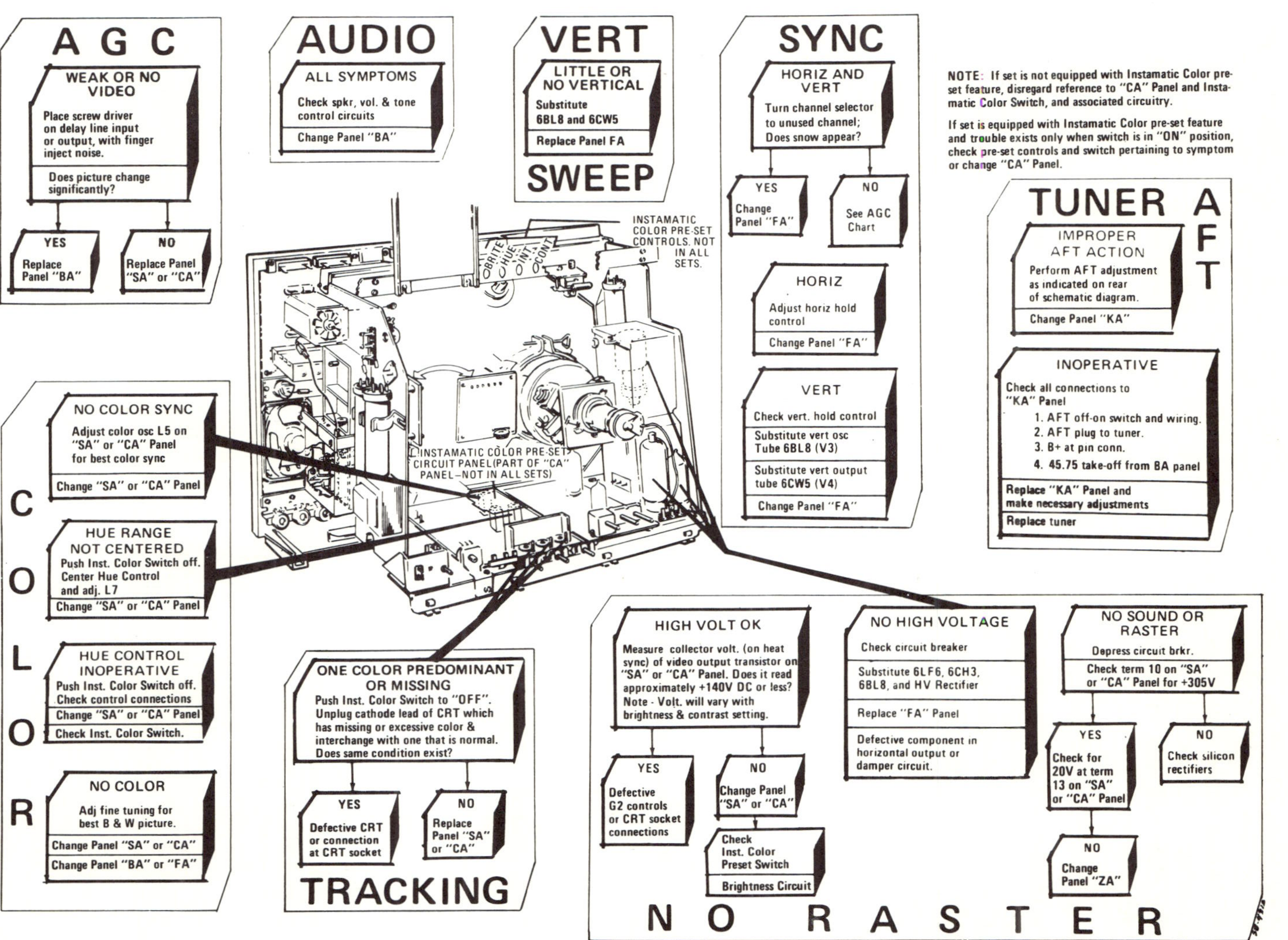

Reprinted with permission of Quasar Electronics Company

FIGURE 17-3

4. If you don't find the trouble, consider other parts of the set that influence that part with control voltages or signals.

5. Use exact replacement parts of equal wattage or voltage rating whenever possible.

6. Use the manufacturer's manual to identify parts, voltages, and waveforms.

Servicing Modules

Panel changing is fast becoming a way of servicing in a hurry; see Figure 17-3. Panel replacement is convenient, fast, gets the customer serviced within a few minutes, and is expensive. Some of the modules cost in excess of $20. Whether you can charge the customer enough to make a profit and recover the cost of the module depends on your location and his willingness to pay. In some shops, if a set is out of warranty, the practice is to replace the module, giving the customer quick service, then to repair his old module to go into another set. In this case you must charge enough to pay for the time it will take you to repair the module at the shop and for the part or parts needed to fix it.

Other repairmen take the whole TV to the shop, repair the customer's bad module, and put it back in his set. This way the repairman knows exactly what to charge in time and parts, but it does not allow the fast service the system is supposed to provide. Actually, if the customer bought a set with the understanding that it could be fixed instantly, so he would never miss programs, he will expect that kind of service. He should be willing to pay more for this service, too, so charging the full module price is not unreasonable. If you replace the module and fix it for another set, you should allow generously for your time, as you have given the customer fast service. Stocking replacement modules requires the shop to tie up considerable money, and general repairmen can't afford to stock modules for all brands.

At this point in laboratory work, if available, you should try actual servicing of sets with supervision.

If this book is the text in a school situation your lab instructor will likely "bug" sets and you will find the faults.

Quiz for Chapter 17

Identify the probable faulty stage for each symptom:

▶ 1. The color is weak, but the colors and video are correct, and the sound is low. The sync tears a little at times. The tuner has been checked, as have the IF and video and they test normally with a signal generator input to the set. Power supply voltages are good. Only when the antenna terminals are connected is the signal poor. The AGC seems to be OK when on signal generator.

▶ 2. The colors constantly drift through the picture. The hue control won't help the condition.

▶ 3. All channels buzz the set, and the picture tears on most of them regardless of the setting of fine tuning.

▶ 4. The picture doesn't fill the screen vertically, and there is a foldover of picture at the bottom.

▶ 5. Flesh tones cannot be brought in properly with the hue control, though the picture and colors are steady.

▶ 6. The picture has a blue cast all the time, even on black-and-white broadcasts.

▶ 7. The picture tube will not light, and there is no light in the neck where the heater should be. The sound is absent, without even a hiss or hum. The set is solid state so no other tubes are to be observed.

▶ 8. The picture tube will not light, and there is a light in the neck of it. You note a sizzling sound and a smell of tar.

18. Service Adjustments for Color TV

PROJECT: Performing Service Adjustments

WHEN THE Radiola III receiver came out in 1923, RCA felt it necessary to issue a 16-page user's manual with it. After all, it had three controls the listener must operate. Today color TV sets contain so many controls that the vast majority of them should be operated only by a skilled serviceman. A glance at Figure 18-1 will reveal what must be set at one time or another in order for the set to work well. Although the service adjustments will vary in name and position from set to set, if you learn the general ones, you will usually be able to adjust any set with a little thought and use of the manufacturer's manual for that set. Let's examine some of the controls and learn how to adjust them.

LOW-VOLTAGE POWER SUPPLY

Not all sets have a regulated power supply, nor will all of those that do have a service adjustment. With the Quasar set you have studied in this book, the 20-volt supply for the transistors is adjustable. You will find it near the middle of Figure 18-1 on the DC regulator panel, "ZA." Schematically, this can be reviewed in Figure 11-3. To set the voltage you would measure the (+) DC voltage from terminal 2 to ground and adjust the control shown for the 20 volts desired at terminal 2. A high-impedance voltmeter should be used, not a 1,000-ohms-per volt type. Another volt-

age adjustment arrangement appears on the RCA TV in Figure 18-3. Since power companies don't all put out the same AC voltage, a switch, S104, is provided. If the power supply voltage of AC is 120 volts, the switch is set in the left or normal position. If the AC is higher, say 128 volts, the switch is put in the right or "HI" setting. This adjusts the set for the higher input of AC voltage.

HIGH-VOLTAGE POWER SUPPLY

Some sets have adjustable focus and grid 1 voltages for the picture tube. The voltage can be from a high-voltage or a low-voltage power supply, but we will consider them under "High-Voltage." Note on the Quasar set, Figure 18-1, the master G1 taps. A connecting wire is put on either the 40- or 80-volt tap. See Figure 16-1, and note that for grid 1 in the picture tube a choice of 75 or 38 volts is listed. In Figure 18-1 the choice is 40 or 80 volts. This biases the grid since both video and color come in on the cathodes, only a bias is needed here. Note that Figure 18-1 says to use the 40-volt tap only if turning down the brightness control won't darken the picture tube with use of the 80-volt position.

The focus voltage is the last set of grids in each gun, as shown in Figure 16-1. We find the setting place in Figure 18-1 at lower right of the set. With the brightness and contrast set for a good picture (these are customer controls, as explained in Chapter 10), the wire is connected to 710, 300, or ground voltage, whichever focuses the picture most sharply. This should be done with insulated pliers, held back a little. Other sets have a focus control that you set as you watch the screen. Figure 18-3, the RCA set, has such a control near the high-voltage cage labeled "R116 Focus." Again, for any adjustments always consult the manufacturer's service manual. Our instructions here are merely to familiarize you with the general adjustments, though often they will be similar to service manual instructions.

IF AGC ADJUSTMENT

Adjustment: Tune in a moderately strong signal. Set brightness to normal; contrast to maximum. Adjust AGC control for slightly above normal contrast. Make operational check. Readjust AGC as necessary.

AFT (AFC) DISCRIMINATOR ADJUSTMENT

VISUAL ADJUSTMENT: Defeat AFT and fine tune properly. Activate AFT. If detuning occurs, retune with secondary core only.

COMPLETE ALIGNMENT:

1. Fine tune in best picture (just off burble) with AFT switch in off position.

2. Connect VTVM to AFT output terminal and ground. Set meter to zero center scale.

3. Set AFT switch to ON.

4. Adjust both cores to extreme outer ends of coil.

5. Adjust primary core for first peak reading.

6. Adjust secondary core for zero center scale.

7. Disconnect VTVM and check AFT action by noting any detuning as AFT is switched in and out. Repeat Steps 4 thru 6 as required.

SOUND IF

1. Adjust fine tuning for best picture.

2. Adjust A.T.O. for best sound with weak signal - (disconnect antenna if necessary).

3. Adjust quad coil (L3) for best sound with normal signal.

INSTAMATIC COLOR SWITCH (Not in all models).

When "ON" (in position) Instamatic indicator will be lighted and pre-set controls will be activated. Customer hue, int, contrast and brite controls are inoperative. See Instamatic color pre-set adjustment.

COLOR OSCILLATOR COIL (L5)
Adjust for stable color sync.

MASTER BRIGHTNESS

1. Set both the contrast and brightness controls to minimum (zero beam current).

2. Using a VTVM or 20K ohms per volt meter, measure and note the boost voltage at 710V focus tap.

3. Turn the brightness control to maximum and adjust the Master Brightness for a boost

RF AGC DELAY

NOTE: Factory preset. Adjust only if overloading or busy background is present.

1. Tune in medium or weak station.

2. Turn RF AGC clockwise (as viewed from panel side) until picture is snowy, then counter-clockwise until picture clears. If interference appears, optimize adjustment.

voltage which is 50 volts lower than the initial reading.

IMPORTANT - If G2's, drive controls, CRT or sweep components are changed, the master brightness must be readjusted.

COMPLETE B & W TRACKING PROCEDURE

PRESET: Intensity and G2 controls to minimum, brightness and drive controls to maximum, hue and contrast to center, master brightness to point of highlight blooming.

REDUCE BRIGHTNESS TO LOW LEVEL - Increase one or two G2 controls as needed for gray raster. Leave G2 control of predominant color at minimum.

VIDEO DRIVE

Adjust for white picture highlights at high brightness level. Also see "Background Adjustment."

INCREASE BRIGHTNESS TO HIGH LEVEL - Reduce one or two drive controls if needed for white picture highlights. Leave control of the weakest color at maximum. Recheck at low and high brightness settings and readjust as needed.

VERTICAL PINCUSHION ADJ. FOR OPTIMUM CORRECTION AT TOP AND BOTTOM

TUNE IN STATION SIGNAL

FOR TOUCH-UP ONLY - set drive controls (at high brightness) for white highlights

- set G2 controls (at low brightness) for gray scale.

INSTAMATIC COLOR PRE-SET ADJ. (Not in all models)

1. Depress Instamatic Color Switch to "IN" position.

2. Adjust controls for normal brite, hue, int, and cont. as with customer controls.

3. Check all available channels. Readjust controls as necessary to obtain optimum hue, int, brite and contrast on all channels.

HORIZONTAL BIAS CONTROL ADJUSTMENT R503

Measure 2nd anode voltage with an accurate high - impedance, high-voltage meter. Adjust bias control R503 for 21KV with zero CRT beam current (raster extinguished) at 122V AC line input.

NOTE: Due to the high impedance of the high voltage section, discharge of high voltage at zero beam current is slow. Turn up brite control momentarily to reduce high voltage rapidly.

ALTERNATE METHOD

If a H.V. meter is not available, adjust bias control for ¼" overscan or monitor boost voltage at focus tap and adjust for 710V boost voltage.

WARNING: The horizontal bias control is a critical component and misadjustment of this control may cause excessive X-radiation.

FOCUS TAPS

1. Set BRIGHTNESS and CONTRAST for normal picture.

2. Connect lead to tap which provides best overall focus.

½A. slow blow fuse in horiz output cathode ckt in chassis B-00 & later. Do not use higher rating

VERTICAL LINEARITY & VERTICAL SIZE

Large excursions may affect dynamic convergence. Adjust before converging set.

HORIZONTAL HOLD

Adjust for most stable horizontal sync. Check by changing channels.

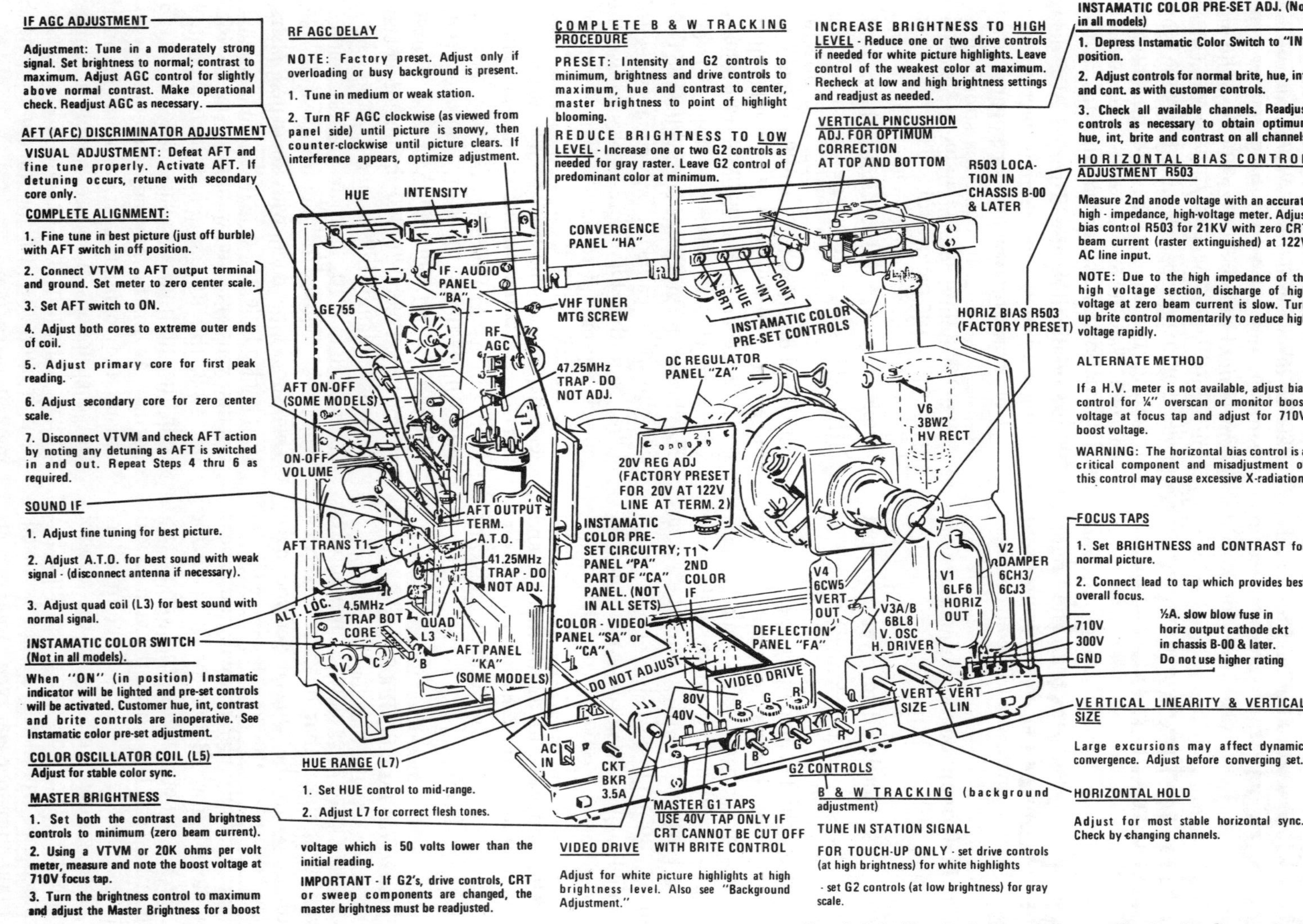

Reprinted with permission of Quasar Electronics Company

FIGURE 18-1

The high-voltage adjustment itself usually requires use of a good voltmeter with a high-voltage probe. Figure 18-3 shows where to measure with an RCA. R415 is set for 26,500 volts. Note that this set has a bleeder to remove the charge on the picture tube after the set is off. Would that they all did. You can get one great shock after the set is off with many picture tubes anodes. The reason the instructions specify that you work with the brightness control at minimum or with zero beam current is that to measure full voltage there must be no current flowing through the picture tube. Always, unless the manual says differently, turn the brightness control all the way down to measure high voltage. The label showing use of the 500-volt scale does not mean you put your regular probe directly to the high voltage. It means you use the high scale, and you also use the *high-voltage probe.* Grip it well to the rear behind the plastic rings. The meter must be grounded to the set, not to the earth.

With the Quasar set, Figure 18-1, note that the point of adjustment is down in the chassis. The high-voltage probe would be attached to the anode on the side of the picture tube. Note the alternate method just under Horizontal Bias Control Adjustment information in Figure 18-1. This would work only if you can focus the set well with the wire set on the 710-volt position. Setting the high voltage in the Quasar set is called "horizontal bias control adjust." This resistor, R503, we noted in Figure 13-3. It sets the bias for the horizontal output tube, but this setting affects the amount of high voltage.

Automatic Gain Control (AGC) and AGC Delay

The AGC and delay adjustments are not even near each other in the Quasar set. In the Zenith set you have been studying they are together at the back of the chassis, to the left. (See the Zenith layout drawing in Chapter 10.) In general the AGC is set for a strong signal. Tune in the strongest channel you have at a time when it comes in at greatest signal strength. Turn up the automatic gain control until there is a buzz in the sound and the picture starts to tear up. Back off until this condition just disappears, and that is the proper setting. You should turn through all strong channels, however, to see if any tear occurs.

The delay adjustment is set for the weakest channel you will receive. Turn the AGC delay setting until snow appears in the picture, then back off until the snow just disappears. This is the proper setting. It requires a strong signal picture without snow normally in order to make this adjustment.

Automatic Fine Tuning (AFT)

Sometimes the automatic fine tuning (AFT) needs a touch-up. If the set has a defeat switch and a manual fine tuning control, defeat the AFT and set the fine tuning control for best picture on a good channel. Then turn on the AFT and adjust the secondary core in the AFT panel so that the best picture occurs with this same setting of the manual fine tuning, though it is out of circuit now. Check between several channels, because slight adjustment may be necessary for the AFT to tune in all channels well. The complete alignment method explained in Figure 18-1 requires a voltmeter that can be centered at midscale. Usually this is a VTVM or transistorized voltmeter. Ordinarily it is not necessary to use this method.

Purity

Purity setting is a process of adjusting for each color to be sure it is the right color and covers the entire viewing area of the screen. The three guns, shown without the glass envelope around them in Figure 18-2, lower left, put out electron beams. There is no color in them, as you know, just electron beams that will strike the proper phosphor dots to make the right color. The amount of beam determines how much will hit the dot and how

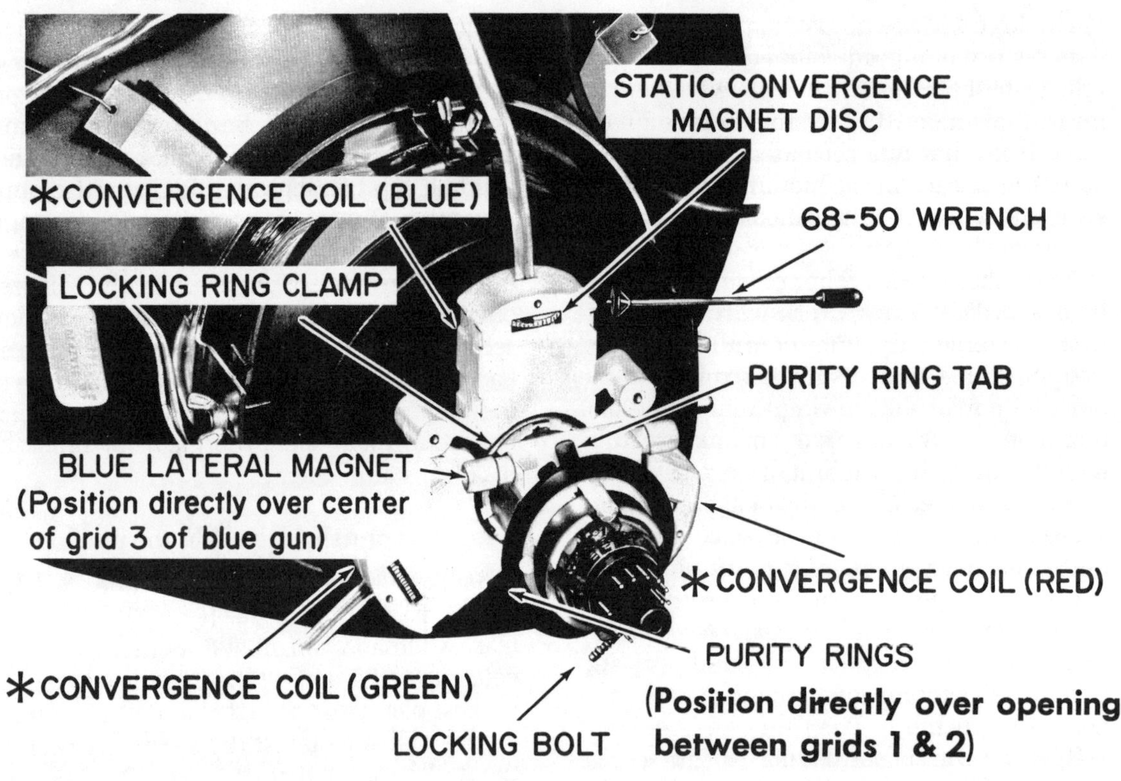

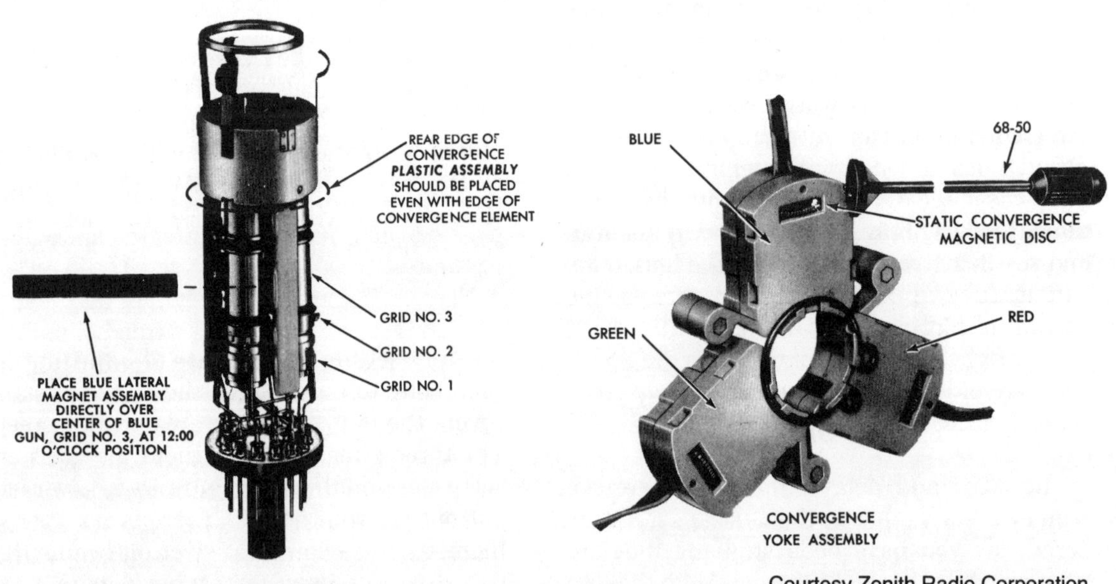

Courtesy Zenith Radio Corporation

FIGURE 18-2

much color will be shown. It is easy to understand that the red gun beam could strike the green phosphor dots somewhere in the picture tube screen, and it would show green, not the red that was broadcast. It is very important that each beam hit only its own set of dots, and purity adjustments help to assure that it does.

To get each beam to hit only its own dots, the set has purity rings, magnets near the picture tube socket. Refer to Figure 18-2 and the upper half of Figure 18-3. The position of the deflection yoke also affects purity, because it affects the outer edges of the picture tube screen, and the adjustment of the purity rings affects purity in the middle of the picture. Convergence, which we will take up presently, also helps make each beam hit only its own dots.

To begin the purity adjustments, usually you loosen the deflection coil wing nuts and bring the deflection coil back against the convergence assembly. Then turn the blue and green screen grid controls, G2 controls, all the way down, leaving a red raster on the screen. Now you rotate the purity rings until the purest red is shown in the center of the screen. Then slide the deflection yoke forward, away from the convergence assembly, until the entire raster is pure red. If you have a crosshatch generator, it would be well to connect it to the antenna and tune in to see if you have the deflection yoke straight instead of tilted. Lacking one, you can tune in a station and see if the picture is straight. Tighten the deflection yoke wing nuts. Now run the red screen grid control down and test the blue, then run the blue down and run up the green. Each should fill the screen with its own color. If not, further tab and yoke positioning may be necessary.

BLACK-AND-WHITE TRACKING OR GRAY SCALE

Tune in a monochrome (black-and-white) picture if you can find it. If not, set the color control to minimum and use a color signal. Turn the G2 controls fully down. If the set has a normal-service switch, put it in service position. Advance each screen control for a thin line of its proper color, just visible. If the lines form on top of each other, adjust for a thin white line. If they don't, adjust them for equal intensity and with each barely visible. Then flip the normal-service switch to normal, and set the drive or gain controls, whichever they are called, to eliminate coloring in the bright areas of the picture. If the set lacks a service position switch use the procedure shown with Figure 18-1.

PINCUSHION

Because of their shape and the positions of the guns, pictures have a tendency to trace with a slight bulge at the top and bottom, rather than straight across at the extremes. The picture shape looks slightly like the bulge of a pincushion; hence the name. With monochrome receivers pincushion can be tolerated or is hardly noticed, but with color it becomes obvious, as the three-color guns magnify the condition.

A circuit that corrects pincushion was noted in the study of vertical sync (see Figure 13-3). It has a connection from the vertical windings of the yoke so that adjustments of the saturable reactor will have a bending effect at the top and bottom of traces to counteract the natural tendency to bulge. The phasing coil is set to straighten the lines at the top and bottom.

Now in Figure 18-1 we find this coil at the upper right. It is merely turned to end the bulges at the top and bottom. Figure 10-2, the Zenith chassis, shows two pincushion adjustments at the lower right back of the chassis. These amp and phase controls are set back and forth until the lines are straight at the top and bottom. Sometimes it helps to use a crosshatch generator to set pincushion rather than just to look at the raster lines. A hand lens can be used to inspect individual lines of the raster.

STATIC AND DYNAMIC CONVERGENCE

Convergence involves bringing the beams of all three color guns into harmony. Not only must the red gun beams hit only red phosphor dots, but each must hit a precise red dot next to precise green and blue dots, not a red dot two spaces away. You must converge or bend the beams so they will do this properly. To converge means to move together.

In the center of the screen the converging is done with permanent magnets. Figure 18-3 shows knobs that are turned to position the magnets for good convergence in the middle of the picture. Figure 18-2 shows discs that are rotated. Zenith recommends the 68-50 wrench for this adjustment. Though adjusting these magnets will affect the entire picture, adjust them only for center convergence. The crosshatch generator should be attached to the antenna terminals and the set should be on and fully warmed up before trying to converge it. The height and vertical linearity controls must be properly set before converging too. Purity, gray scale, and other such adjustments should precede convergence.

A mirror in front of the picture tube helps the work but confuses left and right. After the center of the screen is converged for even crosshatch marks, preferably with each one fully white (red, green, and blue lines on top of each other), you go on to the dynamic convergence, done with the convergence panel. Figure 18-1 shows this panel fastened above the set with screws. This is service position. On the neck of the picture tube are wires to the convergence coils and usually wires to a socket that connect to the horizontal and vertical sync circuits. Figure 13-3 shows connections schematically and Figure 10-5 shows them pictorially.

In some cases the wires permit taking the convergence module around to the front of the TV so you can observe directly as you converge the set dynamically. Dynamic convergence will straighten out the lines at the top, bottom, right, and left sides of the picture. Figure 18-4*a* shows a typical convergence unit. Note that between the top and bottom controls is an illustration to show what each knob or slug does. The knobs are potentiometer adjusts and are hand set. The slug adjusts labeled "Blue Horiz. right," "R & G. Vert. right," and "R & G Horiz. right" must be set with a plastic hexagonal tip tool. Most convergence assembly panels have twelve controls to set. The three static controls at the picture tube neck and an adjustable blue lateral magnet bring the total to 16 settings in all.

The blue lateral magnet is set to merge the blue with the two other colors if nothing else works. Some sets have a blue lateral adjustment that gives a pitch to the blue gun by shifting the convergence yoke assembly slightly.

Here is a typical adjustment pattern.

1. Merge the red and green horizontal lines at bottom center of picture using the "R & G Horiz. bottom" control, illustrated in Figure 18-4*a* at the lower left of the convergence unit. This should be done so the red is on top of the green or green on top of the red across the screen at the very bottom. The crosshatch generator may not make a line exactly at the bottom, but use the lowest available. See Figure 18-4*b* and compare with *a* as we go along.

2. Merge the red and green vertical lines at the bottom center of the picture, using the next control to the right. See "3" in Figure 18-4*b*.

3. Merge the red and green horizontal lines at the top center of the picture, using "R & G Horiz. top," or the upper left-hand knob. On 18-4*b* this is designated as "4".

4. Merge red and green vertical lines at the top center, using the knob called "R & G Vert. top."

5. Repeat the steps thus far, including the static convergence. Each adjustment affects all the others.

6. Merge the blue horizontal line with the

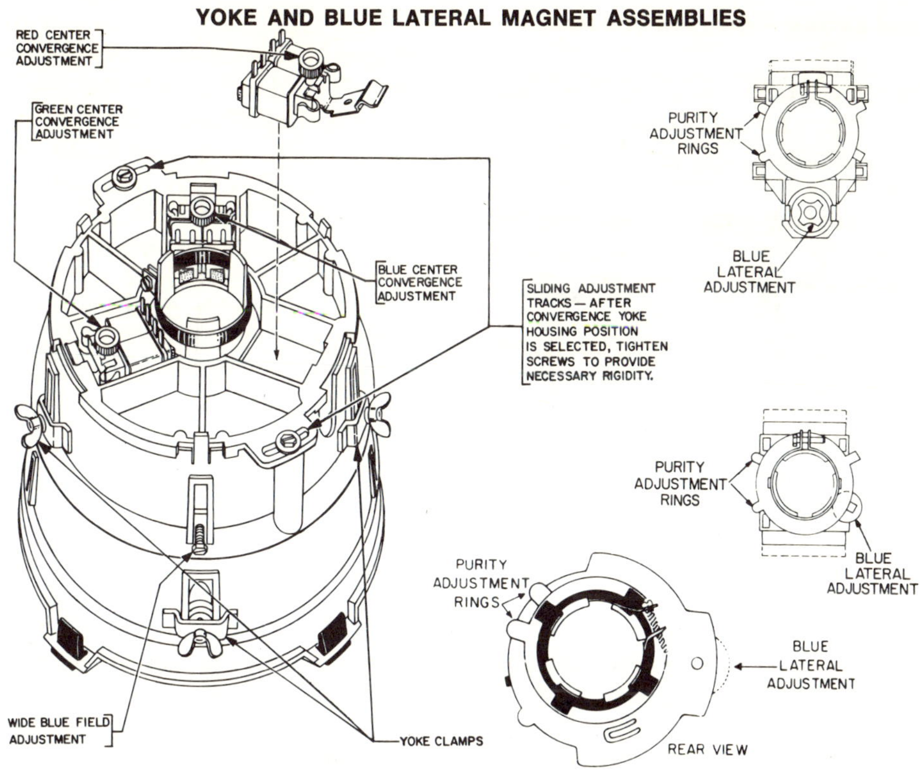

HIGH VOLTAGE ADJUSTMENT

① Preset the High Voltage Adjustment to ⅔ clockwise. Turn the brightness control to minimum.

② Connect the high voltage probe of the VTVM to the kinescope 2nd Anode (use 500 v. scale).

③ Adjust the high voltage control for 26.5KV on the High Voltage Meter.

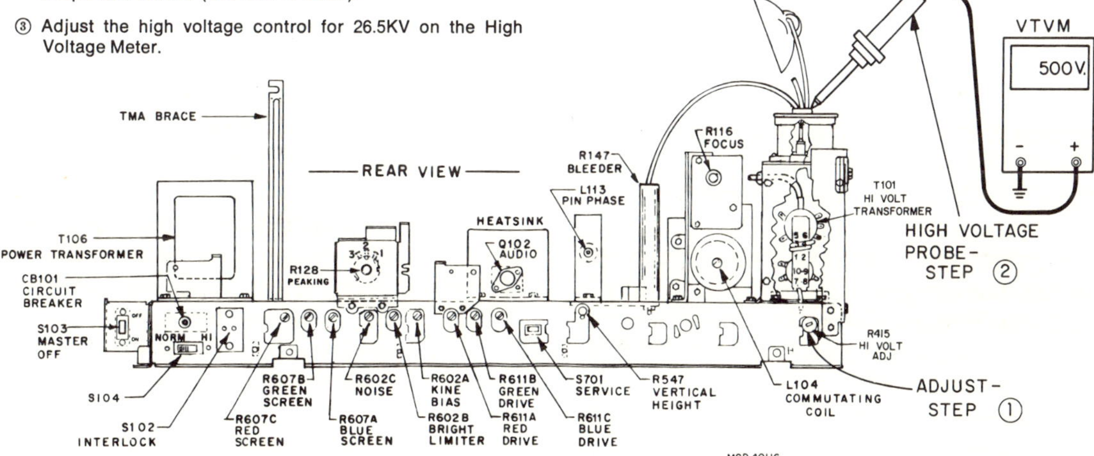

Courtesy RCA

FIGURE 18-3

R & G Horiz. top
R & G Vert. top
Blue Horiz. top
Blue Horiz. left
R & G Vert. left
R & G Horiz. left
A.
B.
(4) (5) (7)
GBR
(11)
(12) G R B
(13)
(1) B
R G
(8)
(9)
(10)
(2) (3) (6)
Convergence Pattern.
R & G Vert. bottom
R & G Horiz. bottom
Blue Horiz. bottom
Blue Horiz. right
R & G Horiz right
R & G Vert. right
OSCILLOSCOPE
SWEEP GENERATOR
MARKER ADDER
MARKER GENERATOR
C.
120Ω
150Ω
C
10K
TP302
VIDEO DETECTOR
LINE SWEEP
D.
MIXER   L 71
SHIELDED LEAD
100 MMF
27 K
TO CIRCUIT UNDER TEST
1N34 OR EQUIV.
10 K
TO OSCILLOSCOPE
Simplified Circuit Diagram of
Detector Probe.
E.
GEN. CABLE
50 Ω
120
56
150
BALANCED 300 Ω
SHIELD
GEN. CABLE
72 Ω
120
82
150
BALANCED 300 Ω
SHIELD
Typical IF and RF Alignment Pads.
Courtesy RCA

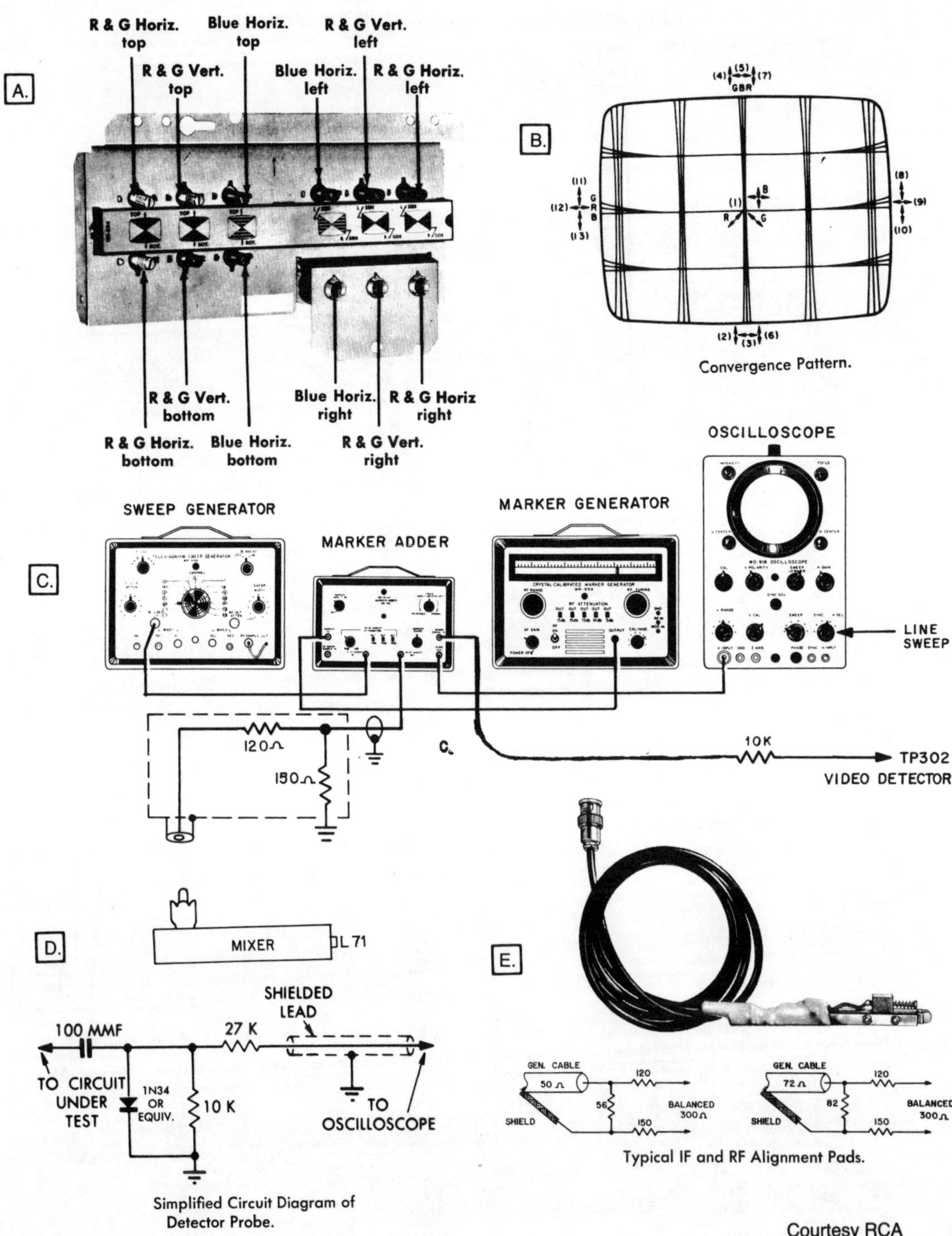

FIGURE 18-4

red and green lines at the bottom center of the picture, using "Blue Horiz. bottom" ("6" in Figure 18-4).

7. Merge the blue horizontal line with the red and green at the top center of the picture, using the "Blue Horiz. top," knob (step "7"). It is good to do steps 6 and 7 several times for straight lines all up and down the center of the picture, preferably with all three colors so merged that they appear white. Touch up static convergence if lines aren't now merged in the middle center.

8. Merge the red and green horizontal lines at the right center, using "R & G Horiz. right."

9. Merge the red and green vertical lines at right center, using the "R & G Vert. right" slug adjust, middle slug. This is shown as "9" in Figure 18-4*b*.

10. Merge the blue horizontal line with the others at right center of picture ("Blue horiz. right" slug).

11. Merge the red and green horizontal lines at left center. Use the "R & G Horiz. left" knob that is situated on panel uppermost right.

12. Merge the red and green vertical lines at left center using the "R & G Vert. left" knob.

13. Merge the blue horizontal line with the others at left center, using the "Blue Horiz. left" knob.

14. Repeat steps 8 through 13 for best left-and-right side convergence.

Every book will have its own convergence steps, and it is best to follow the manufacturer's recommendations. If you have none, then going about it systematically as suggested should do it. Convergence should be done with care, and not just for the fun of it except while learning. If a set doesn't need convergence, leave it alone.

DEGAUSSING

The earth's magnetic field affects deflection and convergence. Any time the TV is moved a magnetic field is likely to build up in the metal of the picture tube and at other places in the set. Modern sets have an automatic degaussing coil that comes on when the set is warming up to demagnetize the picture tube. See Figure 11-3; L801 is this coil. R803 is a thermal resistor and after it heats it changes value so that AC is no longer furnished to L801. If it continued to work, very unusual color would continuously dance over the picture tube screen.

Sets without automatic degaussing and even some with it need to be manually degaussed when the set has been moved or when one is adjusting the purity or the convergence. If there is a color that stays in one corner or over part of the screen all the time, or if the settings of the G2 and drive controls can't remove a tint, or if the purity rings will not produce a complete red or other color, or if the set won't converge, try degaussing.

First remove your watch if it is subject to magnetism. Plug a purchased coil into a wall socket with the coil at least six or more feet from any part of the TV. Take the coil to the face of the picture tube and make circular motions close to it. If the set is on, you will note colors moving all through the picture. Then make circles around the sides and rear of the set. Back the degaussing coil away, still making the circles as you go. When you have taken the coil more than six feet from the set, turn off or unplug the coil. This procedure should remove a color that hangs in one corner of the picture. Since you are using AC and you are going all over, the magnetism of the degaussing coil is constantly changing and it removes the magnetism that has charged the metal of the TV. Once in a while it takes two or three efforts.

A homemade degaussing coil may be made by winding 500 turns of No. 20 enameled wire one foot in diameter, taping it well, and connecting to an extension cord of more than six feet. A switch is optional. I remember well when I was teaching at a tech school that also taught refrigeration. The refrigeration instructor had made the TV boys a degaussing coil with a switch. The TV students came to me and said they couldn't get the coil to work properly. I looked at it and noted a push switch. "Oh," I said, "you have to push the switch to get it to work." They said they had been doing so. Finally I thought of the trouble. It was a light switch operated by a refrigerator door. You would push it in to turn it off, and let it out for the *on* position. I recommend a more standard switch.

Certain controls such as color killer, brightness, contrast, intensity, hue, vertical hold, vertical linearity, and horizontal hold have been covered in previous chapters. They should be adjusted as needed.

QUIZ FOR CHAPTER 18

▶ 1. What kind of picture should be tuned in to set the gray scale properly?

▶ 2. What is the purpose of the blue lateral magnet?

▶ 3. How could you set the high voltage within reason if you lacked a high-voltage probe for your meter?

▶ 4. What does the pincushion reactor do for the set?

▶ 5. ———— convergence is controlled by permanent magnets, and ———— convergence by electromagnets given pulses by the horizontal and vertical sync.

▶ 6. What is meant by purity, and why is its setting important?

▶ 7. What switch is beneficial when setting the gray scale?

▶ 8. What other adjustments should be made before convergence is attempted?

▶ 9. Why is it necessary to repeat convergence steps several times?

▶ 10. How is purity adjusted for the corners and outside portions of the picture tube? For the middle?

▶ 11. What type signal is the AGC set with? The AGC delay?

19. Alignment of a Color TV

Project: Aligning a Color TV

In the last chapter you learned how to adjust the controls throughout a color TV. You learned that they should be adjusted when needed and not every time the set is operated. The same is true of the alignment of a set. A screwdriver mechanic is a real menace around a TV set.

In lesson after lesson you have seen coil after coil with a slug for adjusting the frequency on which it will operate. Ordinarily once the set is in operation, it needs little or no alignment unless a new coil or capacitor that is part of a tuned circuit is replaced. Some of the alignment work done by the repairman is to undo the work of the "electronics expert" who has read a pocket book on tube changing and taken his screwdriver and adjusted every slug in the set to "make it play better." Usually this dangerous person has put the TV way out of alignment. No matter how the set gets out of alignment, however, it must be put back into acceptable order.

Some manufacturers recommend that the repairman not align the tuner, or in fact do much of anything but replace it. Some solid state tuners are very compact, and any changing of position of parts may change frequency. Even so, usually the tuner can be set properly by the slug adjusts mentioned in Chapter 12. With the set tuned to the desired station, adjust the oscillator slugs for that station. This

way is hard to beat. If there is no slug, bending the wire turns of the coil a little will change frequency and align properly. This procedure must be done with a plastic, not metal, device. Some manufacturers recommend that a sweep signal from a generator be run through the tuner for aligning it. It is often advantageous to align the IF first and then adjust the tuner to match it.

Alignment of the IF is critical for several reasons. The sync, video, color, sound, and everything depend on good amplification within the IF system. Since a whole band of frequencies travels through the IF stage there must be sets of coils that accent particular parts of that bandwidth. Again, critical adjustment is necessary. A color TV does not have just one frequency to which every tuner is set. The IF gets rid of adjacent channel interference to a degree. Again, critical adjustment helps tune out other channels that are not quite on IF frequency by heterodyning with the oscillator.

IF alignment should never be done by adjusting the slugs while you watch the picture. With a black-and-white TV you might get by, but with color—never! Often in such adjustment you would turn a slug in a coil and if nothing happened, you would leave it where it is and go to one that shows something. The trouble is that you would have messed up the whole scheme of things by leaving the first slug in a different position. With color TV align only with the proper equipment.

You need a sweep generator that will create a sweep 8 MHz wide and will operate in the VHF ranges for tuner alignment, or up to 50 MHz for IF alignment. If the sweep generator does not have internal marker selection a marker generator must be used with it. It should have attenuations to 0.1 volt so as not to overload circuits. It will need an alignment pad to connect properly to the IF. See Figure 18-4e. A combination of shielded cable and resistors changes it to 300-ohm impedance.

The marker generator, if separate, makes a frequency at some point along the sweep and will make a little blip in the picture on the oscilloscope so you know the amplitude of that frequency. Figure 18-4c shows a setup with a marker generator, which is always a signal generator, and a marker adder that connects up the sweep, marker, and puts the combination sweep with marker into the set at the mixer. At another point the marker adder receives the signal at the detector of the TV and sends it to the oscilloscope. The oscilloscope must have a detector probe to pick up the signal at the output. In c the marker adder takes care of it. D shows a simple type of detector probe that may be made up if the scope has none. The RF or IF must be detected before it goes into the scope.

To get an idea of how a typical IF is aligned, we will consider the Zenith set we have studied throughout the lessons.

Step 1. Set the channel selector to Channel 13. Have the antenna off so you are not receiving the channel.

Step 2. Connect the oscilloscope to test point C1. In Figure 12-6 you will find that this is the output of the emitter resistor of Q105. Figure 12-8 shows it at the left with a violet code. One almost has to have information from the manufacturer or a *Sams Photofact* of the set to find the places and know what to do.

Step 3. Disconnect AGC white/yellow lead from the IF chassis. It is usually necessary to defeat the AGC during alignment. Connect +7 volts to the AGC terminal on the IF chassis. See Figure 12-8. At the +24 volt supply, white/brown, you can connect a potentiometer between that point and ground, and adjust the swinging arm for +7 volts, or you can have a regular Zenith bias box made for the setting of bias. Also batteries in a box and a potentiometer to meter off +7 volts would work. Now you have a steady bias voltage for the IF.

Step 4. Temporarily connect a 470-picafarad capacitor from test point "B" to ground. This also appears on the module layout in Figure 12-8, and in schematic 12-6 you can

see that you are putting any signal that might be at the collector of Q102 to ground. You want to align only the last IF now. Couple the sweep and marker generators to test point "G," using the resistor and capacitor network shown in Figure 19-1 (bottom drawing). Set the sweep generator for a 6-MHz width with a center frequency of 43.5 MHz. Test point "G" is shown on Figures 12-6 and 12-8. It puts the sweep to the base of Q104 through R115. Adjust the sweep generator output for 3.5 volts. Now you are ready to adjust coils.

Step 5. Temporarily detune the 41.25 MHz trap below 39.75 MHz. This is L105; Figure 12-8, right side, shows it. Adjust trap L109B, the fourth IF primary bottom slug on core, for minimum response at approximately 40 MHz. Adjust the fourth IF primary L109A, top core, and the fourth IF secondary, L110, for maximum gain and symmetry of the 42.75 MHz and the 45 MHz marker locations.

Step 6. Vary marker generator frequencies as required to check marker locations at various points on the curve.

Step 7. Disconnect the 470-picafarad capacitor from "B" to ground.

Step 8. Connect sweep and marker signals to test point "A" on the VHF tuner (schematic not in book). With oscilloscope connected to test point C1 (the same place it was), adjust bias for 7 volts again with a bias box or other voltage bias substitution. Adjust the generator for 3.5 volts and note the response. It should resemble Figure 19-1a. Now readjust the bias to 5.5 volts and the waveform will appear as in Figure 19-1b. Adjust the marker for 41.25 MHz and set L105 for minimum. Adjust the marker for 47.25 and set L102A also for minimum. Set the marker for 39.75 and adjust L102B for minimum.

Step 9. Readjust the bias to 7 volts; adjust the sweep generator output for 3.5 volts; adjust the mixer output coil on the tuner for maximum response at 42.75 MHz to approach the response shown in Figure 19-1. Adjust

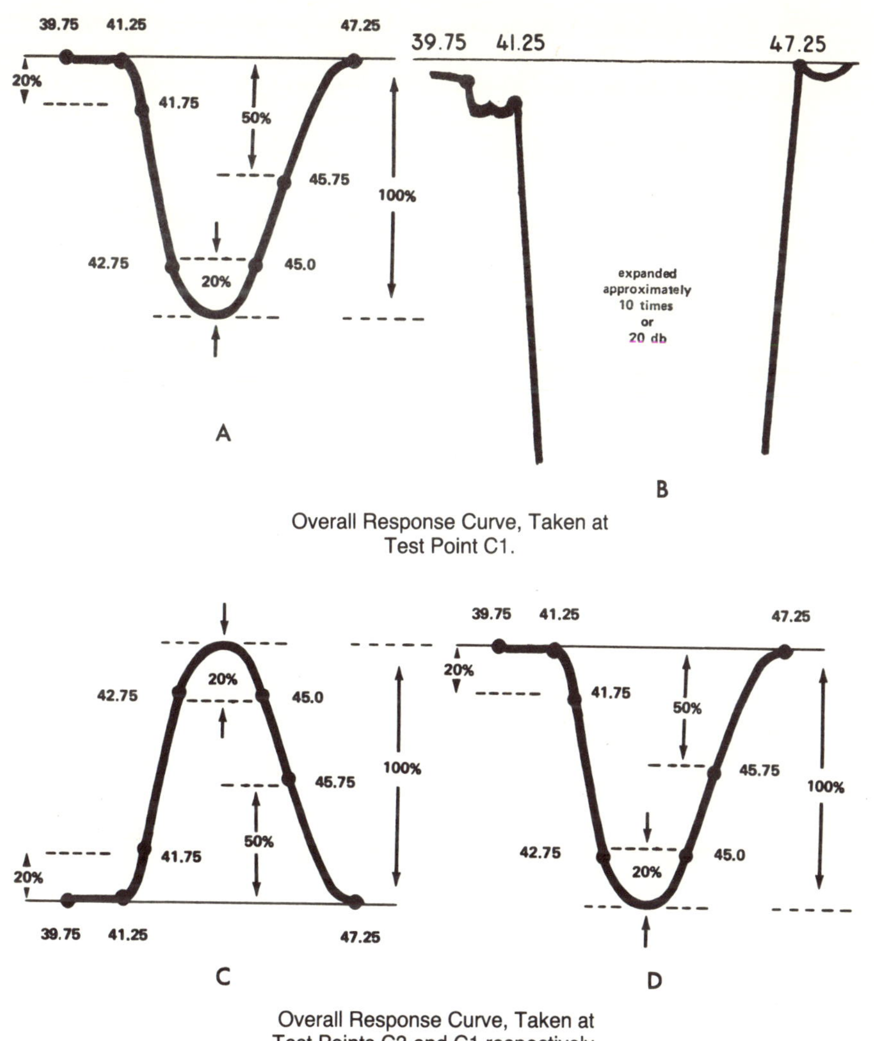

Overall Response Curve, Taken at
Test Point C1.

Overall Response Curve, Taken at
Test Points C3 and C1 respectively.

With oscilloscope connected to test point C1, adjust controls for 3.5 volts peak-to-zero. Connect oscilloscope to test point C3 (oscilloscope should be AC coupled). Waveform at test point C3 should be at least 6 volts peak-to-peak and opposite in polarity to test point C1.

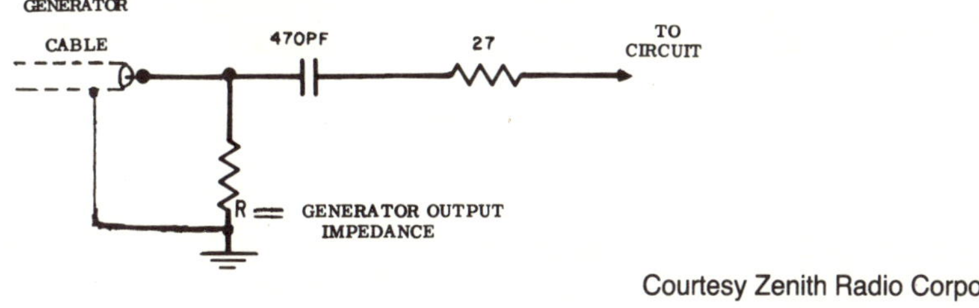

Courtesy Zenith Radio Corporation

FIGURE 19-1

the bottom core of L101B for proper placement of the 41.75 MHz marker. Adjust the top core for maximum amplitude of the curve.

Step 10. Adjust the bottom core farthest from cover, of the second IF coil, L103, for equal amplitude of 42.75 MHz and 45 MHz markers to approach the response in Figure 19-1*d*.

Step 11. Adjust the third IF, L104, for proper placement of the 42.75- and 45-MHz markers to be 20 percent below the top of the curve. It may be necessary to repeat steps 9, 10, and 11.

Step 12. Set the bias for maximum gain of 4.7 volts; adjust the top core of L103 for maximum picture carrier, 45.75-MHz response. Maximum distance from baseline is it. Reset bias for 7 volts. Recheck response curve as shown in Figure 19-1*d*. If necessary, repeat steps 9, 10, and 11 again.

Not all systems are this complicated. For instance, with a certain Sylvania circuit you merely connect the sweep generator and marker to the tuner and the scope to a test point. A 44-MHz center frequency with a 10-MHz sweep is inserted and a +2-volt bias to that test point. Then the marker is moved from 41.25 to 47.25 in steps and all slugs are adjusted for a good response curve. Sound alignment seldom needs sweep or marker equipment. Refer to Figure 14-4. Zenith gives the following information for aligning the module:

"The IC Sound Module can be quickly and effectively adjusted with a strong 'off the air' signal. Using a strong station, tune the quadrature coil, T1101 for best sound. Reduce the signal until the sound becomes noisy. Tune the input coil, T1102 for maximum sound. Re-adjust T1101 for best sound." The term "off the air" as used in the Zenith book refers to a regular station signal. One might prefer to call it "on the air." The important thing to grasp is that sound tuning can be done with a channel tuned in rather than with equipment unless badly off frequency.

The sound alignment of the Quasar set is listed at the center left of Figure 18-1. See Figure 14-2 for the parts affected. The A.T.O. is L13. It is important that L13 be set with a weak signal so that clipping can occur of all amplitude in the IC that follows. Then the strong signal is put in so that the quad coil, L3, will be set for center frequency.

There is an easy way and a hard way to adjust the color alignment. Figure 18-1 gives details. The color oscillator coil, L5, is adjusted for stable color sync. If the colors are drifting through the picture, setting the oscillator may stop it. This puts the reference oscillator on frequency so the chroma will be the correct transmitted phase from the reference for steady, proper color pictures. The hue range control, L7, is used to bring the hue control within the adjustment enough so the customer can set hues at will from the front of the set. First the hue control is set to midrange so there is play either way. Then the range control, L7, on the color panel, is adjusted for good flesh tones. Any drifting now can be compensated for at the front of the set with the hue control.

Now let's do it the harder way, using a color bar generator. Usually this generator is a crosshatch, dot and bar generator. It must put out ten bars. The second bar will correspond to red. The first bar from left is not usually visible, as it is blanked out and the raster overscans. In Figure 16-1, the oscilloscope would be connected to point 24 to observe the green video output. Adjust the hue control so that the tenth bar is maximum negative. Check with the scope at point 23. The sixth bar should be maximum negative on the scope. Now at point 25, the third bar should show maximum negative. Such adjustment should be done only if the customer wants precise setting of the hue control and is not going to fool with it every time he wants to change the color a bit.

Some sets have a cross-talk adjustment. See Figure 10-2, in the main part of the chassis. A color bar generator is connected to the antenna terminals and the bar pattern used. A 25-mfd 25-volt electrolytic capacitor is connected from test point D, to the right of the cross-talk control, to ground during the adjustment. This removes the video signal. The control is adjusted for maximum sharpness and clarity on the leading and trailing edges of the color bars. In other words, it makes for a sharp change from one color to the next. The 25-mfd capacitor is removed after the work.

This Zenith set also has APC or automatic picture control. The color setup switch (between chroma and subcarrier modules, Figure 10-2) is turned to align position. Adjust the APC on subcarrier module for minimum movement of color bars through the picture. Return the color setup switch to normal.

Some sets require chroma bandpass alignment very infrequently. If they have adjustments on the coils in the chroma bandpass circuits, a sweep of from 3 to 4 MHz is usually entered at a specified point. Then the coils are adjusted for a double hump curve on the oscilloscope with the low point between the humps being 3.58 MHz. A marker generator is used to determine where 3.58 is on the curve. The instructions may specify removing

something or jumpering something, but this is peculiar to each set. Follow the manufacturer's recommendation. Ordinarily the chroma bandpass slugs should not be adjusted, but replacement of one of them or other connecting parts might make it necessary.

QUIZ FOR CHAPTER 19

▶ 1. What type of repairs might make alignment of a stage of a TV necessary?

▶ 2. What kind of customer problem might make alignment necessary?

▶ 3. Why is it a mistake to align IF stages while watching the picture with an "on air" program?

▶ 4. Why is IF alignment critical?

▶ 5. How is the sound discriminator ordinarily aligned?

▶ 6. Name the two types of generators needed for IF alignment.

▶ 7. What device interprets the aligning you are doing? What do you watch for in traps? In IF coils?

▶ 8. Explain an easy way to set the color oscillator and hue range controls.

▶ 9. What generator is necessary to adjust cross-talk controls?

▶ 10. How are tuners adjusted for each channel?

20. Professional Practices That Win

CERTAIN FALSEHOODS prevail in every kind of enterprise. Because some auto mechanics install parts they don't need to, and because some auto dealers charge labor for the hours a car sits on the lot or in the garage whether it is being serviced or not, some of the American public has the idea that all auto repairmen are crooks. Because a few mail-order dealers have been dishonest, some people think all mail-order selling is a get-rich-quick scheme.

Because a few TV repairmen have overcharged, or have misrepresented service performed, far too many people think that all TV servicemen are getting rich quickly and less than honestly. To compound the problem, some ignorant folk go around saying that all it takes to repair a TV set is to "change a tube or condenser." These are the same people who want their set fixed for $2, and quickly. Certain tube-changing manuals have contributed to the layman's concept that TV repair is easy and should not cost much.

With the increase of solid state equipment, the tube-changing customer is more at the repairman's mercy. No longer can he try every tube at a free drug store tester and save only the hard-to-find troubles for the repairman. Tube testing never was 95 percent of the needed servicing anyway. Now, most of the parts are soldered into printed boards, and the Saturday mechanic can't cope with them unless he studies a book like this to learn how it all works. Truly, the repairman is getting things more his way. Unfortunately old stories die slowly, and many a customer feels the repairman is out to gyp him.

Certainly courtesy and tact on the part of the repairman are necessary, but just as important is developing understanding on the part of the customer. If you must take the set to the shop, first explain why. Invite the customer to come to the shop and time you on the servicing to see that he gets honest treatment. Naturally, a customer would be a pest, but if he is reminded occasionally that you are charging him time, he will quit talking and let you troubleshoot the set. Of course just suggesting that the customer come to the shop helps build confidence.

It is very hard for the customer to believe that a set could need more than one part. You can show him on a schematic how one part affects another, but he probably won't understand. Trying to explain, however, helps build the customer's confidence in you. In cases such as lightning damage, many parts may be needed, and insurance often covers this servicing. Sometimes the customer had a mediocre set with a weak picture tube, and after the lightning damage he tells you that it worked perfectly before the lightning hit it. Then you are caught between a dishonest customer and an insurance company that doesn't expect to pay for a new picture tube unless the lightning damaged it. Such is a serviceman's lot.

If the set must stay in the shop for a few days the customer is going to demand a loan set in the meantime. I believe that a rental fee should be charged for the loan set, except in cases of extra good customers who buy new sets from you or personal friends. Far too often the loan set will be damaged when you get it back. Of course, some shops loan out 1950 black-and-white sets, so abuse hardly matters.

Every shop has its own credit plan. If you are starting in business, you probably won't be in it long if you grant any credit on service. For selling new sets, you can work through a loan acceptance company and avoid tying up personal funds in purchases. The acceptance company pays you for the new set, and they do all the collecting. You are usually not relieved from service during warranty, but you are relieved from collecting payments or reposession in case of default. For servicing, however, credit is very dangerous. If the person is good for the money, he or she certainly can borrow it at a bank or savings and loan. If he or she is not good for it, you don't want the business.

Often the customer can and will pay, sometimes with a bit of coaxing, bills of $40 and under. It is the larger bills that present a problem. If the set needs a power transformer or a picture tube, chances are trying to get cash for the servicing will be difficult. The hard thing about it is that the serviceman will have paid 60 percent or more of the cost out of his own pocket to get the expensive part. If the customer doesn't have the money when the repaired set is delivered, many things can happen. The customer may make a few payments on the bill and then tell you that you can have the set. Often it is hard to sell a used TV in this kind of situation and come out well. Or the customer will hide from you, tell you the set injured him, claim you didn't fix it, move away, or get divorced—and who can you collect from then?

Probably the only really safe method is to (1) inform the customer that since the bill will be large, a deposit will be required before you order the new picture tube, and/or (2) deliver the set only when the customer assures you that he or she will pay for it when delivered. It is a good idea to call the customer before delivering a set with a substantial bill and say that you will be bringing it by today and that you will set all the controls at that time for the best possible picture. Then ask if he or she will be able to have the money ready to settle the bill.

Many customers think of buying a new set when a color picture tube goes out. The serviceman with a sales shop has at least a chance of selling the new set and making a profit. A problem arises, however, for the person who only does repairs, or for the sales and service representative who wastes an hour or more only to have the customer buy elsewhere. In these cases the serviceman should be entitled to a labor charge, but just try to get it! The doctor is paid whether the patient gets well or not. The schoolteacher is paid whether the child learns or not. Why should the TV technician not be paid if he finds the trouble, whether he is allowed to fix it or not?

One way to avoid this problem is to require the customer to sign a repair order before you examine the set, the way most garages do for cars. In signing such a repair order the customer agrees to pay for labor costs regardless of whether he wishes the set repaired once the trouble is found. In this way the repairman can stay in business. Then if the customer buys a new set from him, the serviceman can, at his option, deduct the labor charged for the old set from the price of the new. Point out this fact to the customer; it provides an incentive to buy from you.

A repair order also should give the repairman the right to sell the set to someone else if the customer does not pay for repairs within ninety days. In this case you keep the set at your shop until the customer has the money to pay for the repairs. You should call the customer frequently, reminding him that you are ready to deliver the set and asking him to be ready to accept it. Finally, after the ninety-days, remind him once more, and if he can't pay, you can sell the set to recover the servicing fee. This must be written out carefully on the repair order form.

Selling new sets need not be an expensive enterprise. If you live in a city where the distributor of a reliable brand of TV is located, sometimes you can arrange to show your potential customer a new set in the distributor's showroom. In this way you don't tie up your money in new sets, display space, or other things, although you might be required to pay a small fee per set sold. At any rate, this is an easy way to sell new sets when the customer doesn't want to pay a lot to have the old one fixed. You would act as salesman at the distributor's house. You take the customer there, demonstrate the set, stress that it is right out of the factory and hasn't sat in a TV shop for months. Chances are the distributor will already have arranged a payment plan. Of course if selling sets appeals to you, in time you might want a sales and service shop with a showroom and all the extras.

Sometimes the distributor may sell you servicing contracts. Such plans vary, but basically they are as follows. You deliver new sets and are paid a fee to install, align, set up controls, converge, and so on. These are sets sold retail by the distributor himself or sold by furniture stores with no TV repairman. Besides the fee for delivery and installation, you are paid a flat fee to keep each set serviced during the warranty period. The distributor will furnish the parts, but the time is on you. Since the flat fee is low, and the service adjustments that need to be set are legion, and customers are critical about color TV, you will really earn the money, and then some.

Yet if you accept a thousand contracts at $7 each, you have a sort of living the first year you are in business. Yes, you will likely sweat a lot for that $7,000, but it will be a good way to get started. It beats sitting there wishing the phone would ring. If your work is good, and if you treat everyone courteously, chances are you will retain a healthy number of the people as customers after the warranty expires. In time, it is not a good deal, as your hourly income will soon surpass the amount you were getting with the service contracts, but it is a way to start. Then, too, in time you won't want to be delivering sales to the other fellows for just a few bucks.

We cannot all be born salesmen, but usually a born salesman is not the nicest person to be around all the time. *Good* selling does not call attention to itself. The customer must think he made the decision without influence.

A good salesman guides; he does not bully.

Selling service is important. The customer must understand that just as there is "no free lunch," there is no absolutely free TV. Because sets are complicated and so many parts depend on other parts, it is natural for the TV to need periodic service. Explain that an auto needs tires every year or so, that a child needs new clothing as the old wears out, and that a TV likewise is always wearing, wearing, wearing. It, too, needs repairs from time to time. Explain that every time the car needs a tire it is not necessary to junk it and get a new car; nor would you junk your child when the clothes wear out. Similarly, it is usually far better to repair a TV when parts wear out than to replace it.

Get the customer to figure the cost of servicing. If he watches 2 hours a night, that amounts to over 700 hours of entertainment a year. At the movies that could cost over a thousand dollars. Now compare this figure with the purchase of a color TV for, say, $500, divided by about 10 years. The cost is $50 a year, and even with another $50 for servicing, the cost is still about 30 cents a night for the entertainment plus a few cents for the electricity. What's more, the whole family can watch for the same low fee. Considered on a cost per hour basis, TV servicing is not much of an expense. Commuters often get a grease-and-oil job for their car each month at a total cost of possibly $75 to $100 a year. They expect the expense. TV is similar. The set needs "lubrication" of overheated or shorted parts. Stress that color TV service costs more than black-and-white and is needed more often, but that it is worth it.

In selling new sets, emphasize the reputation of the manufacturer, the way engineers test and retest the circuits to make them as foolproof as possible. Assure the customer that he is making an investment for many years, so he should get a good set that he will be happy with for that length of time. Offer to look over his living room and suggest a wood style that will harmonize with it; show the customer that you are interested in how the set will look in his home. Put yourself in his shoes. By all means have sets well converged before you demonstrate them. I marvel at the poor convergence and other problems I see in display sets in stores. How do they ever sell any when they look so bad?

Know something about the manufacturer's special features. If the picture tube has a black matrix between color dots, show literature to demonstrate this, then show the set and ask the customer if he can't see the difference. If the set has a certain system that is supposed to hold vertical sync better than others, explain this fact as nontechnically as you can and say that the feature is worth extra money but that it won't cost the customer a cent more. Do your homework and be ready to compare sets with the customer who shops around.

Think of selling points for any season. In summer don't emphasize the reruns; stress the baseball and specials. In fall talk about football and the new season of regular programs. In winter suggest a TV as the perfect Christmas present for the entire family, pointing out that the season is only half over for all their favorite programs. Be adaptable. Talk family shows if that is what the family seems to like. Finally, make a good trade-in offer on the old set.

You may well decide to work for others rather than to organize your own TV shop. Certainly the regular paycheck is nicer at first than waiting around for business. Also, an established shop will have a lot of test equipment, service manuals, and such. You can begin in a shop with an eye on having your own shop later, or on moving into the service manager job when there is one.

Much of what has been said will apply to work in other people's shops, but they will probably have established systems for crediting, selling, and so on. You must comply with them and perhaps you can learn from them. The shop foreman may have a pet method for servicing sets, and you must do it his way. He may be an analyzer fiend, or insist that all servicing be done with a scope, or think that anything more complicated than a voltmeter is a waste. If the service manager has been servicing for a long time, his experience can be worth many dollars to you. Get all the tips you can from him.

If you must start your own shop, here are some organization suggestions that might save money. You can be rather sure that you will not make a living at first. It takes months to become known and trusted in any community. During the infancy of the business, it often pays to service sets in the customer's home and in your spare room, basement, or heated garage. Get a basic supply of test equipment such as a scope, a bar, dot, and crosshatch generator (one unit), a tube tester (or build the one in Appendix B), and a picture tube tester. You have a voltmeter-ohmmeter that will do for quite a while.

You can probably get by without aligning sets for some time, so don't invest in sweep or marker generators at first. Buy tubes and transistors just as needed. Get a supply of resistors, capacitors, solder, and the other things that you are sure to need. A good soldering gun with low wattage is a must. Build a workbench of scrap lumber. After all, you are the only one who will see it at first. Either get an old truck or station wagon to haul the sets in, or hire a local mover to bring the set in and out at first. True, you won't make money at this, but it will get you established. My main advice is not to spend several thousand dollars at the start.

City zoning could prevent you from servicing in your house. If so, you may get a contractor's type license that specifies that you will repair at the customer's home. True, you will have to sneak some home after dark, but this is a temporary arrangement. Sometimes you can make a deal with a department store so that you repair in their shoproom or basement, which is zoned properly, in exchange for giving them a low rate for servicing sets they sell.

Now suppose that you have a shop, and you

want to get some business. First you should get all the free publicity you can. Tell all your relatives to tell all their friends. Visit old friends and tell them about your shop. Urge them to tell others. Put up signs in laundromats and stores that have bulletin boards. You can type these up, hand print them, or have a few duplicated inexpensively. Think of what your shop will have that makes it unique. Maybe you are the only shop on a certain street or in a section of town. Perhaps you are the most recently trained technician, with the most up-to-date knowledge.

You can set a good price to get going, but beware. First, the established shops may lower their service call rate below yours temporarily if you are hurting them at all. Second, the price motive tends to attract the "change a tube or condenser and charge me $2" jerk. Of course, price can attract legitimate customers, too, but be aware that advertising a low price does bring in calls from the people who expect everything for nothing.

With a supply of printed cards you can go door to door, giving them out and suggesting that the customer put the card on the TV so when it does need service he or she will know how to get in touch with you fast. Often a door-to-door campaign will produce customers who tell you they are having trouble with the set, and what would you charge to look at it. You reply, "Since I'm here anyway there will be no service charge. If the set needs shop attention it will be [so many dollars] per hour plus parts."

Normally, you would have a service call charge, and often this covers the first hour of repair time too. With the cost of gasoline today, you must charge to come out even on service calls and try to make a profit too. Often, in larger cities, it takes the better part of an hour to get to the customer's house, look at the set, talk to the customer, and take the set to the shop. More time will be used in delivering the set later. You must establish the service call charge with these facts in mind if the charge must cover an hour of ser-

vicing time too. So many shops have this arrangement that it is hard to do otherwise. Suppose the service call is $10; then it takes four hours to repair the set. You bill at $5 an hour. You would charge on the basis of three hours (the service charge covered the fourth) or a total of $25 for the four hours plus delivery and pickup time. Since delivery and pickup probably add another hour, the $25 just comes out to the $5 an hour you are trying to make. The service call must cover travel plus time spent talking to customer. True, you might have some markup on parts, but this is likely to pay only for gas and truck upkeep.

Newspaper ads seldom pay off immediately. For both sales and service a display newspaper ad can pay, but for servicing alone, a classified ad is the largest you should try. Chances are it won't pay its way, but if it gets a customer, it helps the general cause. Spot announcement commercials on radio can help if the station is small and has low rates. Yellow Pages ads for servicing alone usually should be no more than a column inch. It is hard to track down results on Yellow Pages ads. Some people thumb through the Yellow Pages when they need a serviceman, but far more ask a neighbor for a recommendation. Get the word around; be neat, honorable, and friendly, and in time you will have a reasonable amount of business.

Occasionally you will find a set you can't fix. Then you must either admit defeat to the customer or "farm" the work out. Farming out is a delicate process. You must find another repairman who is willing to do the work, and you must hope he will be discreet. Normally I would suggest that you don't try to make a profit on farmed out work. I have known people who could fix very little, so they farmed out much of their work and made a good living doing so. In time, however, people find out, even in large areas, so farm out as little as you can and charge your expenses to education.

Sometimes the customer refuses to pay—what then? I remember a fellow who worked for a shop. A woman had called saying that she needed her phonograph repaired. When the repairman got there, he found that the only problem was that the needle had dropped out of the tone arm unit. He replaced it and wrote up a bill for $6. She refused to pay. He sat on her sofa and said, "When I get back to the shop one of us is going to have to cough up that $6 and it isn't going to be me!"

She called him a bunch of names, paid him, and told him to get out of her house. OK; let's think about this. First, the customer had called for service and should have expected to pay for it. Since it was a simple repair, however, she felt cheated. How much better it would have been if the man had put in a new needle,

as the fall probably damaged that one anyway (well, it could have), then spent some time looking through and somehow adjusting the rest of the set, then charged her. That way she would have saved face and felt she got $6 worth. Threats and intimidation should be a very last resort. Of course it is possible to get a mechanic's lien on a repaired set if the customer won't pay.

Usually the best policy is to give the customer a ballpark estimate of his costs and be sure he understands. Second, be honest with him. Third, explain why the costs are what they are. And finally, be polite but firm in demanding payment before the set is delivered.

Appendixes

- Parts List
- Testers
- Optional Radio Circuits
- International Code
- Color Codes
- Mathematical Formulas

Appendix A:
Parts List

A PPENDIX A is designed to help you find the parts to construct the laboratory projects in Chapters 1 through 8. Certainly if you find a local radio store that has similar parts, buy from it. Most of the components can be substituted as long as part value and voltage or wattage requirements are met. The audio fre-quency transistors can be of almost any type as long as they are PNP and low-current types. Do not use power transistors. The high-frequency transistor is critical. It must amplify high in the megahertz frequencies. Be sure that any transistor you use for Q3 is PNP and a high-frequency type. In some cases, of course, you can save parts out of used TVs and radios that have been junked. This training method is designed to save you money. Since some of the items you build can actually be used to advantage in a TV service shop, you realize a double economy.

CHAPTER ONE PROJECT PARTS

TRANSISTORS:	Q2 and Q1 PNP Audio 22-15430-6*, 20A1313-2*, or 276-2004
RESISTORS:	R1 220K
(CARBON)	R2 1000
	R3. 330K
CAPACITORS:	
(DISC)	C1, C2, C3 0.01 Mfd.
EARPHONE(S):	1000-2000 ohm single or headset
BATTERY:	6 Volt Lantern Type
MISC.:	Masonite Panel 6″ × 9″ × ¼″

Wood Supports (2) each 6″ × 1″ × ¾″
Fahnestock Clips (2) each ¾″ × 1″
Machine screws and nuts to attach fahnestock clips
Wood screws to connect panel to supports
Hook up wire #20 enameled, 100′ coil will do all projects in book.
Antenna Wire, Stranded TV antenna guy wire 50-100′
Use of soldering gun or iron, screwdrivers, pliers, drill or ice pick, sandpaper, electricians tape.

CHAPTER TWO PROJECT ADDITIONAL PARTS

DIODE:	D1 1N34 or 1N60
CAPACITOR:	C4 Variable 365 Pfd. 18A1651-9*

Coil: L1   Tuning Coil 65 close wound turns enameled wire (from 100′ roll listed above) on a 5″ × 2″ diameter piece PVC water pipe

Misc.:   Masonite Panel 6″ × 9″ × ¼″
Knob with set screw for variable capacitor
Screws ETC. to mount variable capacitor

Chapter Three Project Additional Parts

Transistor: Q3   PNP High Frequency 22-15450-4* or 276-2007

Resistors:
(Carbon)   R4   1300
R6   100K

(Variable)   R5   100K Potentiometer   271-092

Capacitors
(Disc)   C5   0.01 Mfd.
C6   100Pfd.

Coil: L2   Secondary of Tuning Coil 10 turns close wound on same form as L1, part of 100′ wire #20 listed with chapter one materials.

Misc.:   Knob to fit R5

Chapter Four Project Additional Parts

Capacitor:
(Disc)   C7   0.05Mfd.

Chapter Five Additional Part

Use of a Pocket Superheterodyne Circuit Radio

Chapter Six Project Additional Parts

Necessary only if you wish to do the optional radio control room simulation exercises:
Resistors:
(Variable)   R1,R2,R3,R4,R5,RM-100,000 potentiometers

Capacitor:
(Disc)   CM   0.05 Mfd.

SWITCHES: (DOUBLE THROW)	S1,S2,S3,S5	275-403

TRANSFORMERS:	T1 Audio Output Modulation	273-1380 273-1378

TRANSISTOR: QM Audio   22-15430-6*, 276-2004

METER:
(DC TYPE)   M   1 Milliamp   35A6165-4*

MISC.:   Microphones
Turntables (2)
Tape Player
Wood to build Console, Turntable Cabinets, Etc.
Clock
Glass Panel if Using Another Area for Studio

CHAPTER SEVEN PROJECT ADDITIONAL PARTS

TEST LEADS   A Red and a Black Flexible Lead
Alligator Clips
Probe

CHAPTER EIGHT PROJECT ADDITIONAL PARTS

DIODES: D1, D2 1N34 or 1N60

CAPACITOR:
(DISC)   C1 500Pfd.
C2 0.001Mfd.

RESISTORS:
(CARBON)   R1 27K
R2 5 Meg
R3 470K
R4 10Meg
R5 1 Meg
R6 100K
R7 15K

All resistor values are in ohms. K following number means thousand ohms. Example: R1 of chapter eight supplies is 27,000 ohms. Meg is for million. Example, Chapter Eight List-R4 is 10,000,000 ohms. ¼ watt resistors or larger may be used throughout. C3 should have a voltage rating of 400. Others above 6 volts.

Most parts are available at radio supply stores, hardware stores, or lumber yards depending upon the item. Certain electronic parts may not be available in smaller communities. For your convenience these items have been listed with a stock number. If the stock number has an asterisk following it such as Q3 PNP High Frequency (transistor) 22-15450-4* the stock number of the part is for Burstein-Applebee, 3199 Mercier St., Kansas City Missouri 64111. Request a catalog before ordering by mail. If the stock number following the part description has no asterisk such as R5 100K Potentiometer 271-092, the

stock number of the part is from Radio Shack. They have retail stores in all cities and several thousand towns, but you can write for a catalog to Radio Shack, 2617 7th St., Fort Worth, Texas 76107.

Appendix B: Testers

THE TUBE tester shown schematically in Figure A-1 costs little to make and can be adapted to other type tubes later, so it will test nearly every TV tube in use today. It is not designed for 4-, 5-, and 6-pin tubes which are obsolete, and it won't test 50-volt radio tubes, as servicing tube radios is no longer profitable or necessary. The drawing shows the tube sockets as viewed from underneath the set. The 7-pin socket at upper left is used for the miniatures so often used in IF and the RF stages of tuners. The octal socket tests the keyway 8-pin tubes that will be found occasionally in horizontal and vertical sync circuits such as a 6SN7. Next is the 9-pin socket that tests a great number of miniature 6- and 12-volt heater tubes. Finally, the 12-pin compactron socket will test all the compactron tubes that are used in portable and other TV sets. The six wafer switches are at the middle of the schematic, and chances are you will want to mount them in the middle of the chassis too. The arrangement is such that any pin of any tube can be connected to heater, control grid, cathode, screen grid, or plate. It makes no difference what type tube you are testing; the combinations can be made to accommodate it properly.

The wiring can be a real headache, but all it really amounts to is wiring pin 1 of each tube to pin 1 of every other tube, then pin 2 of each tube to pin 2 of every other tube, and so on until all pins of the same number are wired to the pins of the same number on all other tubes. Then you do the same thing with all six switches. Finally, you add wires so that each pin 1 is connected to each switch contact position 1, each pin 2 is connected to each switch contact position 2, and so on. The tester should be wired with as little wire as possible,

not squared off nicely as shown. Good soldering is a must when everything is connected.

On the top of the tester, label each rotary wafer switch so you will know what it connects (heater, heater, cathode, control grid, and so on). You will need to fasten pointer knobs to the shafts of switches on top of the tube tester; then you will need to number positions on top. In this schematic, lines that are connected have a big dot at the connecting point. Wires that overlap but are insulated from each other have no dots. In some cases "loops" are drawn to indicate that wires are not joined but in other places they are not.

The power transformer can be only a heater transformer, but be sure it is a 120-volt primary with a 12-volt center tapped secondary. This way you can test both 6.3- and 12.6-volt heater tubes. In fact, you can test the 14-volt type too if you have patience for the heaters to warm. Use whatever type switch you want for S2 and S1, the on-off switch. For R1, the grid return resistor, a 1-megohm resistor is good. R2 should be a 1,000-ohm potentiometer; a 2-watt unit should suffice. R2 is the bias control, and moving it when the tube is being read in transconductance determines whether the tube will amplify. R3, the screen grid resistor, could do well with 4,700 ohms, and the plate load resistor, R4, should be 1,000 ohms. These should be 1 watt or larger.

This tube tester has been designed to work with your homemade voltmeter (Chapter 8) to save the cost of another meter unit. Voltages are measured across R4 instead of currents being read directly with a milliammeter.

The switch contacts for the shorts tester, S3, are commoned—1 to 1, 2 to 2, and so on—for both sections of the switch. This switch must be arranged so that the *A* swinger can be set independently of the *B* swinger. You may need two complete switches. The letter *N* designates a neon bulb, and it must be the type that works directly from 120 volts of AC. It may have an internal resistor or another way of connecting the voltage properly. Note that pin 1 of each section of the switch goes to the

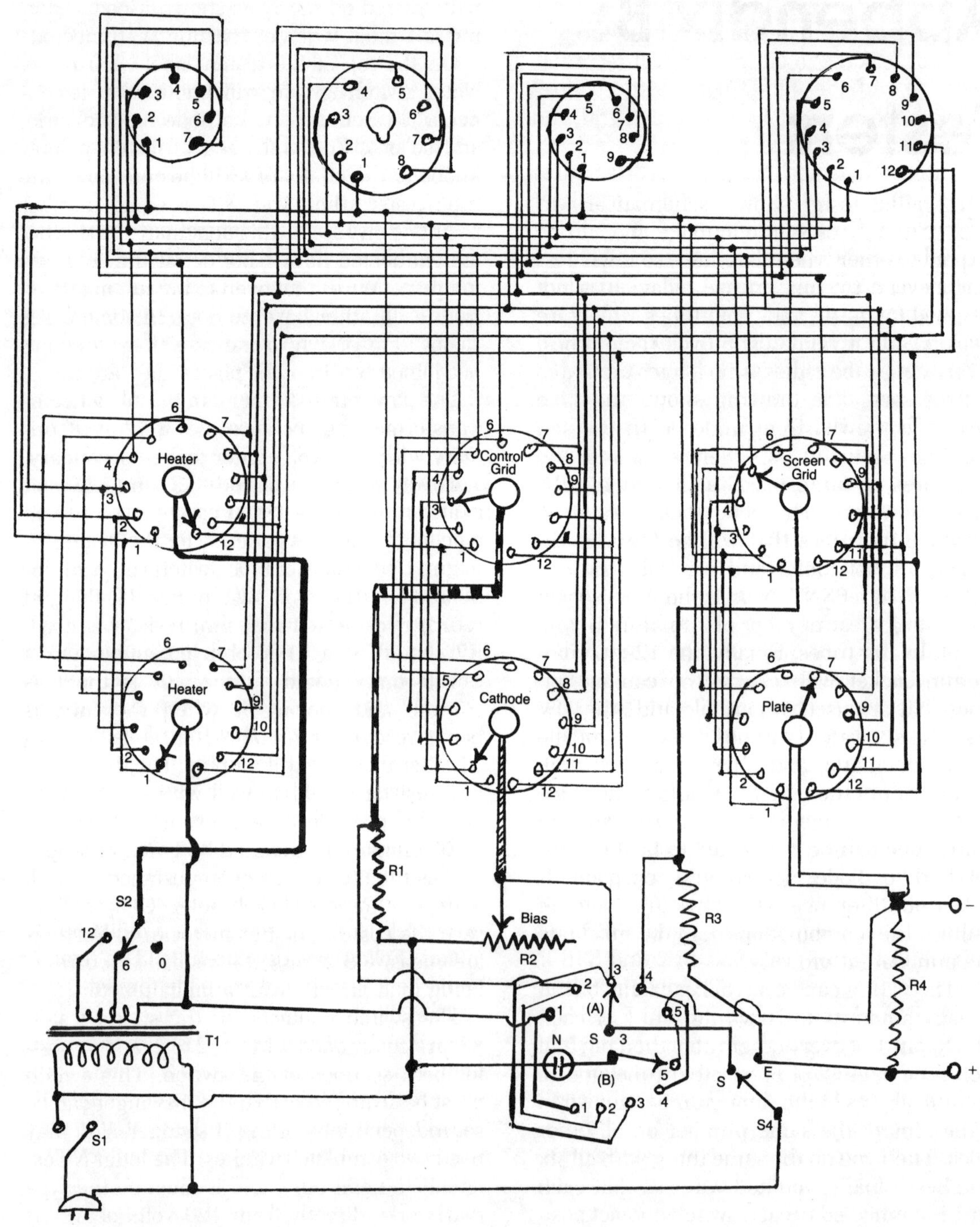

FIGURE A-1
TUBE TESTER

heater. It is only necessary for the shorts tester to connect to one heater of the tube, as shorts or leakage to either side will record as heater to cathode. Pin 2 goes to the control grid, 3 to the cathode, 4 to the screen grid, and 5 to plate.

As an example, let's try testing a 6BE6 tube, a type often used as a converter-oscillator combination. In a tube manual you find that this tube needs 6.3 volts of heater, and that heater terminals are 3 and 4. The cathode is pin 2; the control grid is pin 1. Pin 5 is the plate, 6 is the screen grid of two parts, and 7 is the suppressor grid. You set all the contact switches properly, forgetting about anything for pin 7, and set S2 in 6-volt position. Since this is a 7-pin tube it obviously goes in the 7-pin socket. First let's check the tube for shorts. S2 should be set for neutral position or 0 volts during shorts testing. S4 must be in S or shorts testing position.

Now, with S1 turned on, you put S3A in the 1 position and rotate S3B to all positions. Naturally the neon will come on when S3B is also in the 1 position, but otherwise the neon should not light. Then set S3A in the 2 position and rotate S3B through all positions to check for shorts, disregarding a flash on 2 with S3B. Then set S3A in the 3 position, and rotate S3B. This is done so you can check each element of the tube for shorts to each other element. Normally only one element would short to the next, such as heater to cathode (check with S3A in 1 and S3B in 3); cathode to control grid (S3A in 2 and S3B in 3); control to screen grid (S3A in 2 and S3B in 4); and screen grid to plate (S3A in 5 and S3B in 4). If you find no shorts, you are ready to check for emission. If you do find shorts do not test for emission, as a shorted tube could damage your tester.

Change S4 to *E* or emission test. S2 must remain set for 6-volt operation with this tube. Let the tube warm up. Now with the voltmeter connect as indicated. A tube such as a 6BE6 will have a small plate current so a small voltage will result. A tube manual will

give plate currents, but they will be less than indicated, as the tube tester is using AC and the tube must do its own rectifying. You will need to try several known good tubes of different types to get an idea of the voltage to expect. Power output types with heavy current will need to be measured with a higher voltage on the voltmeter.

Now move the bias control back and forth. The voltage across R4 also should vary, indicating that the tube is capable of amplification. If the voltage will not change, the control grid can't control electron flow from cathode to plate.

Though this tube tester won't check some types, it is very useful for TV tubes, and it is very adaptable. If you wish to make it test 4-pin types, you would put in a 4-pin socket and common up the wires as before.

The picture tube tester and rejuvenator, Figure A-2, is made with a power transformer that has a primary of 120 volts, a secondary of 300 to 400 (maximum) points *A* to *B*, a 6.3-volt secondary at points *B* to *C*, and a 5-volt secondary at points *C* to *D*. You will have to use a voltmeter to be sure you have it connected in such a way that from *B* to *D* is 11.3 volts of AC. Older TV transformers that used a 5U4 rectifier are ideal for this construction. C1 is 4 mfd. R1 is 1,000 ohms. R2 is a 100,000-ohm potentiometer. S2 needs only two positions. S3 is a switch that you push to make contact; it is off until pushed, and off when released. The meter is a 5-milliamp DC meter. Follow polarity carefully. The rectifier must be a 1-amp unit, cathode toward S3.

The alligator clips 1, 2, 3, and 4 are connected to the picture tube pins with the socket removed. Clips 3 and 4 go to the heaters of the picture tube. You test each gun separately. R2 is calibrated with a good tube so that it measures about 4 milliamps on the meter. R2 is 100,000 ohms.

For red you would set clip 1 to the control grid of the red gun and clip 2 to its cathode. After making all tests you would connect clip 1 to the grid of the blue gun and 2 to its

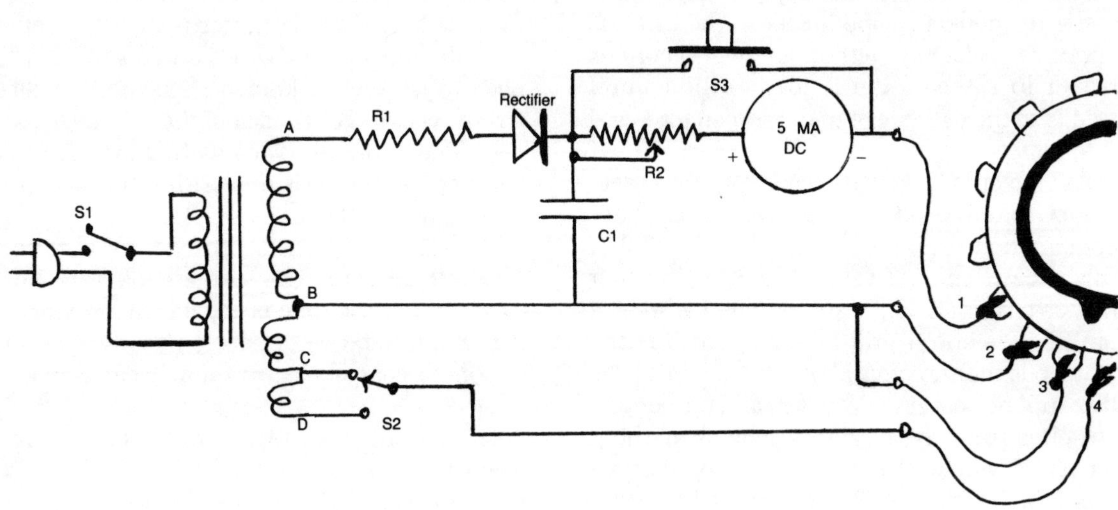

FIGURE A-2
PICTURE TUBE TESTER-REJUVENATOR

cathode. After the test (and rejuvenation if necessary) you would do the same for the green gun. If only one color were missing or weak you would connect clip 1 to its grid and clip 2 to its cathode.

Unplug the TV and turn on S1. Note the reading on the meter. You will have calibrated the meter on a good set so you will know a low reading. For testing, S2 is in the 6-volt position at *C*. If the reading is low touch the rejuvenate button, S3, momentarily. You might see little arcs in the neck of the picture tube or you might not. Just touch the S3 switch for a second or two. Then release. Now if the meter shows an improved reading you have repaired the picture tube. If not, flip the heater boost switch, S2, to boost or 11.3-volt position at *D*. Give the tube plenty of time to warm up the extra amount. Now hit S3 again briefly. Return the heater boost switch, S2, to the 6.3-volt position, and give the rejuvenate button,

S3, two more hits. This should rejuvenate a bad section of a tube. Of course, you can complete this process several times, but you risk burning out the picture tube with the extra heavy heater voltage.

Figure A-3 shows how a heater transformer-oscilloscope combination can test transistors in circuit. The secondary must be 6.3 volts. The resistor is 270 ohms. The vertical and horizontal connections of scope to circuit are shown. First connect with miniclips to emitter and base. If the transistor is good you will get a right angle either upright or inverted as in upper drawings, depending on the type of transistor. If the angle is curved there is some leakage, as in the middle drawings. If the section is shorted a vertical line will show. If open, a horizontal line will show. Then test as shown from base to collector. The same lines will indicate good, leaky, shorted, or open transistors.

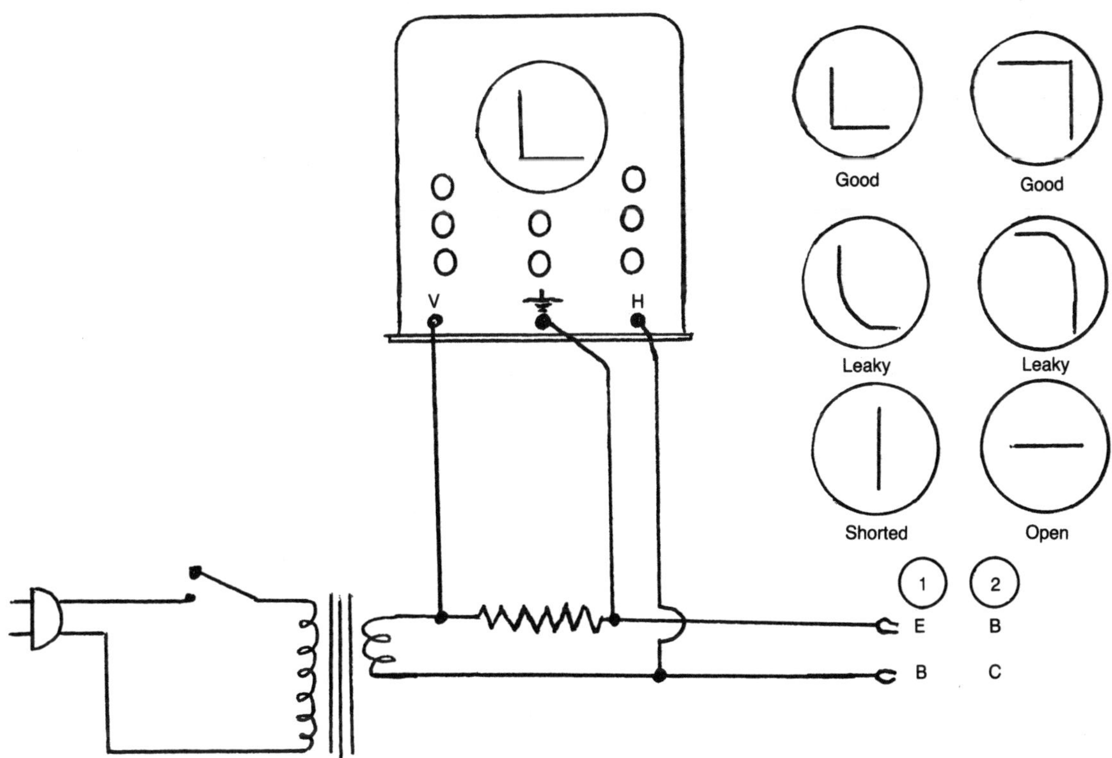

FIGURE A-3
TRANSISTOR TESTER USING SCOPE

Appendix C: Optional Radio Circuits

THE CIRCUITS in Appendix C are intended as extra exercises for those who find building radios a rewarding hobby and who may take up amateur radio as a pastime. The CB receiver, Figure A-4, is just about as simple an outfit as you can construct and hear CB with any degree of reliability. True, now and then a crystal radio or even a phonograph will pick up CB, but normally it takes a radio that works on the CB band of 27 MHz.

The parts values are given. Build the radio on a chassis and panel similar to those used in Chapters 1 through 6. Keep the leads between parts as short as possible, as you are dealing with short waves, and even a long wire affects tuning. This is a grounded base circuit for Q1, although the feedback is actually to ground and up through the 1,000-pfd. capacitor to the base. When the collector and base are tied together in the tuned circuit, the circuit is super-regenerative rather than regenerative. Instead of a whistle as you advance the 100,000-ohm pot, you will hear a thud and then hissing. Tune just above the thud before the hiss gets its loudest, as this is often the most sensitive point and picks up the CB "good old boys" best. About 33 feet of antenna, including the lead-in, will work well.

The code transmitter in Figure A-5 can be put on the air only by someone with at least a novice or technician amateur license. Howev-

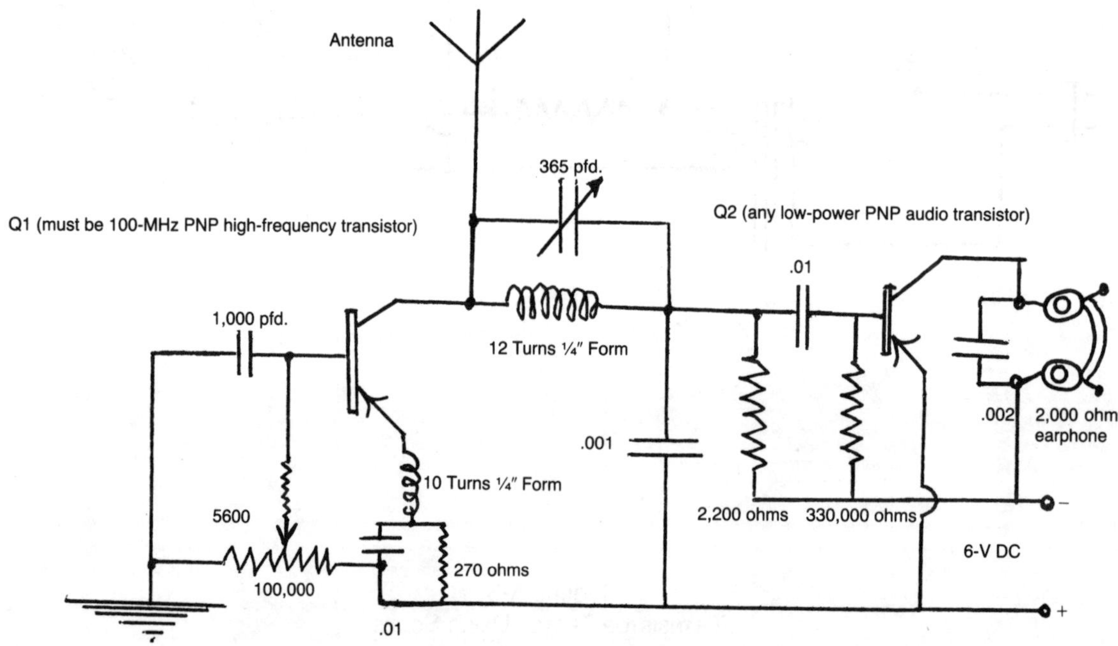

This CB receiver is a super-regenerative type. Run variable resistor just above pop into hiss region.
Dial in CB near top of variable capacitor.

FIGURE A-4
CITIZEN'S BAND RECEIVER

er, if you keep the antenna short or connect up a lightbulb instead of an antenna as a dummy load you can work it without trouble from the FCC, as you are practicing code and listening to it on a nearby shortwave radio. In fact, this transmitter should broadcast code to nearby homes even without antenna and not interfere with other services if you keep it on amateur frequencies.

The crystal must be for the frequency on which you wish to broadcast, or it can be for a lower frequency and you can tune L2 and L3 and the variable capacitor for the second or third harmonic. With high-frequency transmission you won't be able to get crystals that will oscillate that high, so operating on a harmonic is necessary. If you wish to be able to broadcast on various frequencies, a plug-in arrangement for crystal, L2 and L3 is necessary. L3 can be on top of L2 with a layer of plastic tape between coils, or L3 can be wound above L2 on the same coil form.

A junked TV with a good power supply can be used to supply voltages. You could even

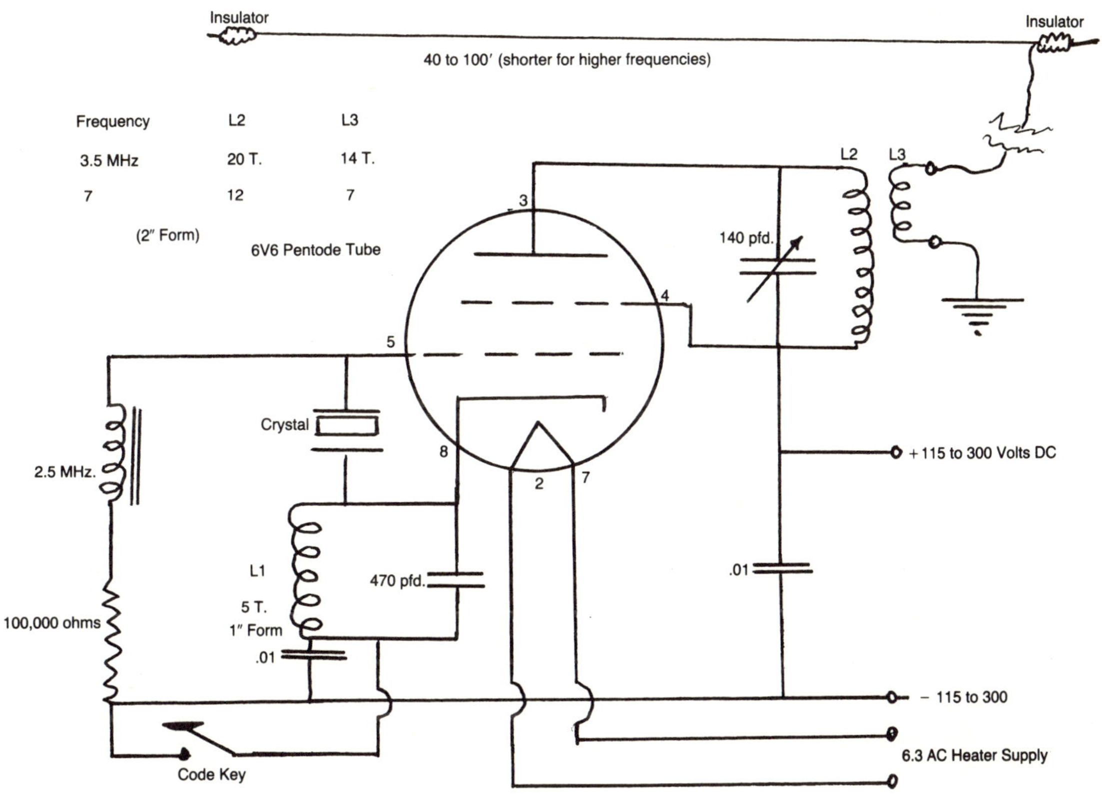

FIGURE A-5
CODE TRANSMITTER

build the transmitter on the TV chassis. The key should be well insulated, because from 115 to 300 volts, depending on the supply you use, will be connected and disconnected by it as you send code.

Higher frequencies than those shown can be broadcast by using the appropriate crystal and fewer turns on L2 and L3. This transmitter definitely could work for CB, so be careful that your code doesn't cover up your neighbor's talk. If you are interested in becoming a ham operator get a copy of the *AMECO Radio Amateur License Guide for Novice thru General Class* from a bookstore or radio parts store. Present cost is less than a dollar. Also see a ham in your area for tips.

Appendix D: International Code

L EARNING THE code is very important. Think of the dots as "di," as that is the sound they will have as they come out of a receiver. The dashes will sound like "daah." The dash should be held down to sound three times as long as the dot. As you practice you will become accustomed to thinking of the letter *P* as "di daah daah dit." *F* would be "di di daah dit." Naturally this information is for those interested in amateur transmissions; it is not required in TV work. In fact, one can even get a professional FCC license and work at a TV station without knowing the code.

A	·—	V	···—
B	—···	W	·——
C	—·—·	X	—··—
D	—··	Y	—·——
E	·	Z	——··
F	··—·	1	·————
G	——·	2	··———
H	····	3	···——
I	··	4	····—
J	·———	5	·····
K	—·—	6	—····
L	·—··	7	——···
M	——	8	———··
N	—·	9	————·
O	———	0	—————
P	·——·	Period	·—·—·—
Q	——·—	Interrogation	··——··
R	·—·	Break	—···—
S	···	Wait	·—···
T	—	End of message	·
U	··—	End of transmission	···—·—

FIGURE A-6
INTERNATIONAL CODE

Appendix E: Color Codes

THE RESISTOR and capacitor color codes must be learned by TV repairmen, radio hobbyists, and just about anyone who has anything to do with electronics. See Figure A-7. Note the orientation of the resistors in *A*, *B*, and *C*. At *A*, you start at the end where the colors are nearest and work across. A fourth color of silver means that the resistor is within 10 percent tolerance of the exact ohms it is supposed to have. Gold means 5 percent.

The first three colors indicate the ohms. The first color at the left tells the first number in the ohms reading. Suppose it were red. Note that red stands for the number 2, so you know that the first number of ohms is 2. If the second color were violet, it would mean 7. So far you have 27. If the third color were brown or 1, you would know to add one zero. The third color always tells how many zeros to add to the other two numbers. In this case, with one zero, you get 270 ohms for this resistor; that is, a red/violet/brown with the red at the extreme left is a 270-ohm resistor. The resistor in a TV may be mounted backward, but look to the end where the colors start and work from left to right. With *B* and *C* you will need to pay attention to the drawings to interpret ohms. Fortunately, these styles are infrequently used today. Remember that the first two colors are actual numbers and the third is zeros.

The same is true of capacitors (Figure A-8), but the number will be in picafarads. The microfarad capacitors usually have the number written on them, i.e. 0.002 or 0.1. Add the zeros as before. It would be easy for the novice repairman to confuse resistors and capacitors with color codes, as they look somewhat alike, especially *C*, *D*, *G*, *H*, and *I* capacitors. The *D* capacitor always has a large white center to show that it is not a resistor. *C* and *I* are usually fatter than a resistor and have a bit more taper at the ends. Note that they also have more bands of color. *G* and *H* might compare to *B* with resistors, as they all have wires looped around their edges, but note that the resistor has a dot in the middle, whereas the capacitors have several color dots bunched toward the left. Disc capacitors such as *A* and *B* often are stamped with the number instead of marked with colors. If it is a 0.01 or a 0.005, it is in microfarads, but if it is a high number such as 1,000 or 670, it is picafarads. Such a small disc unit could not possibly have that many microfarads. The colors always refer to picafarads.

The wiring color codes are not always as shown in Figure A-9, but often they will be. Because of the heat in TV sets color coded wires around a power transformer will often all look black. Careful observation and a little scraping can sometimes reveal the color the wire insulation was.

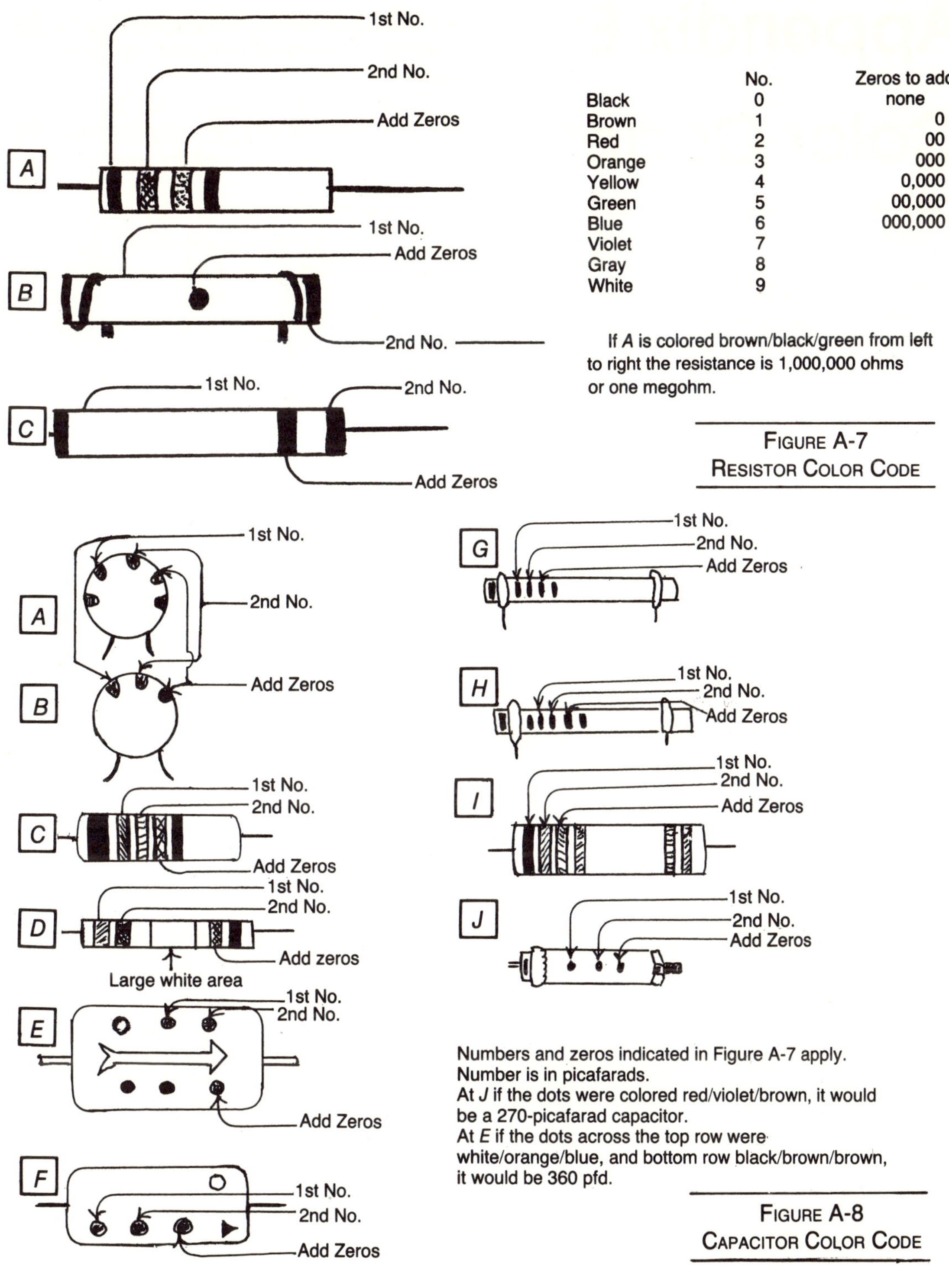

	No.	Zeros to add
Black	0	none
Brown	1	0
Red	2	00
Orange	3	000
Yellow	4	0,000
Green	5	00,000
Blue	6	000,000
Violet	7	
Gray	8	
White	9	

If *A* is colored brown/black/green from left to right the resistance is 1,000,000 ohms or one megohm.

FIGURE A-7
RESISTOR COLOR CODE

Numbers and zeros indicated in Figure A-7 apply.
Number is in picafarads.
At *J* if the dots were colored red/violet/brown, it would be a 270-picafarad capacitor.
At *E* if the dots across the top row were white/orange/blue, and bottom row black/brown/brown, it would be 360 pfd.

FIGURE A-8
CAPACITOR COLOR CODE

Grounds	Black	
Heaters	Brown	
B+ Power Supply	Red	
Screen Grids	Orange	
Cathodes	Yellow	Emitters
Control Grids	Green	Bases
Plates (Anodes)	Blue	Collectors
B− Power Supply	Violet	
AC Power Line	Gray	

IF Transformers:

Plate Lead	Blue	
B+ Lead	Red	
Grid Lead	Green	Diode
Grid Return	Black	Diode Return

AF Transformers:

Plate in Primary	Blue	Collector
Power Supply	Red	
Grid Lead Secondary	Green	Base
Grid Return	Black	Base Return

Power Transformers:

AC Primary	Black
AC Primary Center Tap	Black/Yellow
Secondary to Plates	Red
Secondary to Plates Center Tap	Red/Yellow
Heaters for Rectifier	Yellow
Other Heater Leads	Brown

FIGURE A-9
WIRING COLOR CODES

Appendix F: Mathematical Formulas

To USE electronics math formulas, you must always convert milliamps to amps: 5 milliamps is 0.005 amps. Use the 0.005 in the formulas shown. 300 milliamps is 0.3 amps. For the frequency formulas convert picafarads to farads; for instance, 365 picafarads is 0.000000000365 farads. If a coil offered 100 microhenries it would have to be converted to 0.0001 henries. The answer would come out in Hz. If you want the answer to come out in MHz leave the coil inductance in microhenries and change the picafarads to microfarads. For instance, 365 picafarads is 0.000365 microfarads.

·If you had a coil that had about 12 turns as in Figure A-5 for the transmitter tuning output, it would offer roughly 10 microhenries of magnetic resistance. If the 140-picafarad capacitor were tuned about a third of the way, to 50 picafarads, what frequency would this circuit tune? Leave the 10 microhenries as is and convert the 50 picafarads to microfarads to get 0.000050 microfarads. First multiply the two together and get 0.0005 as the answer. Then take the square root of this and get approximately 0.0224, which you must multiply times 6.28 and get approximately 0.141. Now divide this into 1 and you will get very close to 7 or 7 MHz. Note at the chart in Figure A-5 this is the mentioned frequency.

With resistance formulas use ohms throughout. Perhaps this is the most useful information here. Sometimes you don't have a 10,000-ohm resistor but have two 5,000-ohm resistors. You can hook them in series and get the 10,000 you need. Also if you needed 500 ohms and had none but did have two 1,000-ohm resistors, you could twist them in parallel and the total would be the 500 you need. Mounting two in small spaces is the problem with circuit boards.

$$\text{Current (in amps)} \times \text{Resistance (in ohms)} = \text{Volts}$$

$$\frac{\text{Ohms}}{\text{Current} \mid \text{Volts}} \qquad \frac{\text{Current}}{\text{Ohms} \mid \text{Volts}} \qquad \text{Current} \times \text{Volts} = \text{Watts}$$

Resistors in series
$$R1 + R2 + R3 = \text{total ohms}$$

Resistors in parallel
$$\frac{1}{\dfrac{1}{R1} + \dfrac{1}{R2} + \dfrac{1}{R3}} = \text{Total Resistance}$$

$$\text{Frequency in Hz.} = \frac{1}{6.28 \times \sqrt{L} \times C}$$

$$\text{Wavelength in meters} = \frac{300,000,000}{\text{Frequency in Hz.}}$$

L = Inductance in henries

C = Capacitance in farads

FIGURE A-10
ELECTRONICS MATHEMATICS

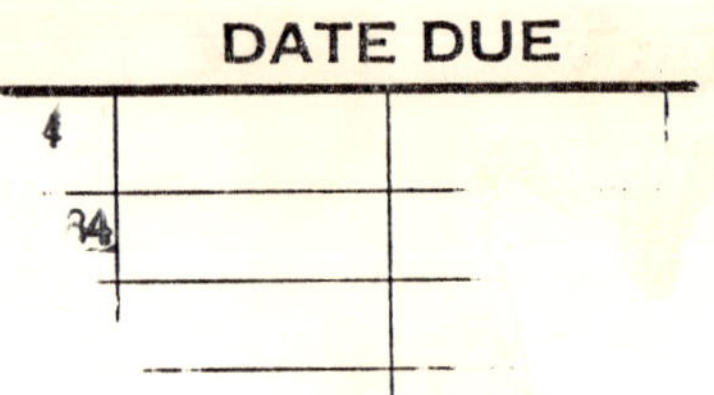
DATE DUE